Takayuki Ito, Minjie Zhang, Valentin Robu, Shaheen Fatima,
and Tokuro Matsuo (Eds.)

New Trends in Agent-Based Complex Automated Negotiations

Studies in Computational Intelligence, Volume 383

Editor-in-Chief

Prof. Janusz Kacprzyk
Systems Research Institute
Polish Academy of Sciences
ul. Newelska 6
01-447 Warsaw
Poland
E-mail: kacprzyk@ibspan.waw.pl

Further volumes of this series can be found on our
homepage: springer.com

Vol. 362. Pascal Bouvry, Horacio González-Vélez, and
Joanna Kolodziej (Eds.)
*Intelligent Decision Systems in Large-Scale Distributed
Environments,* 2011
ISBN 978-3-642-21270-3

Vol. 363. Kishan G. Mehrotra, Chilukuri Mohan, Jae C. Oh,
Pramod K. Varshney, and Moonis Ali (Eds.)
Developing Concepts in Applied Intelligence, 2011
ISBN 978-3-642-21331-1

Vol. 364. Roger Lee (Ed.)
Computer and Information Science, 2011
ISBN 978-3-642-21377-9

Vol. 365. Roger Lee (Ed.)
*Computers, Networks, Systems, and Industrial
Engineering 2011,* 2011
ISBN 978-3-642-21374-8

Vol. 366. Mario Köppen, Gerald Schaefer, and
Ajith Abraham (Eds.)
Intelligent Computational Optimization in Engineering, 2011
ISBN 978-3-642-21704-3

Vol. 367. Gabriel Luque and Enrique Alba
Parallel Genetic Algorithms, 2011
ISBN 978-3-642-22083-8

Vol. 368. Roger Lee (Ed.)
*Software Engineering, Artificial Intelligence, Networking and
Parallel/Distributed Computing 2011,* 2011
ISBN 978-3-642-22287-0

Vol. 369. Dominik Ryzko, Piotr Gawrysiak, Henryk Rybinski,
and Marzena Kryszkiewicz (Eds.)
Emerging Intelligent Technologies in Industry, 2011
ISBN 978-3-642-22731-8

Vol. 370. Alexander Mehler, Kai-Uwe Kühnberger,
Henning Lobin, Harald Lüngen, Angelika Storrer, and
Andreas Witt (Eds.)
*Modeling, Learning, and Processing of Text Technological
Data Structures,* 2011
ISBN 978-3-642-22612-0

Vol. 371. Leonid Perlovsky, Ross Deming, and Roman Ilin
(Eds.)
*Emotional Cognitive Neural Algorithms with Engineering
Applications,* 2011
ISBN 978-3-642-22829-2

Vol. 372. António E. Ruano and
Annamária R. Várkonyi-Kóczy (Eds.)
New Advances in Intelligent Signal Processing, 2011
ISBN 978-3-642-11738-1

Vol. 373. Oleg Okun, Giorgio Valentini, and Matteo Re (Eds.)
Ensembles in Machine Learning Applications, 2011
ISBN 978-3-642-22909-1

Vol. 374. Dimitri Plemenos and Georgios Miaoulis (Eds.)
Intelligent Computer Graphics 2011, 2011
ISBN 978-3-642-22906-0

Vol. 375. Marenglen Biba and Fatos Xhafa (Eds.)
Learning Structure and Schemas from Documents, 2011
ISBN 978-3-642-22912-1

Vol. 376. Toyohide Watanabe and Lakhmi C. Jain (Eds.)
Innovations in Intelligent Machines – 2, 2011
ISBN 978-3-642-23189-6

Vol. 377. Roger Lee (Ed.)
*Software Engineering Research, Management
and Applications 2011,* 2011
ISBN 978-3-642-23201-5

Vol. 378. János Fodor, Ryszard Klempous, and
Carmen Paz Suárez Araujo (Eds.)
Recent Advances in Intelligent Engineering Systems, 2011
ISBN 978-3-642-23228-2

Vol. 379. Ferrante Neri, Carlos Cotta, and
Pablo Moscato (Eds.)
Handbook of Memetic Algorithms, 2011
ISBN 978-3-642-23246-6

Vol. 380. Anthony Brabazon, Michael O'Neill, and
Dietmar Maringer (Eds.)
Natural Computing in Computational Finance, 2011
ISBN 978-3-642-23335-7

Vol. 381. Radosław Katarzyniak, Tzu-Fu Chiu,
Chao-Fu Hong, and Ngoc Thanh Nguyen (Eds.)
*Semantic Methods for Knowledge Management and
Communication,* 2011
ISBN 978-3-642-23417-0

Vol. 382. F.M.T. Brazier, Kees Nieuwenhuis, Gregor Pavlin,
Martijn Warnier, and Costin Badica (Eds.)
Intelligent Distributed Computing V, 2011
ISBN 978-3-642-24012-6

Vol. 383. Takayuki Ito, Minjie Zhang, Valentin Robu,
Shaheen Fatima, and Tokuro Matsuo (Eds.)
*New Trends in Agent-Based Complex Automated
Negotiations,* 2012
ISBN 978-3-642-24695-1

Takayuki Ito, Minjie Zhang, Valentin Robu,
Shaheen Fatima, and Tokuro Matsuo (Eds.)

New Trends in Agent-Based Complex Automated Negotiations

 Springer

Editors

Takayuki Ito
Nagoya Institute of Technology
Graduate School of Engineering
Gokiso-cho, 466-8555 Nagoya, Showa
Japan
E-mail: itota@ics.nitech.ac.jp

Dr. Shaheen Fatima
Loughborough University
Dept. Computer Studies
Loughborough, Leics.
United Kingdom
E-mail: S.S.Fatima@lboro.ac.uk

Minjie Zhang
University of Wollongong
School of Information
Technology & Computer Science
Northfields Avenue, 2522
Wollongong New South
Wales
Australia
E-mail: minjie@uow.edu.au

Tokuro Matsuo
Yamagata University
School of Engineering
Dept. of Mechanical Systems
Engineering, Jonan 4-3-16
992-8510 Yonezawa
Japan
E-mail: matsuo@yz.yamagata-u.ac.jp

Valentin Robu
University of Southampton
School of Electronics and
Computer Science, Southampton
United Kingdom
E-mail: vr2@ecs.soton.ac.uk

ISBN 978-3-642-26968-4   ISBN 978-3-642-24696-8 (eBook)

DOI 10.1007/978-3-642-24696-8

Studies in Computational Intelligence   ISSN 1860-949X

Typeset & Cover Design: Scientific Publishing Services Pvt. Ltd., Chennai, India.

Printed on acid-free paper

9 8 7 6 5 4 3 2 1

springer.com

New Trends in Agent-Based Complex Automated Negotiations

Preface

Complex Automated Negotiations have been widely studied and are becoming an important, emerging area in the field of Autonomous Agents and Multi-Agent Systems. In general, automated negotiations can be complex, since there are a lot of factors that characterize such negotiations. These factors include the number of issues, dependency between issues, representation of utility, negotiation protocol, negotiation form (bilateral or multi-party), time constraints, etc. Software agents can support automation or simulation of such complex negotiations on the behalf of their owners, and can provide them with adequate bargaining strategies. In many multi-issue bargaining settings, negotiation becomes more than a zero-sum game, so bargaining agents have an incentive to cooperate in order to achieve efficient win-win agreements. Also, in a complex negotiation, there could be multiple issues that are interdependent. Thus, agent's utility will become more complex than simple utility functions. Further, negotiation forms and protocols could be different between bilateral situations and multi-party situations. To realize such a complex automated negotiation, we have to incorporate advanced Artificial Intelligence technologies includes search, CSP, graphical utility models, Bays nets, auctions, utility graphs, predicting and learning methods. Applications could include e-commerce tools, decision-making support tools, negotiation support tools, collaboration tools, etc. In this book, we solicit papers on all aspects of such complex automated negotiations in the field of Autonomous Agents and Multi-Agent Systems.

In addition, this book includes papers on the ANAC 2010 (Automated Negotiating Agents Competition), in which automated agents who have different negotiation strategies and implemented by different developers are automatically negotiate in the several negotiation domains. ANAC is the one of the typical competition in which strategies for automated negotiating agents are evaluated in a tournament style.

Finally, we would like to extend our sincere thanks to all authors. This book would not have been possible without valuable supports and contributions of the cooperators. Especially, we would like to appreciate *Katsuhide Fujita* in Nagoya Institute of Technology making a contribution to editing this book.

Japan, 28th, February 2011   Takayuki Ito
                                                             Minjie Zhang
                                                             Valentin Robu
                                                             Shaheen Fatima
                                                             Tokuro Matsuo

Contents

List of Contributors

Reyhan Aydoğan
Department of Computer Engineering,
Boğaziçi University, Bebek, 34342, Istanbul,
Turkey
e-mail: reyhan.aydogan@gmail.com

Pınar Yolum
Department of Computer Engineering,
Boğaziçi University, Bebek, 34342,
Istanbul,Turkey
e-mail: pinar.yolum@boun.edu.tr

Fenghui Ren
School of Computer Science and Software
Engineering, University of Wollongong,
Australia
e-mail: fr510@uow.edu.au

Minjie Zhang
School of Computer Science and Software
Engineering, University of Wollongong,
Australia
e-mail: minjie,john@uow.edu.au

John Fulcher
School of Computer Science and Software
Engineering, University of Wollongong,
Australia
e-mail: john@uow.edu.au

Katsuhide Fujita
Department of Computer Science and Engineering, Nagoya Institute of Technology,

Nagoya, Aichi, Japan
e-mail:
fujita@itolab.mta.nitech.ac.jp

Takayuki Ito
School of Techno-Business Administration,
Nagoya Institute of Technology, Nagoya,
Aichi, Japan
e-mail:
ito.takayuki@nitech.ac.jp

Mark Klein
Sloan School of Management, Massachusetts
Institute of Technology, Cambridge, MA,
U.S.A.
e-mail: m_klein@mit.edu

Simone A. Ludwig
Department of Computer Science, North
Dakota State University, Fargo, North
Dakota, U.S.A.
e-mail: simone.ludwig@ndsu.edu

Thomas Schoene
Department of Computer Science, University
of Saskatchewan, Saskatoon, Saskatchewan,
Canada
e-mail: ths299@mail.usask.ca

Raz Lin
Department of Computer Science, Bar-Ilan
University, Ramat-Gan, Israel 52900
e-mail: linraz@cs.biu.ac.il

Sarit Kraus
Department of Computer Science, Bar-Ilan
University, Ramat-Gan, Israel 52900, and
Institute for Advanced Computer Studies,
University of Maryland, College Park, MD
20742 USA
e-mail: sarit@cs.biu.ac.il

Satoshi Takahashi
Graduate School of System & Information
Engineering, University of Tsukuba,
Tsukuba, Ibaraki, Japan
e-mail:
takahashi2007@e-activity.org

Tokuro Matsuo
Graduate School of Science & Engineering,
Yamagata University, Yonezawa, Yamagata,
Japan
e-mail:
matsuo@yz.yamagata-u.ac.jp

Tim Baarslag
Man Machine Interaction Group, Delft
University of Technology, Delft, the
Netherlands
e-mail: T.Baarslag@tudelft.nl

Koen Hindriks
Man Machine Interaction Group, Delft
University of Technology, Delft, the
Netherlands
e-mail: T.Baarslag@tudelft.nl

Catholijn Jonker
Man Machine Interaction Group, Delft
University of Technology, Delft, the
Netherlands
e-mail: T.Baarslag@tudelft.nl

Shogo Kawaguchi
Department of Computer Science,
Nagoya Institute of Technology, Nagoya,
Aichi, Japan
e-mail: kawaguchi@itolab.mta.
nitech.ac.jp

Bo An
Department of Computer Science, University
of Massachusetts, Amherst, USA
e-mail: ban@cs.umass.edu

Victor Lesser
Department of Computer Science,
University of Massachusetts, Amherst, USA
e-mail: lesser@cs.umass.edu

Colin R. Williams
School of Electronics and Computer
Science, University of Southampton,
University Road, Southampton, SO17 1BJ
e-mail: crw104@ecs.soton.ac.uk

Valentin Robu
School of Electronics and Computer
Science, University of Southampton,
University Road, Southampton, SO17 1BJ
e-mail: vr2@ecs.soton.ac.uk

Enrico H. Gerding
School of Electronics and Computer
Science, University of Southampton,
University Road, Southampton, SO17 1BJ,
e-mail: eg@ecs.soton.ac.uk

Nicholas R. Jennings
School of Electronics and Computer
Science, University of Southampton,
University Road, Southampton, SO17 1BJ
e-mail: nrj@ecs.soton.ac.uk

Liviu Dan Şerban
Babeş-Bolyai University Str. Theodor Mihali
58-60, 400591, Cluj-Napoca, Romania
e-mail:
Liviu.Serban@econ.ubbcluj.ro

Gheorghe Cosmin Silaghi
Babeş-Bolyai University Str. Theodor Mihali
58-60, 400591, Cluj-Napoca, Romania
e-mail:Gheorghe.Silaghi@econ.
ubbcluj.ro

Cristian Marius Litan
Babeş-Bolyai University Str. Theodor Mihali
58-60, 400591, Cluj-Napoca, Romania
e-mail:
Cristian.Litan@econ.ubbcluj.ro

Niels van Galen Last
Man-Machine Interaction,
Faculty of EEMCS, Delft University of
Technology, Mekelweg 4, 2628CD, Delft,
The Netherlands
e-mail:
n.a.vangalenlast@student.
tudelft.nl

Part I
Agent-Based Complex Automated Negotiations

The Effect of Preference Representation on Learning Preferences in Negotiation

Reyhan Aydoğan and Pınar Yolum

Abstract. In online and dynamic e-commerce environments, it is beneficial for parties to consider each other's preferences in carrying out transactions. This is especially important when parties are negotiating, since considering preferences will lead to faster closing of deals. However, in general may not be possible to know other participants' preferences. Thus, learning others' preferences from the bids exchanged during the negotiation becomes an important task. To achieve this, the producer agent may need to make assumptions about the consumer's preferences and even its negotiation strategy. Nevertheless, these assumptions may become inconsistent with a variety of preference representations. Therefore, it is more desired to develop a learning algorithm, which is independent from the participants' preference representations and negotiation strategies. This study presents a negotiation framework in which the producer agent learns an approximate model of the consumer's preferences regardless of the consumer's preference representation. For this purpose, we study our previously proposed inductive learning algorithm, namely Revisable Candidate Elimination Algorithm (RCEA). Our experimental results show that a producer agent can learn the consumer's preferences via RCEA when the consumer represents its preferences using constraints or CP-nets. Further, in both cases, learning speeds up the negotiation considerably.

1 Introduction

In agent-mediated service negotiations, a service consumer and a producer try to reach an agreement on a service to achieve their own goals. Of course, there may be some conflicts in participants' preferences, which may sometimes delay and sometimes make it impossible to reach a consensus. However, in many settings, if there

Reyhan Aydoğan · Pınar Yolum
Department of Computer Engineering, Boğaziçi University, Bebek, 34342, Istanbul,Turkey
e-mail: reyhan.aydogan@gmail.com, pinar.yolum@boun.edu.tr

T. Ito et al. (Eds.): New Trends in Agent-Based Complex Automated Negotiations, SCI 383, pp. 3–20.
springerlink.com   © Springer-Verlag Berlin Heidelberg 2012

are services satisfying both the consumer and the producer, agents can come to an
agreement [11]. To reach a consensus, they make an offer to each other in accor-
dance with their own preferences. If it is possible, after a number of interaction,
they agree on a *service*. During this negotiation process, users' preferences play a
key role in both generation of the offers and determination of the acceptability of
the counter offers.

Understanding and responding to other participant's needs are also important to
be able to negotiate effectively. However, preferences of participants are almost always private. On one hand, participants may try to learn other's preferences through
interactions over time. But learning other's preferences in negotiation is a very complicated task since the participant does not know how the other agent keeps its preferences. For instance, the agent may use constraints or utility functions to represent
its preferences. Alternatively, the agent may use an ordering of the alternatives for its
preferences. Further, the agent does not know whether preferential interdependencies among issues exist. This uncertainty leads the agent to make some assumptions
about its opponent's preferences or negotiation strategy. In open and dynamic environments, these assumptions may not work as expected. Consider a producer agent
that tries to learn a consumer's preferences and assumes that the issues are independent. This assumption may be consistent with some consumer's preferences but it
will fail in others.

Furthermore, the number of training instances (bids exchanged during the negotiation) may be inadequate to learn the opponent's preferences especially if the
agent meets the opponent for the first time. It is difficult to find the exact model in
a few interactions. Therefore, the aim of the learning in negotiation should not be
finding the exact preferences. On the contrary, the aim should be acquiring a model
leading the producer to understand the consumer's need and revise its counter offers
accordingly.

This paper studies an automated service negotiation in which the producer agent
tries to learn the consumer's preferences via concept learning. To achieve this, the
producer agent uses Revisable Candidate Elimination Algorithm (RCEA) [2], which
is designed for learning preferences in the form of conjunctive and disjunctive rules.
The producer has only one assumption about the consumer's preferences; that is
the consumer's preferences are stable during the entire negotiation. To see the performance of the producer with different preference models, we develop two kinds
of negotiation strategies for the consumer. First, the consumer represents its user's
preferences with a set of constraints in the form of conjunctives and disjunctives.
Second, the consumer expresses its user's preferences via Conditional Preference
Networks (CP-Nets) [3]. This preference structure is different from the former in a
way that it keeps a partial preference ordering. To see the impact of learning on different preference model, we also develop a producer, which generates its offers according to the producer's preferences without considering consumer's preferences.
We show that even when the preference structure is different, the producer using
RCEA concludes negotiations faster than the producer considering only its own
preferences.

2 Background

We briefly explain the structure of CP-nets and how we interpret a given CP-net under ceteris paribus semantics in Section 2.1. Further, Section 2.2 gives the background for the learning algorithm, which is used in this study in order to learn the consumer's preferences during the negotiation, namely Revisable Candidate Elimination Algorithm (RCEA).

2.1 Conditional Preference Networks

Conditional Preference Network (CP-net) is a graphical preference model for representing qualitative preference orderings in a compact and efficient way [3]. It is a convenient way to elicit the user's preferences since people have a tendency to express their preferences in a qualitative way rather than in a quantitative way. Consider the apartment renting domain explained in Example 1.

Example 1. For simplicity, we have only three attributes in our apartment renting domain: *Price*, *Neighborhood* and *Parking Area*. There are three neighborhoods: *Etiler*, *Kadikoy* and *Kartal* whereas the valid values for the price are categorized as *High*, *Medium* and *Low*. A parking area may exist or not. Thus, the domain for parking area has two values: *Yes* and *No*.

In CP-net terminology, by using the preference statement such as "I prefer *Etiler* to *Kartal*" we mean that if all other attributes are the same, I prefer *Etiler* to *Kartal*. Such an interpretation is called the preference ceteris paribus, "everything else being equal" [6]. Here, other possible attributes of the item such as price do not affect the preference for the neighborhood attribute (preferential independence). Moreover, CP-net is able to represent conditional preferences such as "If the apartment is at *Kartal*, I prefer an apartment having a parking area". Here the preference for the parking area depends on the value of the neighborhood attribute (conditionally preferential independent). When two apartment outcomes are compared, if remaining attributes are the same value and the neighborhood is assigned to *Kartal*, the user prefers an apartment having parking area.

   A CP-net is a graphical representation in which each node represents an attribute (issue) and the edges show the preferential dependency. Here, if there is an edge from node X to Y, the variable X is called the parent node whereas Y is called the child node. The preference ordering for the child node depends on the parent node's instantiation. In other words, the value of parent attribute affects the user's preferences over the values of child attribute. Consequently, we can express the conditional preferences.

   Moreover, each node is associated with a conditional preference table (CPT) including the preference statements indicating strictly more preferred relation, $\succ$. In other words, a preference statement indicates which attribute value is dominant over others. For example, *Low* $\succ$ *Medium* means that *Low* is preferred over *Medium*.

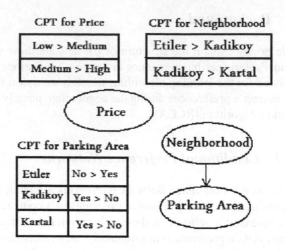

Fig. 1 A sample CP-net for apartment renting domain

Figure 1 depicts a sample CP-net for this domain. According to this CP-net, *Neighborhood* is the parent of *Parking Area* because the user's preferences on "parking area" depends on the neighborhood. The user specifies that if the neighborhood is Etiler, then she does not need a parking area but for other locations she does. By using the preference statements specified in the CPTs, we can infer some conclusions such as if the price is the same for two apartments and neighborhood is Kartal, the user will prefer an apartment having a parking area. In the case that the neighborhood is Etiler, she will prefer an apartment not having a parking area. When there are two apartments that have a parking area and are at the same neighborhood, the user prefers the apartment having with a low price to medium price and the apartment with a medium price is better for the user over that with a high price. According to the user, Etiler is preferred over Kadikoy and Kadikoy is preferred over Kartal.

2.2 RCEA

Revisable Candidate Elimination Algorithm (RCEA) [2] is a concept learning algorithm based on Candidate Elimination Algorithm (CEA) [10]. There are two hypothesis sets: the most general set (G) and the most specific set (S). The general set involves the hypothesis as general as possible including all positive samples while excluding negatives. Contrary, S consists of specific hypotheses minimally covering the positive samples. Different from CEA, this algorithm also keeps the history of all negative and positive samples.

During the training process, when a positive sample comes, the algorithm first checks whether any hypotheses in S cover this sample. If none of the hypotheses covers this sample, the most similar specific hypothesis to the current sample is generalized in a minimal way that it covers the current sample but does not cover any negative samples. If this cannot be possible, that current sample is added as a separate hypothesis into S. Also, each positive sample should be covered by at least one general hypothesis in G. If none of the hypotheses in G covers the current sample, a revision process is invoked to generate new general hypotheses covering this sample but not covering the negative samples.

When a negative sample comes, the algorithm minimally specializes all hypotheses in G covering that sample. If there are any hypotheses in S covering this negative sample, these hypotheses are removed from S. Since each positive sample should be covered by at least one of the hypotheses in S, each positive sample in the positive sample set is checked whether it is covered by S. If none of the specific hypotheses in S covers a positive sample, this positive sample is added as a separate hypothesis to S.

Table 1 Semantic Similarities

Attribute	Value-1	Value-2	Similarity
Neighborhood	Etiler	Kartal	0.2
Neighborhood	Kadikoy	Etiler	0.5
Neighborhood	Kartal	Kadikoy	0.4
Price	Low	Medium	0.6
Price	Medium	High	0.6
Price	Low	High	0.3

Furthermore, this algorithm uses ontology reasoning in generalization and specialization of the hypotheses. A threshold value determines whether a generalization will be performed for each issue in the hypothesis, which will be generalized to cover the current positive sample. Generalization of a hypothesis is different for each issue in the hypothesis, in fact it depends on the property of the issues. If there is an ontological information such that a hierarchy exists on the values of the issue, the generalization depends on the ratio of the covered branches in the hierarchical tree. Otherwise, we apply a heuristic that favors generalization to Thing (?) with a higher probability when more different values for the issue exist. For instance, Table 1 shows the semantic similarities for the domain values in Example 1. Note that these similarities are decided by a domain expert and agent has an access to these similarities via an ontology. According to these similarities, if the values for the price in a hypothesis is *Low* and *High*, it has more probability to generalize Thing (?) concept rather than *Low* and *Medium*. Note that ? means that any value is admissible for that issue.

3   Negotiation Framework

In our negotiation framework, a consumer and a producer negotiate over a service. This process is performed in a turn-taking way: the consumer starts by requesting a service and then the producer offers an alternative service until either a consensus or a deadline is reached. According to our proposed approach, the common understanding between agents is provided via a shared service ontology involving the description of the issues constructing the service and their domain information such as semantic similarities for the domain values. Both agents use this ontology to describe the services that are of interest to them [2].

During negotiation, the consumer agent uses its own preferences to generate its offer and to decide whether the producer's counter offer is acceptable or not. Similarly, the producer tries to offer a service that will be acceptable by the consumer (while respecting its own goals). Hence, to understand negotiation better, we need to study both the consumer's aspect and the producer's aspect separately.

From a consumer's point of view, she is represented by a consumer agent that captures her preferences. A consumer agent's preference representation is instrumental in defining the range of preferences it can capture. Without doubt, the agent's preference representation plays an important role in defining how the agent negotiates. We study two different preference representations:

- Conjunctive and disjunctive constraints (Section 3.1)
- Conditional Preference Networks (Section 3.2)

From a producer's point of view, the consumer's preferences are private so the producer agent does not know the consumer's preferences. However, over interactions, the producer may attempt to learn what the consumer prefers. Overall, when generating an offer, the producer agent can use three pieces of information: the possible services it can offer (based on its service repository), its own preferences (so that it will not offer unrealistic services), and the predicted model of consumer's preferences. This last information may simply contain the consumer's last request or may be a more subtle model that is built over interactions using machine learning. Ideally, the learning algorithm should collect the beneficial information for negotiating effectively regardless of the preference representation of the consumer. In the proposed approach the producer applies inductive learning, particularly RCEA.

3.1   Constraint-Based Consumer Strategy

This negotiation strategy is used by the consumer and is developed for the case that the consumer represents its user's preferences via constraints in the form of conjunctives and disjunctives. For example, if the user prefers either an apartment at Etiler or an apartment having a parking area at Kadikoy, it is represented as *(Neighborhood="Etiler")* ∨ *(Neighborhood="Kadikoy"* ∧ *Parking Area="Yes")*.

According to these preferences, it is straightforward to generate consumer's requests. First, the agent generates values according to the existing constraints and for unconstrained issues it arbitrarily chooses the values from their domain. For instance, consider that the consumer has the preference previously described. The agent chooses a constraint; for example it can choose either (Neighborhod="Etiler") or (Neighborhood="Kadikoy" ∧ Parking Area="Yes"). Assume that the former constraint is chosen. It initializes the neighborhood as Etiler. For other issues such as Parking Area and Price, it chooses values arbitrarily such as "Yes" and "Low" respectively. Consequently, the consumer requests the service *(Yes, Etiler, Low)*.

Note that as long as the constraints are fulfilled, none of the service requests is better than another: They are equally preferred by the consumer. According to this representation, a service can be either acceptable or rejectable. Moreover, an alternative service offered by the producer is accepted when it satisfies one of the constraints. If the offer does not satisfy any constraints, the consumer will reject that offer. At the end of the negotiation, the utility that the consumer gains is either one or zero. If it completes the negotiation successfully, it gains a utility of one. Otherwise, it gains zero utility.

3.2 CPnet-Based Consumer Strategy

This strategy is developed for the consumers representing their preferences in the form of CP-nets. The consumer first induces a preference graph from a given CP-net. To do this, it applies *improving flips* on the CP-net. An improving flip is changing the value of a single issue with a more desired value by using the CPT of that issue. For example, according to the given CP-net in Figure 1, Etiler is preferred rather than Kadikoy for the neighborhood. When we apply improving flip to *(Yes, Kadikoy, Low)*, we derive the following service *(Yes, Etiler, Low)*. To construct a preference graph, the agent starts with the worst choice(s). By applying improving flips, more desired services are obtained. The agent draws an edge from less desired service to more desired one. Consequently, the leaf node(s) keep the most desired service(s) and the root contains the least desired service. Figure 2 draws the induced preference graph for the given CP-net in Figure 1.

The desirability of the nodes increase from top to bottom. For intermediate nodes, we only compare the nodes having a path from others. The nodes having no path to each other cannot be compared under ceteris paribus interpretation. Since CP-nets only capture partial information, not all alternatives are comparable. However, the consumer agent needs to compare each service to be able to negotiate effectively. To achieve this, we develop a heuristic based on the idea of capturing the depth of a service node in the preference graph [1]. The depth of a service node stands for the length of the longest path from the root node to that service node.

If there is an edge from x to y, we ensure that the depth of y is higher than that of x. We expect that the depth of a service node plays a key role in measurement of the desirability for the consumer. According to this approach, the higher depth a

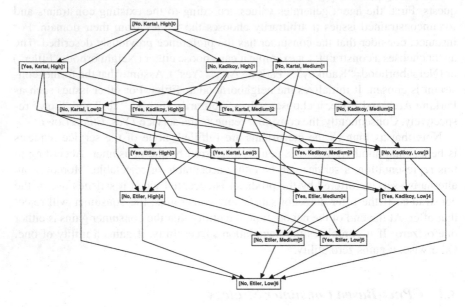

Fig. 2 Induced Preference Graph from the CP-net in Figure 1

service has, the more likely it to be preferred by the user. The services having the
same depth are equally preferable. For example, the desirability of the services *(No,
Etiler, Medium)* and *(Yes, Etiler, Low)* are the same since their depth is equal to five.
The service *(No, Etiler, Low)* is assumed to be more desirable than the previous two
services because its depth is higher than five. This approach enables the consumer
agent to compare any two services services and decide which one is more preferred.

When applying this strategy, the consumer agent first generates the most desired
service (whose depth is the highest) and then it concedes over time. Concession
is performed in a way that the depth of the current offer is equal to or less than
that of its previous request. To illustrate this, assume that our consumer agent uses
the preference graph drawn in Figure 2. The consumer first requests the service
(No, Etiler, Low). If the producer does not supply this service and the depth of the
producer's offer is less than that of the consumer's request, the consumer asks for
either *(No, Etiler, Medium)* or *(Yes, Etiler, Low)*. As seen from this example, the
consumer concedes in terms of the depth of the request over time.

Moreover, the consumer keeps the producer's best offer up to now. Consequently,
it can check if the depth of its current request is less than or equal to that of the
producer's best offer. If it is the case, it can request the producer's best offer and
take the service from the producer instead of prolonging the process unnecessarily.
If the producer's offer is equally desired or more desired than the consumer's current
request, that offer will be accepted by the consumer. Otherwise, it will be rejected
and the consumer will generate a new request.

3.3 Selfish Producer

This producer agent considers its own preferences, service repository and the consumer's last request to generate an alternative service offer during the negotiation. The producer's service repository consists of available services, which are ranked according to the producer's desirability. To generate a counter offer, the producer first selects the services having the highest rank for itself. Among these candidate services, the producer offers the most similar service to the consumer's last request. This agent is named as selfish because it first considers its own preferences and then takes the consumer's last offer into account. This agent does not attempt to learn the consumer's preferences but it assumes that the consumer's last request reflects some part of the consumer's unknown preferences.

3.4 Collaborative Producer

This producer aims to get a long term profit and it cares for the consumer's preferences as much as its own preferences. This producer assumes that if a consumer is satisfied at the end of the negotiation, it may interact with the same producer again in future. Thus, the aim is to satisfy the consumer's need as well as to satisfy its own preferences. However, the producer does not know the consumer's preferences. Therefore, the producer tries to learn the consumer's preferences. Learning a consumer's preference during the negotiation is a complicated task since the producer may not know the consumer's preference representation. The preference representation may vary for the consumers. For instance, one consumer may use utility functions whereas another may represent its preferences via constraints or CP-nets. The method of learning the consumer's preferences is deeply affected by the preference representation.

In this study, our producer agent applies an inductive learning method to learn the preferences. To achieve this, Revisable Candidate Elimination Algorithm (RCEA) is used. The consumer's current requests are considered as positive training instances where the producer's counter offer rejected by the consumer are the negative training samples. Regardless of the consumer's preference representation, the producer uses this learning method to learn some useful information.

The main components of RCEA are the most general hypothesis set (G) and the most specific hypothesis set (S). According to RCEA, since G keeps the most general hypotheses covering the positive samples, if a service offer is not covered by G, that offer will possibly rejected by the consumer. Thus, the producer does not prefer the services not covered by at least one hypothesis in G even though offering that service seems to be beneficial as far as the producer's preferences are concerned. Since S keeps minimal generalization of the consumer's all previous requests, a service that is similar to S is an ideal offer.

According to the proposed negotiation strategy, the producer selects a service that has not been offered before and whose rank is the highest for the producer among the candidates. All candidate services are tested if they are covered by at least one hypothesis in *G*. The services that are not covered by *G* are eliminated from the current candidate service list. Among the remaining candidate services, the most similar service to *S* is offered by the producer. If there are not any remaining candidate services, the producer checks for the services whose rank is the next highest for the producer. These services are also checked to see if they are covered by *G*. The same process goes on until finding a service covered by *G* or exhausting all services in the producer's service repository. In the case that none of the services in the producer's repository is covered by *G*, the producer prefers to offer an unoffered service whose rank for the producer is the highest. Let us examine this process with an example scenario. Assume that our producer has a service repository whose content is shown in Table 2 and our consumer's preference is represented as { { (Neighborhood="Etiler") } ∨ { (Parking Area="Yes") ∧ (Neighborhood="Kartal" ∨ "Kadikoy") ∧ (Price="Low") } }.

Table 2 Producer's Sample Service Repository

Service ID	[Parking, Price]	Neighborhood, Rank
Service 1	[No, Kartal, Low]	3
Service 2	[Yes, Kartal, High]	3
Service 3	[No, Kartal, Medium]	3
Service 4	[Yes, Kadikoy, High]	2
Service 5	[Yes, Etiler, High]	2
Service 6	[No, Kadikoy, High]	1
Service 7	[Yes, Kartal, Low]	1
Service 8	[No, Kadikoy, Medium]	1
Service 9	[Yes, Kartal, Medium]	1

* *Rank* corresponds the utility of the service for the producer.

Table 3 illustrates how the consumer and producer negotiate over an apartment service. According to the consumer's preferences, the consumer first requests the service *(Yes, Etiler, Low)*. The producer agent starts the learning process in a way that *G* involves the most general hypothesis *(?, ?, ?)* and the specific hypothesis is assigned with this first request *(Yes, Etiler, Low)*. Since all services in the repository are covered by *G*, the producer offers the most similar service to *S*, whose rank is the maximum value, in this case three. As a result, *(Yes, Kartal, High)* is offered by the producer. Since this offer does not satisfy the consumer's constraints, it is rejected by the consumer. Note that the rejected offers are our negative training samples. According to RCEA, the hypotheses in *G* covering this negative sample are specialized not to cover it any more. To do this, the agent compares the values of each issue in the specific hypothesis *(Yes, Etiler, Low)* covered by the current general hypothesis *(?, ?, ?)* with those in the negative example. Different values are used to specialize the hypothesis. For example, the neighborhood and the price are different. By changing the value of the neighborhood issue in the general hypothesis with that in specific hypothesis we obtain the general hypothesis *(?, Etiler, ?)*. Further, this is checked to see if it covers any negative samples. Since it does not cover any negative samples, this specialization becomes valid. At the end of the specialization process, G contains new two hypotheses: *(?, Etiler, ?)* and *(?, ?, Low)*.

Second, the consumer requests *(Yes, Kadikoy, Low)*. Among the services having the highest rank for the producer, only *(No, Kartal, Low)* is covered by G. Thus, this service is offered to the consumer. Note that, we also generalize the specific hypothesis to cover this new positive example. Current S is equal to {*(Yes, [Etiler, Kadikoy], Low)*}. Since the current offer is rejected by the consumer, the general hypotheses covering this sample are specialized minimally. Thus, the hypothesis *(?, ?, Low)* is specialized. As a result of this process, the producer generates the following hypotheses: *(?, Kadikoy, Low)* and *(Yes, ?, Low)*.

Last, the consumer requests *(Yes, Kartal, Low)*. The producer looks for a service having the highest rank that has not been offered yet. Only *(No, Kartal, Medium)* remains in the candidate list involving the most preferable services for the producer. However, this service is not covered by any hypotheses in G. Thus, the producer does not offer this service. The producer checks for the services having the second highest rank. Service 4 and Service 5 are the candidate services. Since Service 4, *(Yes, Kadikoy, High)* is not covered by G, it is also eliminated. Therefore, Service 5 is offered by the producer and the consumer accepts this service at the end.

As a result, it is seen that the current version space is consistent with the consumer's preferences. Of course, the performance of this learning algorithm depends on the order and content of the service request-offer pairs. The producer makes only one assumption about the consumer that the preference of the consumer does not change during the negotiation. This learning approach is independent from the consumer's preference structure and its negotiation strategy. It can be used when the consumer represents its preferences via CP-nets.

4 Evaluation and Experiments

In our experimentation set-up, a consumer and a producer negotiate on an apartment renting service, which is explained in Example 1. To observe the impact of our learning approach in negotiation, the same consumer agent negotiates with both the selfish producer (Section 3.3) and the collaborative producer (Section 3.4) having the same service repository with the same ranks for the services. Since the service repository and the services' ranks affect the negotiation outcome significantly, we generate 1000 difference service repositories with different ranks randomly. Each consumer agent negotiates with 1000 different producers having same negotiation strategy but different service repository including nine varied services.

Table 3 Interaction between consumer and producer

Request or offer	Parking	Neighborhood	Price
First request R_1 (+) :	Yes	Etiler	Low
Current sets: $S_0 = \{R_1\}$ $G_0 = \{(?,?,?)\}$			
First counter offer (-):	Yes	Kartal	High
Current sets: $S_1 = \{R_1\}$ $G_1=\{(?, \text{Etiler}, ?), (?, ?, \text{Low})\}$			
Second request R_2(+):	Yes	Kadikoy	Low
Current sets: $S_2 = \{(\text{Yes}, [\text{Etiler}, \text{Kadikoy}], \text{Low})\}$ $G_2=G_1$			
Second counter offer (-):	No	Kartal	Low
Current sets: $S_3 = \{R_2\}$ $G_3=\{(?, \text{Etiler}, ?), (\text{Yes}, ?, \text{Low}), (?, \text{Kadikoy}, \text{Low})\}$			
Third request R_3 (+)	Yes	Kartal	Low
Current sets: $S_4= \{(\text{Yes}, ?, \text{Low})\}$ $G_4= G_3$			
Third counter offer:	Yes	Etiler	High

To study the effect of the consumer preference representations, the same producer agent negotiates with different consumers that have varying preference representations, namely constraint-based and CPnet-based. The first five consumers use conjunctive and disjunctive constraints for their preferences whereas the next five consumers use CP-nets to represent their preferences. Our expectation is that our collaborative producer agent would negotiate with the consumer using CP-nets as well as that using constraints and we believe that learning the consumer's preferences with RCEA would expedite the negotiation process regardless of consumer's strategy and preference model.

Our first evaluation metric is the utility of the outcome for the consumer and the producer at the end of each negotiation (for a total of 1000 negotiations). Note that the utility of the outcome for the consumer using CP-nets is between zero and the maximum depth of the induced preference graph. For the consumer agent using constraints, the utility is estimated between zero and one. If the consumer completes the negotiation successfully, the consumer receives a utility of one. Otherwise, it gains zero utility. The service's rank value represents the utility for its producer and it is between one and three.

Table 4 shows the negotiation outcomes for both consumers and producers out of 1000 negotiations per consumer when the consumers' preferences are represented by conjunctive and disjunctive constraints. For the first, second and fifth preferences, all producer agents (collaborative and selfish) satisfy the consumer's need. Therefore, the average utility of the outcomes for the consumer in those cases is equal to one. For the third and forth preferences, the average utility of the outcomes for the consumer is equal to 0.80 and 0.42 respectively when the producer is collaborative whereas those are equal to 0.58 and 0.37 when the producer is selfish. To sum up, when the producer agent uses RCEA to learn the consumer's preferences and generates its counter offer accordingly (collaborative), the average utility of the outcome for the consumer is up to approximately 1.4 times higher than that in the case when the producer takes care first its own preferences, then the consumer's last offer (selfish). According to these results, it can be said that when the producer is collaborative, the consumer gains more or at worst equally compared to the case when the producer is selfish.

From the point of view of the producer, when the producer is collaborative and the consumer's average utility is higher with respect to the case when the producer is selfish (see the average utilities for the preference 3 and preference 4), the producer's gain is also higher. This stems from the fact that in some cases only the collaborative producer satisfies the consumer's needs because it only considers the learned preferences whereas the selfish producer also considers its own preferences. When this is the case, the collaborative agent obtains the utility of the service that has been accepted by the consumer whereas the selfish agent gains zero utility because the participants do not come to an agreement. According to the other preference profiles, the average utility of outcomes for both producers is approximately the same.

Table 4 Outcomes When Constraint-based Strategy

Preference	Avg. utility for consumer		Avg. utility for producer	
	Selfish	Collaborative	Selfish	Collaborative
Preference-1	1.0	1.0	2.77	2.77
Preference-2	1.0	1.0	2.88	2.86
Preference-3	0.58	0.80	1.61	1.96
Preference-4	0.37	0.42	1.07	1.16
Preference-5	1.0	1.0	2.86	2.86

Since the consumer's preferences are in the form conjunctives and disjunctives and RCEA is designed for learning such kind of preferences, the increase in the average utility for the consumer when the producer is collaborative is an expected result. The main matter is whether the average utility for the consumer will increase by learning consumer's preferences with RCEA when the consumer represents its preferences with CP-nets.

Table 5 shows the overall negotiation outcomes for both participants when the consumer uses CPnet-based strategy. As far as the consumer's gain is concerned, when the producer is collaborative, the consumer gains up to 8 per cent more than that the case when the producer is selfish. On the other hand, from the point of view of the producer, the selfish agent earns up to 11 per cent more than the collaborative agent. There is a tradeoff between satisfying the consumer's needs well and gaining ground itself. The producer may wish a long term profit, which means if the consumer is satisfied, it may have a tendency to carry on business with the same producer. Thus, the customer satisfaction is important for long-term businesses.

Table 5 Outcomes When CP-Net Based Strategy

Preference	Avg. utility for consumer		Avg. utility for producer	
	Selfish	Collaborative	Selfish	Collaborative
Preference-6	4.85	5.23	2.57	2.32
Preference-7	5.62	5.89	2.56	2.38
Preference-8	4.86	5.15	2.56	2.33
Preference-9	4.06	4.24	2.58	2.32
Preference-10	5.90	6.16	2.55	2.35

To evaluate and compare the performances of the producer strategies over 5000 negotiations in terms of the utility of the outcomes for the consumer using CP-nets, we also use distribution functions called performance profiles [4]. Figure 3 shows that in 96 per cent of the negotiations collaborative producer's negotiation strategy does as well as the best possible strategy in terms of the utility of the outcome for the consumer while selfish producer's strategy does as well as the best possible strategy in 78 per cent of the negotiation.

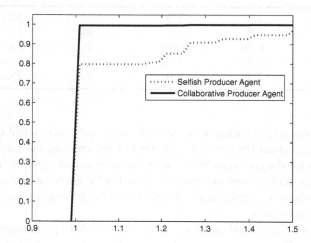

Fig. 3 Performance Profile for the Utility of Outcome

Our next evaluation criterion is the number of interactions between agents to complete a negotiation session. It is desirable to complete negotiation as early as possible. Table 6 shows the average number of interactions between agents to reach a consensus when the consumers preferences are in the form of conjunctives

Table 6 Average Number of Interactions When Constraint Based Strategy

Preference	Selfish	Collaborative
Preference-1	1.61	1.33
Preference-2	1.37	1.30
Preference-3	2.59	2.24
Preference-4	2.40	2.35
Preference-5	1.31	1.19

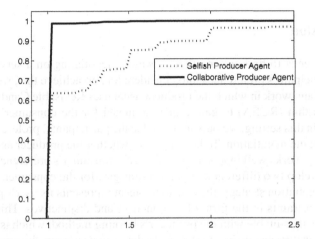

Fig. 4 Performance Profile for # of Interactions

and disjunctives. The difference is most significant in Preference 1. Overall, we see that when the producer is collaborative, the agents negotiate up to 21 per cent faster than those do when the producer is selfish. As a result, the collaborative producer negotiates faster when the consumer uses constraints for its preferences.

Table 7 shows the number of the interactions between agents to complete the negotiation process when the consumers represent their preferences with CP-nets. The collaborative producer negotiates up to 30 per cent faster than the selfish producer. According to the performance profile in Figure 4, in 97 per cent of the negotiations collaborative producer negotiates as fast as the fastest possible agent while selfish producer negotiates as fast as the fastest possible agent in 64 per cent of the negotiations when the consumer uses CP-nets to represent its preferences. The results support our expectation that the agents negotiate faster when the producer learns the consumer's preferences with RCEA and generates offers accordingly (when the producer is collaborative) for both preference representations.

Table 7 Average Number of Interactions When CP-net Based Strategy

Preference	Selfish	Collaborative
Preference-1	3.24	2.5
Preference-2	3.29	2.69
Preference-3	3.24	2.69
Preference-4	3.22	2.56
Preference-5	3.18	2.67

5 Discussion

This study intends to find out an efficient way of negotiating on a service by taking other participant's preferences into consideration. To achieve this, we propose a negotiation framework in which the producer agent uses Revisable Candidate Elimination Algorithm (RCEA) to gain a generic model for the consumer's unknown preferences. In this setting, we only assume that the participants' preferences do not change during the negotiation. To be able to see whether the producer agent's negotiation strategy works well independently from the consumer's preference representation, we develop two different negotiation strategies for the consumer. According to the first negotiation strategy, the consumer agent represents its user's preferences with some constraints in the form of conjunctives and disjunctives. This representation is perfectly suitable with the producer's learning method, which is capable of learning conjunctive and disjunctive rules. In the second strategy, the consumer uses CP-nets for its preferences. This is different from the first one in many ways. First, it keeps a partial preference ordering instead of strict constraints. Second, the consumer also needs to develop some heuristic to be able to negotiate with these preferences. Last, a service that has not been accepted before may become acceptable over time since there is a preference ordering instead of strict constraints. Although the proposed learning algorithm is not designed for learning a preference ordering, it is shown that it learns generic information, which also enables the negotiation faster. Our experimental results also support our expectation that if the producer uses RCEA to learn the consumer's preferences and generate service offers accordingly (when the producer is collaborative), the negotiation ends early. Also, the gain of the consumer is increased so that the consumer may have a tendency to interact with the same producer again rather than another producer.

Learning opponent agent's preferences and modeling of these preferences during the negotiation is a complicated task. Faratin, Sierra and Jennings propose not to model opponent's preferences; instead, to reach agreements with a trade-off strategy by using some heuristics in order to increase the social welfare [5]. By means of trade-off, an agent may pay higher price in order to gain an earlier delivery time. Agents use their own previous proposals and opponents' last offer in order to generate a new offer to their opponents. This newly generated offer should have the same utility value with its previous counter-offer and it is closer to the opponent's last offer. Their heuristic model for trade-offs includes fuzzy similarity estimation and a hill-climbing exploration for possibly acceptable offers. Although we address a similar problem, we use inductive learning for the producer to learn consumer's preferences and revise producer's counter-offers by using these learned preferences. Further, we do not consider only the consumer's last offer but also previous offers.

Luo *et al.* propose a prioritized fuzzy constraint based negotiation model in which the buyer agent specifies its offers by using constraints [9]. According to the proposed model, the agent reveals its partial preference information (some parts of its constraints) in a minimal fashion. That is, the buyer agent sends a constraints to the seller. The seller agent checks the constraints and if it cannot satisfy this constraints, it requests the buyer to relax its constraints. In contrast, all offers in our negotiation model represent a single service, a vector of value rather than constraints. In our model, the constraints in the form of conjunctives and disjunctives are used for representing consumer's preferences. Moreover, the consumer does not reveal its partial preferences. The challenge in our study is to find out a model for consumer's private preferences.

Tykhonov and Hindriks purpose to take the opponent agent's preferences into account to improve the negotiation process [7]. Since opponent's preferences are not known by the agent, a Bayesian learning is applied to learn the opponent's preferences from bid exchanges. They make some assumptions about preference structure and opponent's negotiation strategy. The preferences are represented as a weighted sum of utility functions (linearly additive functions). Moreover, Tykhonov and Hindriks assume issues to be independent. In our study, we do not make any assumption about the consumer's preference model or its negotiation strategy. We only assume that the consumer's preferences do not change during the negotiation. Further, our learning algorithm is not specific to learn a particular preference representation. We test our learning algorithm with two different preference representations and the results show that learning with RCEA enables the negotiation process faster in both cases.

Jonker, Robu and Treur propose a negotiation architecture in which two agents negotiate over multiple attributes under incomplete preference information [8]. After determining the concession step, the agent estimates a target utility for its current bid. This target utility is distributed over the attributes. In the case of determination of the attribute values, the agent does not only consider its own preference weights but also opponent's preference weights. Thus, the agent tries to predict opponent's unknown preferences. In our study, we focus on learning a general preference model to decide which services may possibly be rejected by the consumer and which services are more likely to be acceptable for the consumer. Contrary to that study, we do not take the preference weights into account but it would be interesting to extend our approach with preference weights. Further, we deal with the case where the agent has partial preference information about its own users whereas in their study the agent has partial preference information about its opponent's preferences and the agent tries to predict opponent's unknown preference parameters.

6 Conclusion

This study pursues learning a generic model of the opponent agent's preferences during negotiation without considering the opponent's preference representation. To achieve this, we suggest to use a concept learning algorithm, namely RCEA. The fact that such kind of learning process enables the negotiation faster, is supported by the experimental results.

As a future work, we plan to develop a consumer agent whose preferences are represented by utility functions and to set up a negotiation framework in which our collaborative producer agent applying RCEA learning algorithm negotiates with that agent. Thus, we can observe whether our proposed approach also works well with utility-based preference models.

Acknowledgements. This research is supported by Boğaziçi University Research Fund under grant BAP09A106P and Turkish Research and Technology Council CAREER Award under grant 105E073.

References

1. Aydoğan, R., Taşdemir, N., Yolum, P.: Reasoning and Negotiating with Complex Preferences Using CP-nets. In: Ketter, W., La Poutré, H., Sadeh, N., Shehory, O., Walsh, W. (eds.) AMEC 2008. LNBIP, vol. 44, pp. 15–28. Springer, Heidelberg (2010)
2. Aydoğan, R., Yolum, P.: Learning opponent s preferences for effective negotiation: an approach based on concept learning. Autonomous Agents and Multi-Agent Systems (in Press)
3. Boutilier, C., Brafman, R.I., Domshlak, C., Hoos, H.H., Poole, D.: Cp-nets: A tool for representing and reasoning with conditional ceteris paribus preference statements. J. Artif. Intell. Res. (JAIR) 21, 135–191 (2004)
4. Dolan, E.D., Moré, J.J.: Benchmarking optimization software with performance profiles. Mathematical Programming 91, 201–213 (2002)
5. Faratin, P., Sierra, C., Jennings, N.R.: Using similarity criteria to make issue trade-offs in automated negotiations. Artificial Intelligence 142, 205–237 (2002)
6. Hansson, S.O.: What is ceteris paribus preference? Journal of Philosophical Logic 25(3), 307–332 (1996)
7. Hindriks, K., Tykhonov, D.: Opponent modelling in automated multi-issue negotiation using bayesian learning. In: 7th International Joint Conference on Autonomous Agents and Multiagent Systems (AAMAS), pp. 331–338 (2008)
8. Jonker, C.M., Robu, V., Treur, J.: An agent architecture for multi-attribute negotiation using incomplete preference information. Autonomous Agents and Multi-Agent Systems 15(2), 221–252 (2007)
9. Luo, X., Jennings, N.R., Shadbolt, N., Fung Leung, H., Man Lee, J.H.: A fuzzy constraint based model for bilateral, multi-issue negotiations in semi-competitive environments. Artifical Intelligence 148(1-2), 53–102 (2003)
10. Mitchell, T.M.: Machine Learning. McGraw-Hill, New York (1997)
11. Raiffa, H.: The Art and Science of Negotiation. Harvard University Press, Cambridge (1982)

Bilateral Single-Issue Negotiation Model Considering Nonlinear Utility and Time Constraint

Fenghui Ren, Minjie Zhang and John Fulcher

Abstract. Bilateral single-issue negotiation is studied a lot by researchers as a fundamental research issue in agent negotiation. During a negotiation with time constraint, a negotiation decision function is usually predefined by negotiators to express their expectations on negotiation outcomes in different rounds. By combining the negotiation decision function with negotiators' utility functions, offers can be generated accurately and efficiently to satisfy negotiators expectations in each round. However, such a negotiation procedure may not work well when negotiators' utility functions are nonlinear. For example, if negotiators' utility functions are non-monotonic, negotiators may find several offers that come with the same utility; and if negotiators' utility functions are discrete, negotiators may not find an offer to satisfy their expected utility exactly. In order to solve such a problem caused by nonlinear utility functions, we propose a novel negotiation approach in this paper. Firstly, a 3D model is introduced to illustrate the relationships among utility functions, time constraints and counter-offers. Then two negotiation mechanisms are proposed to handle two types of nonlinear utility functions respectively, ie. a multiple offers mechanism is introduced to handle non-monotonic utility functions, and an approximating offer mechanism is introduced to handle discrete utility functions. Lastly, a combined negotiation mechanism is proposed to handle nonlinear utility functions in general situations. The experimental results demonstrate the success of the proposed approach. By employing the proposed approach, negotiators with nonlinear utility functions can also perform negotiations efficiently.

1 Introduction

Agent negotiation is one of the most significant research issues in multi-agent systems (MASs). Lots of previous work has been done to solve the challenges in agent

Fenghui Ren · Minjie Zhang · John Fulcher
School of Computer Science and Software Engineering, University of Wollongong, Australia
e-mail: {fr510,minjie,john}@uow.edu.au

T. Ito et al. (Eds.): New Trends in Agent-Based Complex Automated Negotiations, SCI 383, pp. 21–37.
springerlink.com   © Springer-Verlag Berlin Heidelberg 2012

negotiation. To list a few of them, Narayanan et al. [9, 10] adopted a Markov chain framework to model bilateral negotiation and employed Bayesian learning to enable agents to learn an optimal strategy in incomplete information settings. Fatima et al. [3] investigated the negotiation outcomes in incomplete information settings through the comparison of the difference between two agent's negotiation deadlines, and proposed an agenda-based framework to help self-interested agents to maximize their utilities. Brzostowski et al. [1] proposed an approach to predict the opponent's behaviors based only on the historical offers of the current negotiation. They claimed that time and imitation are two main factors which influence an agent's behaviors during negotiation. However, most existing approaches are based on an assumption that all negotiators employ monotonic continuous linear utility functions, and not much work has been done on nonlinear utility functions, i.e. non-monotonic, and/or discrete nonlinear utility functions. According to our studies, agents may also employ a nonlinear utility function in many real-world negotiations. For example, as shown in Figure 1, in a scheduling problem for task allocation, an employee feels happy to be assigned work between $9AM - 12AM$ and $1PM - 3PM$, but feels unhappy to work hard during the first one or last two hours of a day. The employee's temper in a working day is a non-monotonic function. In Figure 2, a potential car purchaser may have different preferences for a car's color. Because each model of car only has limited colors, so the car purchaser's preferences on a car's color is a discrete function.

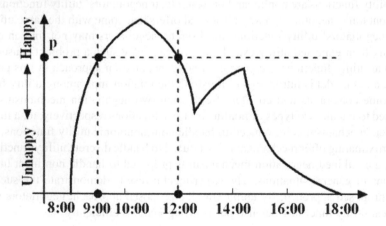

Fig. 1 An employee's temper for a day

In a negotiation with time constraint, an agent usually defines a negotiation strategy to make concessions throughout a negotiation. Firstly, according to the negotiation strategy, agents can calculate the possible maximal utility they can gain at a certain moment. Then, agents can find a particular offer for the expected utility according to their utility functions. Because most negotiation models assume that agents employ monotonic continuous utility functions, a particular offer can always be found to satisfy agents expected utilities at any negotiation round. However, when agents employ non-monotonic and/or discrete utility functions, it cannot be guaranteed that a particular offer will be found. For example, when the utility function is non-monotonic, agents may have multiple options on offers in order to reach the expected utility. As shown in Figure 1, in order to ensure that an employee's happiness is p ($p \in [0, 1]$), a job can be assigned to the employee at either 9AM or 12AM. Also, when the utility function is discrete, an agent perhaps cannot find an offer to satisfy the expected utility exactly. As shown in Figure 2, none of the available colors can make a car purchaser's happiness equal q ($q \in [0, 1]$) exactly.

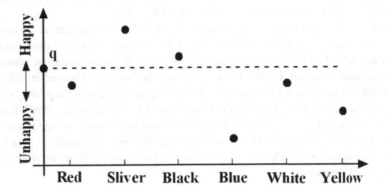

Fig. 2 A customer's flavour on a car's color

In order to solve the offer generation problem when agents employ nonlinear utility functions in time constraint negotiations, we propose a novel negotiation model in this paper, which contains two mechanisms to handle the situation when agents have non-monotonic and/or the discrete utility functions, respectively. For the non-monotonic utility functions, a multiple offer mechanism will be introduced to allow agents to generate equivalent offers in a single negotiation round. For the discrete utility functions, an approximating offer mechanism will be introduced to allow agents to generate an approximating offer. Eventually, the two mechanisms will be combined together to solve general situations in nonlinear utility functions.

The rest of this paper is organized as follows. Section 2 briefly introduces a general bilateral single issue negotiation model with linear utility functions. Section 3 introduces a 3D negotiation model and two negotiation mechanisms to handle nonlinear utility functions in bilateral single issue negotiation. Section 4 extends Rubinstein's alternating offers protocol based on the proposed negotiation model and mechanisms. Section 5 demonstrates a negotiation proceeding between two nonlinear utility agents by employing the proposed mechanisms. Section 6 compares our work with some related work on nonlinear utility agents. Section 5 concludes this paper and explores our future work.

2   A General Negotiation Model for Linear Utility Function

Before we introduce the negotiation model for nonlinear utility functions, we would like to briefly introduce a general negotiation model for linear utility functions by considering time constraints. A general bilateral single issue negotiation is performed between two agents on a good's price. Let b denote the buyer, s denote the seller, and let $[IP^a, RP^a]$ denote the range of values for price that are acceptable to Agent a, where $a \in \{b, s\}$. Usually, for the buyer $IP^b \leq RP^b$, and for the seller $IP^s \geq RP^a$. We use $\hat{a}$ to denote Agent a's opponent. Obviously, an agreement can be reached between Agent b and Agent s only when there is an intersection between their price ranges. If the buyer's reserved price is smaller than the seller's reserved price $(RP^b < RP^s)$, an agreement usually can not be achieved.

Usually, each agent has a negotiation deadline. If an agreement cannot be achieved before the deadline, then the agent has to quit the negotiation, and the negotiation fails. Let T^a denote Agent a's deadline. A negotiation can be started by either the buyer or seller. During the negotiation, the buyer and the seller will send alternating offers to each other until both sides agree on an offer together, or one side quits the negotiation. This negotiation protocol is known as the alternating offers protocol [11]. Let $p^t_{\hat{a} \to a}$ denote the price offered by Agent $\hat{a}$ to Agent a at time t. Once Agent a receives the offer, it will map the offer $p^t_{\hat{a} \to a}$ to a utility value based on its utility function U^a. A general formula of a linear utility function for Agent a is defined in Equation 1.

$$U^a(p^t_{\hat{a} \to a}) = \frac{p^t_{\hat{a} \to a} - RP^a}{IP^a - RP^a} \tag{1}$$

Usually, a utility value is a real number between $[0, 1]$, and it indicates an agent's profit for a given offer. In general, for Agent a, if the value of U^a for $p^t_{\hat{a} \to a}$ at time t is greater than the value of the counter-offer Agent a is ready to send in the next time period, t', ie., $U^a(p^t_{\hat{a} \to a}) \geq U^a(p^{t'}_{a \to \hat{a}})$ for $t' = t + 1$, then Agent a accepts the offer at time t and the negotiation ends successfully in the agreement $p^t_{\hat{a} \to a}$. Otherwise, a counter-offer $p^{t'}_{a \to \hat{a}}$ will be sent by Agent a to Agent $\hat{a}$ in the next round, t'. Thus, the action, A^a, that Agent a takes at time t, in response to the offer $p^t_{\hat{a} \to a}$ is defined as:

$$A^a(p^t_{\hat{a} \to a}) = \begin{cases} Quit & \text{if } t > T^a, \\ Accept & \text{if } U^a(p^t_{\hat{a} \to a}) \geq U^a(p^{t'}_{a \to \hat{a}}), \\ Offer\ p^{t'}_{a \to \hat{a}} & \text{otherwise.} \end{cases} \tag{2}$$

If Agent a does not accept the price $p^t_{\hat{a} \to a}$ at time t and $t < T^a$, then it will send a counter-offer at time t' to Agent $\hat{a}$ as a response. Usually, agents may employ different negotiation tactics [2] based on different criteria, such as time, resources and previous offers and counter offers. The time-dependent tactic is the most popular criteria when agents generate their counter-offers by considering time. In this tactic, the predominant factor used to decide which value to offer in time t and vary the value of offer is depending on t and the deadline T^a. Usually, the counter-offer made by Agent a to Agent $\hat{a}$ at time t $(0 \leq t \leq T^a)$ is modeled as a function ϕ^a depending on time as follows:

$$
p_{a \to \hat{a}}^t = \begin{cases} IP^a + \phi^a(t)(RP^a - IP^a) & \text{for } a = b, \\ RP^a + (1 - \phi^a(t))(IP^a - RP^a) & \text{for } a = s. \end{cases} \tag{3}
$$

where function $\phi^a(t)$ is called the Negotiation Decision Function (NDF) [2] and is defined as follows:

$$
\phi^a(t) = k^a + (1 - k^a)\left(\frac{t}{T^a}\right)^{1/\psi} \tag{4}
$$

where k^a is the initial utility and $k^a \in [0, 1]$. If $k^a = 0$, then at the beginning of a negotiation the initial price IP^a will be delivered, and when the deadline is reached the reserved price RP^a will be delivered.

Theoretically, the NDF has an infinite number of possible tactics when ψ ($\psi \geq 0$) has different values. However, in Figure 3 three extreme cases show clearly different patterns of behaviors for ψ located in different ranges.

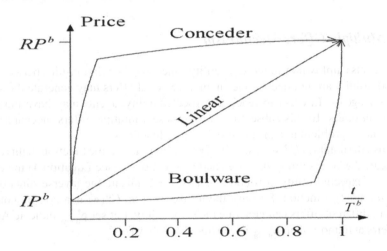

Fig. 3 Negotiation decision functions for the buyer [3]

- Conceder: When $\psi > 1$, the rate of change in the slope is decreasing. It represents a negotiation behavior where the agent will give larger concessions during the early stages of the negotiation, but smaller concessions during the later stages of the negotiation.
- Linear: When $\psi = 1$, the rate of change in the slope is zero. It represents a negotiation behavior where the agent will give a constant concession throughout the negotiation.

- Boulware: When $0 < \psi < 1$, the rate of change in the slope is increasing. It represents a negotiation behavior where the agent will give smaller concession during the early stages of the negotiation, but larger concessions during the later stages of the negotiation.

3   Negotiation Model for Nonlinear Utility Function

In the previous section, we briefly introduced a general bilateral single-issue negotiation model for linear utility functions. However, such a model is based on an assumption that all negotiators employ linear utility functions, i.e. Equation 1 is monotonic and continuous. Therefore, during the negotiation, a unique offer can be generated to satisfy agents expectations on utility in each round. However, if the utility functions employed by agents are non-monotonic and/or discrete, this general model cannot guarantee to produce correct offer/s to satisfy agents expectations. In this section, we introduce two negotiation mechanisms for non-monotonic utility functions, and discrete utility functions, respectively. Finally, the two mechanisms are combined together to handle nonlinear utility functions in general situations.

3.1   Multiple Offers Mechanism

When agents employ non-monotonic utility functions, the relationships between offers and utilities are not one-to-one anymore. Several offers may generate the same utility for agents. In order to reach an expected utility, agents may have multiple options on offers. In this subsection, we propose a multiple offers mechanism to handle such a problem in non-monotonic utility functions.

We use the notation $U^a(\bullet)$, where $U^a(\bullet)$ is a non-monotonic function in this case, to indicate Agent a's utility function, and notation $\phi^a(t)$ (see Equation 4) indicates Agent a's expected utility at round t. Let $U_a^{-1}(\bullet)$ indicate the inverse function of Agent a's utility function. Because the utility function $U^a(\bullet)$ is a non-monotonic function, different offers can generate the same utility. Let set $\mathbf{p}_{a \rightarrow \hat{a}}^t$ indicate Agent a's offer/s at round t, then $\mathbf{p}_{a \rightarrow \hat{a}}^t$ is generated as follows:

$$\mathbf{p}_{a \rightarrow \hat{a}}^t = U_a^{-1}(\phi^a(t)) \tag{5}$$

where $\phi^a(t)$ is Agent a's expected utility at round t, and defined in Equation 4.

In Figure 4, we introduce a 3D model to illustrate the relationships among negotiation time, utility and negotiated issue. Let the x-axis indicate the negotiation time, y-axis indicate negotiated issue, and z-axis indicate the utility. As shown in Figure 4, Line X on Plane A is the negotiation decision function, Line Y on Plane B is the utility function. Both the negotiation decision function and the utility function are known by Agent a. This 3D model indicates how to produce the offer generation function (Line Z on Plane C) based on these two known functions when the utility function is non-monotonic. For example, in round t, according to the negotiation decision function, Agent a expects a utility of 0.8, and according to the utility function (Line Y), two points (Points m, n) can reach the expectation. Then, by employing Equation 5, two points (Points b, p) can be found as Agent a's offers. Finally, Agent a will generate two counter-offers (Points e, f) at round t. By employing such a multiple offer mechanism, Agent a's counter-offer generation functions can be drawn on Surface C.

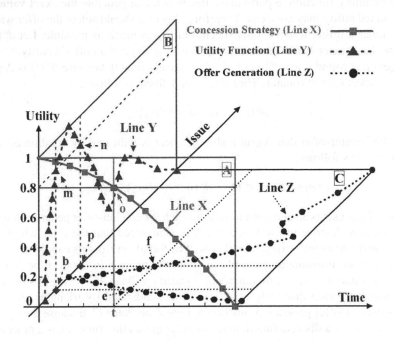

Fig. 4 Negotiation model for non-monotonic continuous utility function

Based on the above example, the counter-offer generation function by considering the non-monotonic utility function is defined by the following three steps.

Step 1:   According to the agents' negotiation decision function, agents calculate their expected utilities at negotiation round t;

Step 2:   According to the agents' utility function, agents discover offers which can bring the expected utility exactly;

Step 3:   Agents send the found offers at round t as counter-offers.

3.2   Approximating Offer Mechanism

In some situations, agents may employ discrete utility functions. Because the number of possible offers is limited to discrete functions, agents may not find offers to satisfy their expected utilities exactly. In order to handle such a problem in discrete utility functions, we propose an approximating offer mechanism in this subsection.

Let set $\mathbf{O} = \{o_i | i \in [1, I]\}$ be the all possible offers that an agent can choose during a negotiation, and notation $U^a(o_i)$ be the utility of Offer o_i. At negotiation round t, Agent a firstly employs the negotiation decision function to calculate its expected utility, i.e. $\phi^a(t)$ (see Equation 4). The value of $\phi^a(t)$ indicates the maximal utility that Agent a can gain in the remaining negotiation rounds. Because Agent a employ a discrete utility function, a particular offer which can generate the exact value as the expected utility may not exist. Therefore, Agent a should select the offer which can generate a utility to close the expected utility as much as possible. Let $d_t^a(o_i)$ indicate the distance between the expected utility $\phi^a(t)$ and an offer's utility $U^a(o_i)$ for Agent a at round t, and $d_t^a(o_i)$ should be greater than 0, because $\phi^a(t)$ is Agent a's best expectation at round t. Then $d_t^a(o_i)$ is defined as follows.

$$d_t^a(o_i) = \phi^a(t) - U^a(o_i) \tag{6}$$

Then, the counter-offer that Agent a should select is indicated by notation o_c, and o_c is decided as follows.

$$o_c = \forall o_i \in \mathbf{O}, \exists o_c \in \mathbf{O} \Rightarrow d_t^a(o_c) < d_t^a(o_i) \cap d_t^a(o_c) \geq 0 \tag{7}$$

Equation 7 represents an approximating approach for counter-offer generation when agents employ discrete utility functions. It is guaranteed that the offer which can mostly satisfy an agent's expected utility will be found out from its candidate offers. In Figure 5, we illustrate an example to show the process in detail. Line X is an agent's negotiation decision function, and located on Plane A. The agent's possible offers are distributed discretely on Plane B, and indicated by triangles. And the agent's counter-offer generation function is Line Z on Plane C. Because the agent's utility function is a discrete function, so the offer generation function is a piecewise function. For example, as shown in Figure 5, at round t, an agent's expected utility is 0.8. However, none of the agent's offers can bring 0.8 utility to the agent exactly. Then, by employing the approximating offer mechanism, Point h is found as the closest utility to 0.8, and Point q is selected as the counter-offer. Finally, the agent sends the counter-offer at round t, which is indicated by Point d on Line Z.

Based on the above example, the counter-offer generation function by considering the discrete feature of utility functions is defined by the following three steps.

Step 1: According to the agents' negotiation decision function, agents calculate their expected utilities at the negotiation round t;

Step 2: According to the agents' utility function, agents find out the offer which can reach the expected utility as much as possible from their offer candidates;

Step 3: Agents send the found offer at round t as counter-offers.

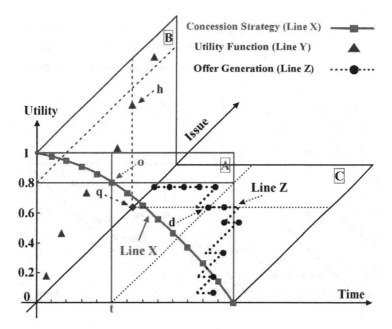

Fig. 5 Negotiation model for discrete utility function

3.3 Combined Mechanism

Subsection 3.1 proposed a multiple offers mechanism for agents employing non-monotonic utility functions, and subsection 3.2 proposed an approximating offer mechanism for agents employing discrete utility functions. In a more general situation, agents' utility functions may be a mixture of non-monotonic and discrete functions.

For example, as shown in Figure 6, an agent employs a non-monotonic and discrete utility function. At round t, the agent's expected utility is indicated by Point o. However, according to the agent's utility function, none of the agent's offers can satisfy its expectation exactly. So the agent should employ the approximating offer mechanism to find the closest offer/s to its expectation. Because the agent's utility function is non-monotonic, the two closest utilities (Points m, n) to the agent's expectation are found. So by employing the multiple offers mechanism, two counter-offers (Points e, f) are sent at round t.

Let set $\mathbf{O} = \{o_i | i \in [1, I]\}$ be all possible offers that an agent can choose during a negotiation, set $\mathbf{O_c} = \{o_j^c | j \in [1, J]\}$ contains Agent a's all possible counter-offers at round t, and $\mathbf{O_c} \subseteq \mathbf{O}$. Then the counter-offer o_j^c is selected as follows.

$$o_j^c = \forall o_i \in \mathbf{O}, \exists o_j^c \in \mathbf{O} \Rightarrow d_t^a(o_j^c) < d_t^a(o_i) \cap d_t^a(o_j^c) \geq 0 \qquad (8)$$

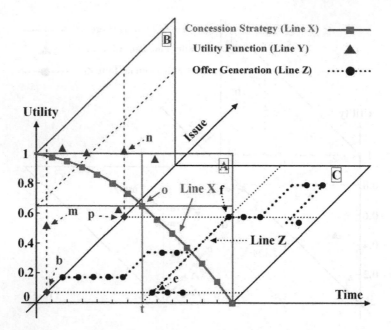

Fig. 6 Negotiation model for general nonlinear utility function

and

$$\forall o_k^c \in \mathbf{O_c}, \forall o_j^c \in \mathbf{O_c}, k \neq j \Rightarrow d_t^a(o_k^c) = d_t^a(o_j^c) \qquad (9)$$

where function $d_t^a(\bullet)$ is defined in Equation 6.

Based on the above example, the counter-offer generation function by considering both the non-monotonic and discrete utility function is defined by the following three steps.

Step 1: According to the agents' negotiation decision functions, agents calculate their expected utilities at the negotiation round t;

Step 2: According to the agents' utility functions, agents find out the counter-offer/s which can reach the expected utilities as much as possible from their offer candidates;

Step 3: Agents send the found counter-offer/s at round t as counter-offers.

4   Protocol and Equilibrium

In this subsection, a negotiation protocol for agents with nonlinear utility functions is proposed based on Rubinstein's alternating offers protocol [2].

Step 1: A negotiator assigns negotiation parameters before a negotiation starts, i.e., the initial offer, reservation offer, utility function, negotiation deadline, and negotiation decision function.

Step 2: The negotiator calculates its best expected offer according to the negotiation decision by considering time constraints, and sends counter-offer/s to its opponent.

Step 3: If the opponent accepts any counter-offer, then the negotiation is completed. Otherwise, the opponent will send back an offer/s, and wait for the negotiator's response. If the current negotiation round is the deadline, the procedure goes to Step 4. Otherwise, the procedure goes to Step 5.

Step 4: The negotiator will evaluate the opponent's offer/s. If the best evaluation result among the opponent's offer/s can bring any benefit to the negotiator (i.e. the biggest utility is greater than zero), then the negotiator will accept the best offer from the opponent, and the negotiation succeeds with an agreement. If none of the opponent's offer/s can bring any benefit to the negotiator (i.e. the biggest utility is less than zero), then the negotiator will reject all the opponent's offer/s, and the negotiation fails.

Step 5: The negotiator calculates its best expected utility for the next negotiation round based on its negotiation decision function. If the best evaluation result among the opponent's offer/s can reach or exceed the negotiator's expectation, then the negotiator will accept the best offer from the opponent, and the negotiation succeeds with an agreement. If none of the opponent's offer/s can reach the negotiator's expectation, then the negotiator will calculate its counter-offer/s and send it/them to the opponent. The negotiation procedure goes back to Step 3.

Based on the above protocol, the negotiator a's action at round t is defined as follows :

$$
Act^a(t) = \begin{cases}
\textbf{Quit}, when \ t = \tau^a \wedge \max(U^a(\mathbf{p^t_{\hat{a} \to a}})) < 0, \\[2mm]
\textbf{Accept} \ p^{*t}_{\hat{a} \to a}, when \ t \leq \tau^a \wedge U^a(p^{*t}_{\hat{a} \to a}) \geq \phi^a(t), \\[2mm]
\textbf{Offer} \ \mathbf{p^t_{a \to \hat{a}}}, when \ t < \tau^a \wedge U^a(p^{*t}_{\hat{a} \to a}) < \phi^a(t).
\end{cases}
\tag{10}
$$

where τ^a is the negotiator a's negotiation deadline, $U^a(\bullet)$ is the negotiator a's utility function, set $\mathbf{p^t_{\hat{a} \to a}}$ contains all offers from the opponent $\hat{a}$ to the negotiator a, $p^{*t}_{\hat{a} \to a}$ is the best offer from the opponent ($p^{*t}_{\hat{a} \to a} \in \mathbf{p^t_{\hat{a} \to a}}$, $U^a(p^{*t}_{\hat{a} \to a}) = \max(U^a(\mathbf{p^t_{\hat{a} \to a}}))$), and $\mathbf{p^t_{a \to \hat{a}}}$ (see Equation 5) contains all counter-offer/s from the negotiator a to the opponent $\hat{a}$ at round t.

5 Experiment

In this section, we demonstrate the negotiation procedure by employing the proposed negotiation approach between two agents with nonlinear utility functions.

5.1   Setting

A female patient wants to make an appointment with her dentist. The patient usually needs to deliver her children to school in the morning (9 : 00 - 9 : 30) and picks up them from school in the afternoon (16 : 00 - 16 : 30). So she prefers not to make an appointment during school hours. However, if her dentist's timetable is full, and she has no other options, she will also accept an appointment during the school hours, and ask her neighbor to give a lift to her children. Also, she prefers to avoid lunch time (12 : 30-1 : 30), and the time during her favourite TV program (14 : 00 - 14 : 30). Therefore, this patient's preference for an appointment time is a nonlinear function, which is illustrated in Figure 7. During the negotiation, the patient is represented by Agent p.

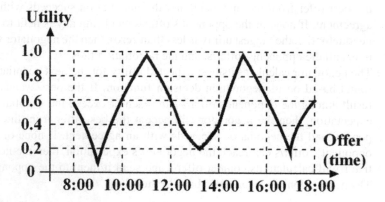

Fig. 7 The patient's preference on her timetable

On the other hand, the dentist's working hours are between 8 : 00 and 18 : 00. But he tries to avoid appointments in the early morning (8 : 00 - 9 : 00), lunch time (13 : 00 - 14 : 00), and late afternoon (17 : 30 - 18 : 00). Also, he already had two appointments at 10 : 00 - 11 : 00 and 14 : 00 - 16 : 00. Therefore, the dentist's preference for the appointment time is a nonlinear and discrete function, which is illustrated in Figure 8. During the negotiation, the dentist is represented by Agent d.

In order to simply the negotiation procedure, both Agents p and d employ linear negotiation decision functions (the correctness of the proposed mechanism is independent on the selection of the negotiation decision function), and set negotiation deadline to the $10th$ round. By employing the proposed multiple offers mechanism and the approximating offer mechanism, Agents p and d can efficiently generate and exchange their offers during the negotiation. Both the patient's and the dentist's offers in each negotiation round are generated and displayed in

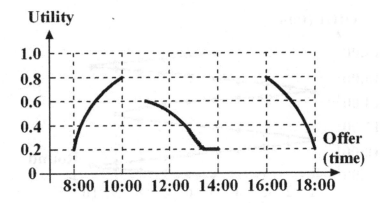

Fig. 8 The dentist's preference on his timetable

Figures 9 and 10, respectively. It can be seen that because both negotiators' utility functions are non-monotonic, they send multiple offers in each negotiation round. The detailed negotiation procedure is displayed in Figure 11. In Figure 11, the x-axis indicates the patient's utility, and the y-axis indicates the dentist's utility. The solid line indicates the dentist's offers in each round, and the negotiation round is marked by the Greek numerals ($I, II, III, ...$). The broken line indicates the patient's offers, and the negotiation round is marked by the Roman numerals ($1, 2, 3, ...$). For example, according to the patient's offer generation function (Figure 9), the patient sends two offers ($11 : 00$ and $15 : 00$) in the first negotiation round. These two offers bring the same utility (i.e. utility equals 1.0) to the patient, but different utilities to the dentist. According to the dentist's utility function (Figure 8), the first offer ($11 : 00$) can bring him a utility value as 0.6, while the utility value of the second offer ($15 : 00$) is zero. According to Figure 11, it can be seen that an agreement was reached at the $5th$ negotiation round. In the $5th$ negotiation round, the patient sent six equivalent offers to the dentist (see Figure 9). All these six equivalent offers bring the same utility (0.6) to the patient, but different utilities to the dentist. According to the dentist's utility function (see Figure 8), two of the patient's offers ($10 : 00$ and $16 : 00$) will bring the utility value 0.8 to the dentist. However, the dentist's expected utility in the $6th$ round is only 0.5. Therefore, the dentist would take either one of these two offers as the agreement.

In this section, we demonstrate the negotiation procedure between two nonlinear utility agents by employing the proposed negotiation approach. It was shown that the proposed negotiation model and mechanisms can efficiently handle nonlinear utility agents, and increase the utilities for all negotiators.

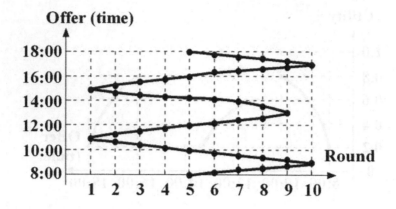

Fig. 9 The patient's offers in each negotiation round

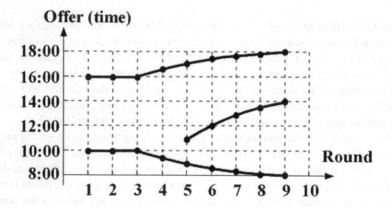

Fig. 10 The dentist's offers in each negotiation round

6   Related Work

There has been a lot of previous work on agent negotiation with nonlinear utility domains. In [4], Fatima et al. studied bilateral multi-issue negotiation between self-interested agents whose utility functions are nonlinear. The authors argued that even though the *package deal procedure* leads multiple negotiation to Pareto optimal, however computing the equilibrium for the *package deal procedure* is not always easy, especially for nonlinear utility functions. In order solve such a problem, the authors introduced two approaches: (i) to approximate nonlinear utility functions by linear functions, then an approximate equilibrium can easily be reached based on the approximate linear function; and (ii) to use the *simultaneous procedure* to negotiate issues in parallel but in dependently. By employing these two approaches, the

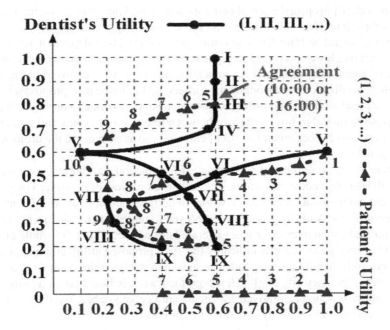

Fig. 11 The comparison between the dentist's utilities and the patient's utilities

approximate equilibrium can be found in polynomial time. This paper also showed that although the *package deal procedure* is known to generate Pareto optimal outcomes, the *simultaneous procedure* outperforms in some cases by considering economic properties. The differences between Fatima's work and our work are that we discussed more detail about the types of nonlinear function, and proposed a multiple offers mechanism for non-monotonic utility functions and an approximating offer mechanism for discrete utility functions. Also, Fatima's work pays more attention to the Pareto optimal agreement searching.

Ito et al. [6] proposed a negotiation protocol for multi-issue negotiation and extended it for nonlinear utility domains. The protocol is based on a combinatorial auction protocol. However, the results of this research are preliminary. In [5, 7], Ito et al. proposed an auction-based negotiation protocol among nonlinear utility agents. This model considered nonlinear utility functions, and interdependency among multi-issue. In order to reach Pareto efficiency, a sampling process is firstly performed on each negotiator's utility space, and these samples adjusted in order to get more feasible contracts. Then each negotiator makes bids, and a mediator is employed to find combinations of bids and to maximize the total value of the bids. The experimental results indicated good performance compared with other negotiation methods. The difference between Ito's work and our work is that we focus on the negotiation model for nonlinear utility functions in single issue negotiation, and Ito pays more attention to searching for Pareto efficiency for nonlinear utility agents in multi-issue negotiation.

Klein et al. [8] proposed a simulated annealing based negotiation model to consider multiple inter-dependent issues and large contract spaces. In order to achieve near-optimal social welfare for negotiation with binary issue dependencies, a negotiation approach was proposed to make substantial progress towards achieving a "win-win" negotiation outcome. An annealing mediator was employed to analyze negotiators' feedback on their opponents' offers, and to make new proposals to increase the social welfare. The difference between Klein's work and our work is that we proposed a model to bridge the gap between linear utility agents and nonlinear utility agents and a mechanism for offer generating, so as to increase the utilities for both negotiators.

Robu et al. [12] proposed an agent strategy for bilateral negotiations over multi-issue with inter-dependent valuations. The authors employed ideas inspired by graph theory and probabilistic influence networks to derive efficient heuristics for multi-issue negotiations. A utility graph was employed to represent complex utility functions, and the best counter-offer was generated through searching the whole utility graph. The experimental results indicated that their proposed utility graph approach can successfully handle interdependent issues, and navigate the contract space to reach Pareto efficiency efficiently. However, their work is based on an assumption that the utility functions employed by agents are decomposable. We do not make any assumption on agent utility functions in this paper.

7 Conclusion and Future Work

In this paper, a bilateral single-issue negotiation model was proposed to handle non-linear utility functions. A 3D model was proposed to illustrate the relationships between an agent's utility function, negotiation decision function, and time constraint. A multiple offer mechanism was introduced to handle non-monotonic utility functions, and an approximating offer mechanism was introduced to handle discrete utility functions. Finally, these two mechanisms were combined to handle nonlinear utility functions in more general situations. The procedure of how an agent generated its counter offers by employing the proposed 3D model and negotiation mechanisms was also introduced. The experimental results also indicated that the proposed negotiation model and mechanisms can efficiently handle nonlinear utility agents, and increase the utilities for all negotiators.

The future work of this research will pay attention to multiple issue negotiations with nonlinear utility functions. The negotiation model and mechanisms presented in this paper will be further extended by considering an agent's nonlinear preference on negotiated issues. Lastly, an extended negotiation protocol will be developed to search for Pareto efficiency in multi-issue negotiation between nonlinear utility agents.

References

1. Brzostowski, J., Kowalczyk, R.: Predicting Partner's Behaviour in Agent Negotiation. In: Proc. of 5th Int. Conf. on Autonomous Agents and Multiagent Systems (AAMAS 2006), pp. 355–361 (2006)
2. Faratin, P., Sierra, C., Jennings, N.: Negotiation Decision Functions for Autonomous Agents. Journal of Robotics and Autonomous Systems 24(3-4), 159–182 (1998)
3. Fatima, S., Wooldridge, M., Jennings, N.: An Agenda-Based Framework for Multi-Issue Negotiation. Artificial Intelligence 152(1), 1–45 (2004)
4. Fatima, S., Wooldridge, M., Jennings, N.: An Analysis of Feasible Solutions for Multi-Issue Negotiation Involving Nonlinear Utility Functions. In: Proc. of 8th Int. Conf. on Autonomous Agents and Multiagent Systems (AAMAS 2009), pp. 1041–1048 (2009)
5. Ito, T., Hattori, H., Klein, M.: Multi-issue negotiation protocol for agents: Exploring nonlinear utility spaces. In: Proceedings of the 20th Int. Conf. on Artificial Intelligence, pp. 1347–1352 (2007)
6. Ito, T., Klein, M.: A Multi-Issue Negotiation Protocol among Competitive Agents and Its Extension to A Nonlinear Utility Negotiation Protocol. In: Proceedings of the 5th Int. Conf. on Autonomous Agents and Multiagent Systems (AAMAS 2006), pp. 435–437 (2006)
7. Ito, T., Klein, M., Hattori, H.: A Multi-Issue Negotiation Protocol among Agents with Nonlinear Utility Functions. Multiagent and Grid Systems 4(1), 67–83 (2008)
8. Klein, M., Faratin, P., Sayama, H., Bar-Yam, Y.: Negotiating Complex Contracts. Group Decision and Negotiation 12(2), 111–125 (2003)
9. Narayanan, V., Jennings, N.: An Adaptive Bilateral Negotiation Model for E-Commerce Settings. In: Proceedings of the Seventh IEEE Int. Conf. on E-Commerce Technology, pp. 34–41 (2005)
10. Narayanan, V., Jennings, N.R.: Learning to Negotiate Optimally in Non-stationary Environments. In: Klusch, M., Rovatsos, M., Payne, T.R. (eds.) CIA 2006. LNCS (LNAI), vol. 4149, pp. 288–300. Springer, Heidelberg (2006)
11. Osborne, M., Rubenstein, A.: A Course in Game Theory. The MIT Press, Cambridge (1994)
12. Robu, V., Somefun, D., Poutré, J.: Modeling Complex Multi-Issue Negotiations Using Utility Graphs. In: Proceedings of the 4th Int. Conf. on Autonomous Agents and Multiagent Systems (AAMAS 2005), p. 287 (2005)

The Effect of Grouping Issues in Multiple Interdependent Issues Negotiation Based on Cone-Constraints

Katsuhide Fujita, Takayuki Ito, and Mark Klein

Abstract. Most real-world negotiation involves multiple interdependent issues, which create agent utility functions that are nonlinear. In this paper, we employ utility functions based on "cone-constraints," which is more realistic than previous formulations. Cone-constraints capture the intuition that agents' utilities for a contract usually decline gradually, rather than step-wise, with distance from their ideal contract. In addition, one of the main challenges in developing effective nonlinear negotiation protocols is scalability; they can produce excessively high failure rates, when there are many issues, due to computational intractability. In this paper, we propose the scalable and efficient protocols by grouping Issues. Our protocols can reduce computational cost, while maintaining good quality outcomes, with decomposing the utility space into several largely independent sub-spaces. We also demonstrate that our proposed protocol is highly scalable when compared to previous efforts in a realistic experimental setting.

1 Introduction

Multi-issue negotiation protocols represent an important field in the study of multi-agent systems. In fact, negotiation, which covers many aspects of our lives, has led to extensive research in the area of automated negotiators; that is, automated

Katsuhide Fujita
Department of Computer Science and Engineering, Nagoya Institute of Technology,
Nagoya, Aichi, Japan
e-mail: fujita@itolab.mta.nitech.ac.jp

Takayuki Ito
School of Techno-Business Administration, Nagoya Institute of Technology,
Nagoya, Aichi, Japan
e-mail: ito.takayuki@nitech.ac.jp

Mark Klein
Sloan School of Management, Massachusetts Institute of Technology,
Cambridge, MA, U.S.
e-mail: m_klein@mit.edu

T. Ito et al. (Eds.): New Trends in Agent-Based Complex Automated Negotiations, SCI 383, pp. 39–55.
springerlink.com   © Springer-Verlag Berlin Heidelberg 2012

agents capable of negotiating with other agents in a specific environment include e-commerce [9], large scale argument [14], and collaborative design. Many real-world negotiations are complex and involve interdependent issues. When designers work together to design a car, for example, the value of a given carburetor is highly dependent on which engine is chosen. Negotiation protocols well-suited for linear utility spaces, unfortunately, work poorly when applied to nonlinear problems [8].

In this paper, we employ a model of highly complex utility spaces based on cone-shaped constraints that can represent more realistic and flexible constraint shapes than existing works like common block-shaped constraints ([7] etc.). Cone-constraints capture the intuition that agents' utilities for a contract usually decline gradually, rather than step-wise, with distance from their ideal contract. Negotiation protocols that are not scalable also work poorly when applied to such highly complex utility space as one based on cone-based constraints as shown in this paper[3].

Recently, some studies have focused on negotiation with nonlinear utility functions. The following are the representative studies on multiple issues negotiations for complex utility spaces: A bidding-based protocol was proposed in [7]. Agents generate bids by finding high regions in their own utility functions, and the mediator finds the optimum combination of submitted bids from the agents. In [18], utility graphs were used to model issue dependencies for binary-valued issues. [6] proposed an approach based on a weighted approximation technique to simplify the utility space. [1] proposed bilateral multi-issue negotiations with time constraints. [15] proposed an auction-based protocol for nonlinear utility spaces generated using weighted constraints, and [16] extended this work to address highly-rugged utility spaces. However, unsolved problem is the scalability of the protocols against the number of issues. Thus, reducing this computational cost has been a key focus in this research.

We propose a new protocol in which a mediator tries to decompose a highly complex utility space into several tractable utility spaces in order to reduce the computational cost. In this protocol, the mediator ignores the weak relations of issues and leaves the strong relations of issues. First, agents generate the interdependency graph, which is a weighted non-directed interdependency graph, by analyzing all their constraints. Next, a mediator can identify issue subgroups by analyzing the interdependency graph. Then, agents generate bids for each issue-group and set the evaluation value of bids for each issue-group. We introduce two ways of setting evaluation values in generating bids for each issue-group. One is the maximum value for the possible utility of the divided bids, and the other is the average for the possible utility of the divided bids. Finally, the mediator finds the optimum combination of submitted bids from agents in every issue-groups, and combines the contracts generated in each issue-group.

The experimental results show that the negotiation protocol based on issue-groups has high scalability compared with existing works. Our proposed protocol can find a more optimal contract in the case where the number of edges becomes small as the interdependency rate increases (Exponent distribution case). The exponent distribution case is realistic, and there are examples in the real world. Also, we analyze the optimality for each setting of the evaluation values.

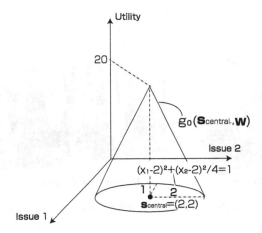

Fig. 1 Example of cone-constraint

The remainder of this paper is organized as follows. First, we describe a model of nonlinear multi-issue negotiation and utility function based on cone-constraints. Second, we describe analyzing protocol to get the interdependency rate, and propose negotiation protocol based on issue-groups. Then we present the experimental results. Finally, we describe related works and draw conclusions.

2 Highly Complex Utility Space

We consider the situation where N agents $(a_1, \ldots, a_N)$ want to reach an agreement with a mediator who manages the negotiation from the man-in-the-middle position. There is a set of issues $S = \{s_1, \ldots, s_M\}$, M issues, $s_j \in S$, to be negotiated. The number of issues represents the number of dimensions of the utility space. For example, if there are three issues[1] the utility space has three dimensions. Issue s_j has a value drawn from the domain of integers $[0, X]$, $i.e.$, $s_j \in \{0, 1,, \ldots, X\}(1 \leq j \leq M)$.[2] A contract is represented by a vector of issue values $\mathbf{s} = (s_1, \ldots, s_M)$. We assume that agents have incentive to cooperate to achieve win-win agreements because the situation is not a zero-sum game. We also assume that agents don't tell a falsehood.

An agent's utility function is represented by constraints. The existing works[7, 15, 2] assume "cube"-shaped constraints. In this paper, we assume "cone"-shaped constraints (called "cone-constraints") that can represent more realistic and flexible constraint shapes in utility space. In fact, "cone"-constraints capture the intuition

---

[1] The issues are shared. They are not "separated" into agents, who are all negotiating a contract that has N (e.g., 10) issues in it. All agents are potentially interested in the values for all M issues.

[2] A discrete domain can come arbitrarily close to a real domain by increasing its size. As a practical matter, many real-world issues that are theoretically real (delivery date, cost) are discretized during negotiations.

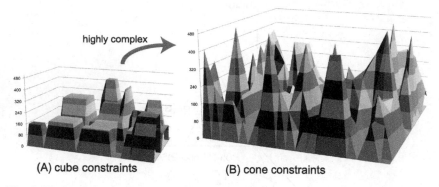

Fig. 2 Nonlinear utility function and complex utility space

that agents' utilities for a contract usually decline gradually, rather than step-wise, with distance from their ideal contract. Figure 1 shows an example of a binary cone-constraint between issues 1 and 2. This cone-constraint has a value of 20, which is maximum if the situation is $s_{central} = [2,2]$. The effective region is $\mathbf{w} = [1,2]$. The formula of the segment of the base is $(x_1 - 2)^2 + (x_2 - 2)^2/4 = 1$.[3] This representation is completely new in studies of multi-issue negotiations.

There are l cone-constraints, $C = \{c_k | 1 \leq k \leq l\}$. Cone-constraint c_k has a gradient function $g_k(s_{central}, \mathbf{w})$, which is defined by two values: central value $s_{central}$, which is the highest utility in c_k, and effective region $\mathbf{w}$, which represents the region where c_k is affected. We assume not only circle-based but also ellipse-based cones. Thus constraint c_k has value $u_a(c_k, s)$ if and only if it is satisfied by contract s. In this paper, effective region $\mathbf{w}$ is not a value but a vector. These formulas can represent utility spaces if they are in an n-dimensional space. Function $\delta_a(c_k, i_j)$ is a effective region of i_j in c_k. $\delta_a(c_k, i_j)$ is $\emptyset$ if c_k has no region regarded as i_j. Every agent has its own, typically unique, set of constraints.

An agent's utility for contract s is defined as $u_a(s) = \sum_{c_k \in C} u_a(c_k, s)$, where $u_a(c_k, s)$ is the utility value of s in c_k. This represents a crucial departure from previous efforts on "simple" multi-issue utility spaces, where contract utility is calculated as the weighted sum of the utilities for individual issues, producing utility functions shaped like flat hyperplanes with a single optimum.

Figure 2 shows an example of a nonlinear utility space with two issues. This utility space is highly nonlinear with many hills and valleys. [7] proposed a utility function based on "cube"-constraints. However, cube-constraints don't sufficiently capture the intuition important in real life. Compared with cube-constraints, the utility function with cone-constraints is highly complex because its highest point is narrower. Therefore, the protocols for making agreements must be analyzed in highly complex utility space. In particular, a simple simulated annealing method

---

[3] The general expression is $\sum_{i=1}^{m} x_i^2/w_i^2 = 1$.

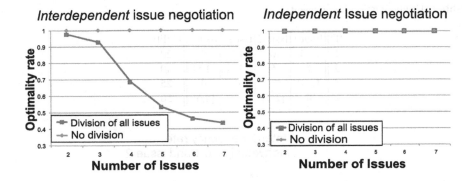

Fig. 3 Relationship of interdependency and optimality rate with nonlinear utility function

to directly find optimal contracts is insufficient in a utility function based on cone-constraints[3].

The objective function for our protocol can be described as follows:

$$\arg\max_{\mathbf{s}} \sum_{a \in N} u_a(\mathbf{s})$$

Our protocol, in other words, tries to find contracts that maximize social welfare, *i.e.*, the total utilities for all agents. Such contracts, by definition, will also be Pareto-optimal.

It is possible to gather and aggregate all individual agents' utility spaces into one central place and then to find the entire optimal contract by using such well-known nonlinear optimization techniques as simulated annealing or evolutionary algorithms. We cannot employ such centralized methods for negotiation purposes, however, because agents typically don't want to reveal too much about their utility functions. Therefore, in this paper, agents don't completely reveal their utility information to third parties.

Note that, in negotiations with multiple *independent* issues, we can find the optimal value for each issue in isolation to quickly find a globally optimal negotiation outcome. In negotiation with multiple *interdependent* issues, however, the mediator can't treat issues independently because the utility of a choice for one issue is potentially influenced by the choices made for other issues. Figure 3 shows the relationship between issue interdependency and negotiation optimality in an example with interdependent issues. In figure 3, we ran an exhaustive social welfare optimizer for each issue independently, as well as for all possible issue combinations. The number of agents is four, and the domain of per issue is five. The linear utility function (independent cases) is generated by $u_a(x) = k * x + c$ (where x is the value for that issue, k and c are constants, and a is the agent). The nonlinear function is generated by some multi-dimensional constraints. If the mediator ignores the issue interdependencies (i.e. finds optima for each issue in isolation), optimality declines

Table 1 Utility function for an agent

ID	Issue1	Issue2	Issue3	Issue4	Utility
1	[2,4]	∅	[4,6]	∅	20
2	∅	5	[3,7]	[1,5]	40
3	[3,9]	∅	∅	∅	25
4	4	[2,7]	9	[4,6]	50

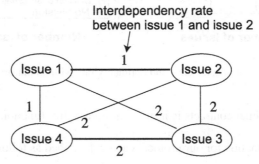

Fig. 4 Interdependency Graph

rapidly as the number of issues increases. This means that the mediator must account for issue interdependencies to find high quality solutions.

However if the negotiation protocol tries to do so by exhaustively considering all issue-value combinations, it quickly encounters intractable computational costs. If we have, for example, only 10 issues with 10 possible values per issue, this produces a space of 10^{10} (10 billion) possible contracts, which is too large to evaluate exhaustively. Negotiation with multiple interdependent issues thus introduces a difficult tradeoff between optimality and computational cost.

3 Issue Interdependency and Interdependency Graph

A issue interdependency for multi-issue negotiations is defined as follows: If there is a constraint between issue X and issue Y, then we assume X and Y are interdependent. If, for example, an agent has a binary constraint between issue 1 and issue 3, issue 1 and issue 3 are interdependent for that agent - see Table 1.

The strength of issue interdependency is measured by *interdependency rate*. We define a measure for the interdependency between *issue* i_j and *issue* i_{jj} for agent a:

$$D_a(i_j, i_{jj}) = \sharp\{c_k | \delta_a(c_k, i_j) \neq \emptyset \cap \delta_a(c_k, i_{jj}) \neq \emptyset\}$$

This measures the number of constraints that inter-relate the two issues.

The agents capture issue interdependency information as an interdependency graph. An interdependency graph is represented as a weighted non-directed graph, in which a node represents an issue, an edge represents the interdependency between issues, and the weight of an edge represents the interdependency rate between the issues. An interdependency graph is thus formally defined as: $G(P,E,w)$: $P = \{1,2,\ldots,|I|\}(finite\ set)$, $E \subset \{\{x,y\}|x,y \in P\}$, $w : E \rightarrow R$. Figure 4 shows the interdependency graph for the constraints listed in Table 1.

4   A Negotiation Protocol Based on Issue Interdependency

Our proposed negotiation protocol works as follows. A mediator gathers private issue interdependency graphs from each agent, generates a social interdependency graph, identifies issue sub-groups, and then uses that information to guide the search for a final agreement. The suitable number of issue-groups is fixed previously. We describe the details below:

[Step1: Analyzing the issue interdependency]. Each agent analyzes issue interdependency in its own utility space, and generates an interdependency graph. Then, each agent sends the interdependency graph to the mediator. Algorithm 1 can generate interdependency graphs. The computation time increases in proportion to the number of constraints.

Listing 1. get_Interdependency(C)

C: a set of constraints

```
1: for c ∈ C do
2:    for i := 0 to Number of issues do
3:       for j := i + 1 to Number of issues do
4:          if Issue i and Issue j are interdependent in c then
5:             interdependencyGraph[i][j]++
6:          end if
7:       end for
8:    end for
9: end for
```

[Step 2: Grouping issues based on the social interdependency graph]. In this step, the mediator identifies the issue-groups. First, the mediator generates a social interdependency graph from the private interdependency graphs submitted by the agents. A social interdependency graph is almost same as a private interdependency graph. The only difference is that the weight of an edge represents the *social* interdependency rate. The social interdependency rate between *issue i_j* and *issue i_{jj}* is defined as: $\sum_{a \in N} D_a(i_j, i_{jj})$. ($D_a(i_j, i_{jj})$: Interdependency rate between *issue i_j* and *issue i_{jj}* by agent a).

Next, the mediator identifies the issue-groups based on the social interdependency graph. In this protocol, the mediator tries to find optimal issue-grouping using simulated annealing (SA) [19]. The evaluation function for the simulated annealing is the sum of the weights of the edges that do not span separate issue-groups. The

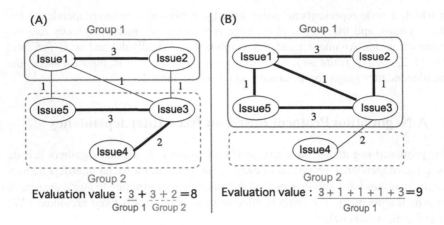

Fig. 5 Evaluation value in identifying issue-groups

goal is to maximize this value. Figure 5 shows an example of evaluation values for
two issue-groups. In Figure 5 (A), the evaluation value is 8 because there are non-
spanning edges between issue 1 and issue 2, issue 3 and issue 4, issue 3 and issue
5, and issue 4 and issue 5. In Figure 5 (B), the evaluation value is 9 because there
are non-spanning edges among issue 1, issue 2, issue 3, and issue 5. The number of
issue-groups is decided before the protocol begins.

[Step3: Generating bids]. Each agent samples its utility space in order to find high-
utility contract regions. A fixed number of samples are taken from a range of random
points, drawing from a uniform distribution. Each agent uses a nonlinear optimizer
based on simulated annealing [19] to try to find the local optimum in its neighborhood. For each contract **s** found by the adjusted sampling, an agent evaluates its
utility by summation of values of satisfied constraints. If that utility is larger than
the reservation value δ (threshold), then the agent defines a bid that covers all the
contracts in the region that have that utility value. The agent needs merely to find
the intersection of all the constraints satisfied by that **s**.

Next, agents divide the bids into some bids for each issue-group, and set the evaluation values for these bids. In this paper, agents set the evaluation value by the
following two types of methods: **(1) Max:** The maximum value for the possible
utility of the divided bids; and **(2) Average** The average of the possible utility of the
divided bids. Figure 6 shows an example in which there are two issues and agents
divide two issue-groups. In Figure 6, an agent generates $B_{all} = [1,2]$ for all issues,
and divides B_{all} into $B_1 = [1,X]$ for issue 1 and $B_2 = [X,2]$ for issue 2 (X: any value).
If an agent sets the evaluation value with the type 1 method, both evaluation values
of B_1 and B_2 are set to 9. This is because the possible utility value of B_1 and B_2 is
the same as the utility value of B_{all}. On the other hand, the evaluation value of B_1 is

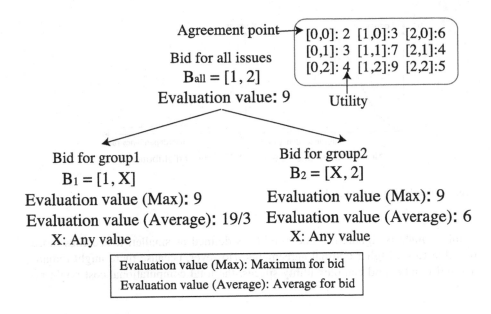

Fig. 6 Division for the bid by agents

set to $19/3$ by analyzing $[1,0]$, $[1,1]$, $[1,2]$ and the evaluation value of B_2 is set to 6 by analyzing $[0,2]$, $[1,2]$, $[2,2]$ in the type 2 method. We demonstrate the effect of employing type 1 or type 2 methods in the experiments.

[Step 4: Finding the Solutions]. The mediator identifies the final contract by finding all the combinations of bids, one from each agent, that are mutually consistent, *i.e.*, that specify overlapping contract regions[4]. If there is more than one such overlap, the mediator selects the one with the highest social welfare (i.e. the highest summed bid value). The mediator employs breadth-first search with branch cutting to find the social-welfare-maximizing bid overlaps. After that, the mediator finds the final contract by consolidating the winning sub-contracts from each issue-group.

Agents are at risk for making an agreement that is not optimal for themselves by dividing the interdependent issues. In other words, there is the possibility of making a low utility agreement by ignoring the interdependency of some issues. However, agents can make a better agreement in this protocol because the mediator identifies the issue-groups based on the rate of interdependency.

There is a trade-off between the optimality of agreement and the computational cost when setting the number of issue-groups. If the number is defined as larger, it becomes difficult to get higher optimal agreements because a large interdependency

---

[4] A bid can specify not just a specific contract but an entire region. For example, if a bid covers the region $[0,2]$ for issue 1 and $[3,5]$ for issue 2, the bid is satisfied by the contract where issue 1 has value 1 and issue 2 has value 4. For a combination of bids to be consistent, the bids must all overlap.

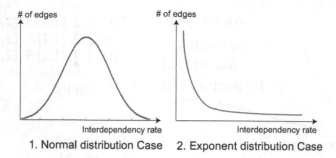

1. Normal distribution Case   2. Exponent distribution Case

Fig. 7 Distribution of the number of edges

number might be ignored. If the number is defined as smaller, it becomes more possible to get higher optimal agreements. However, because there might remain a small number and complex utility spaces, the total computational cost might be large.

5   Experimental Results

5.1   Setting

We conducted several experiments to evaluate the effectiveness of our approach. In each experiment, we ran 100 negotiations between agents. The followings are the parameters in our experiments. The domain for the issue values: $[0,9]$. The number of constraints: 10 unary constraints, 5 binary constraints, 5 trinary constraints, etc. (a unary constraint relates to one issue, a binary constraint relates to two issues, etc). The maximum value for a constraint: *100 × (Number of Issues)*. Constraints that satisfy many issues have, on average, larger utility, which seems reasonable for many domains. In the meeting scheduling domain, for example, higher order constraints concern more people than lower order constraints, so they are more important. The maximum width for a constraint: 7. The following constraints would all be valid: Issue 1 = $[2,6]$, Issue 3 = $[2,9]$. The maximum effective region for a constraint is 7.

In the experiments, interdependency rates are generated by the following two types of the distributions: "1) Normal Distribution Case" and "2) Exponent Distribution Case." In these cases, the distribution of the number of edges is shown as in Figure 7. In general, "1) Normal Distribution Case" is used for evaluating the negotiation protocol like [7].

We compare the following four methods: "(A-1) Issue-groups (Max)," "(A-2) Issue-groups (Average)," "(B) Basic Bidding," and "(C) Q-Factor." "(A-1) Issue-groups (Max)" and "(A-2) Issue-groups (Average)" are protocols based on issue-groups proposed in this paper. In "(A-1) Issue-groups (Max)," the evaluation value for divided bids is set as the maximum value for the possible utility of the divided bid. In "(A-2) Issue-groups (Average)," the evaluation value for divided bids is set as the average of the possible utility of the divided bids. They are defined in Section 4. "(B) Basic Bidding" is the bidding-based protocol proposed in [7], which does not employ issue-grouping. In this protocol, agents generate bids by finding the highest utility regions in their utility functions, and the mediator finds the optimum combination of bids submitted from agents. "(C) Q-Factor" is the Maximum Weight Interdependent Set (MWIS) protocol proposed in [15, 16]. MWIS is a variant of bidding protocol where agents use the Q-factor, a combination of region and utility, to decide which bids to submit. This reduces the failure rate because agents are less likely to submit low-volume bids that do not overlap across agents.

The parameters for generating bids in (A)-(C) are as follows [7]. The number of samples taken during random samplings: *(Number of Issues)* × 200. Also, the initial temperature is 30 degrees for adjusting the sampling point using SA. For each iteration, the temperature is decreased 1 degree. Thus, it is decreased to 0 degrees by 30 iterations. Note that it is important that the annealer does not run too long or too hot because then each sample will tend to find the global optimum instead of the peak of the optimum nearest the sampling point. The threshold used to cut out contract points that have low utility: 100. The limitation on the number of bids per agent: $\sqrt[N]{6,400,000}$ for N agents. It was only practical to run the deal identification algorithm if it explored no more than about 6,400,000 bid combinations.

In "(A-1) Issue-groups (Max)" and "(A-2) Issue-groups (Average)," the initial temperature is 30 degrees for identifying the issue-groups. For each iteration, the temperature is decreased 3 degrees. Thus, it is decreased to 0 degrees by 10 iterations. The annealer should not run too long because the computational cost is at most $_{\sharp\ of\ issues}C_{\sharp\ of\ groups}$. In addition, the number of issue-groups is three. In "(C) Q-Factor," Q (Q-Factor) is defined as $Q = u^{\alpha} * v^{\beta}$ (u: utility value, v: volume of the bid or constraint), $\alpha = 0.5$, $\beta = 0.5$.

To find the optimum contract, we use simulated annealing (SA)[19] because the exhaustive search becomes intractable as the number of issues grows too large. The SA initial temperature is 50.0 and decreases linearly to 0 over the course of 2,500 iterations. The initial contract for each SA run is randomly selected. This simulated annealing method is generally unrealistic in negotiation contexts because they require that agents fully reveal their utility functions to a third party. If the exhaustive search is employed in our complex negotiation setting, the computational complexity is too high with increasing the number of issues. Therefore, we regard a contract by this SA method as an approximate optimal contract. The optimality rate in this experiment is defined as *(The maximum contract calculated by each method) / (The maximum contract calculated by SA)*.

Our code is implemented in Java 2 (1.5) and run on a core 2-duo processor iMac with 1.0 GB memory on a Mac OS X 10.6 operating system.

5.2 Experimental Results

Figure 8 compares the failure rate in a normal distribution case. The number of agents is four. "(A-1) Issue-groups (Max)" and "(A-2) Issue-groups (Average)"

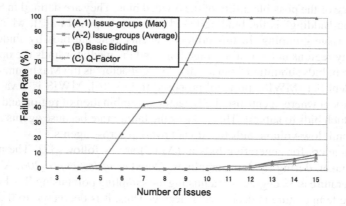

Fig. 8 Failure rate (Normal distribution case)

proposed in this paper have a lower failure rate than "(B) Basic Bidding." This is because "(A-1) Issue-groups (Max)" and "(A-2) Issue-groups (Average)" reduce the computational cost by issue-grouping. Also, "(A-1) Issue-groups (Max)" and "(A-2) Issue-groups (Average)" have almost the same failure rate as "(B) Q-Factor." Our proposed method and Q-Factor can reduce the computational cost by adjusting the number of issue-groups or quality factor, respectively.

Figure 9 compares box-plots of the optimality rate in the normal distribution case. The lines represent the min and max values, the boxes represent +/- *1* standard deviation, and the middle line represents the average. "(A-1) Issue-groups (Max)" and "(A-2) Issue-groups (Average)" achieve higher optimality than "(B) Basic Bidding." This is because "(B) Basic Bidding" fails to make agreements more than "(A-1) Issue-groups (Max)" and "(A-2) Issue-groups (Average)". In addition, the optimality rate of "(A-1) Issue-groups (Max)" can change widely compared with that of "(A-2) Issue-groups (Average)." Therefore, "(A-1) Issue-groups (Max)" can be seen as risk-seeking for making low optimal agreements. "(A-2) Issue-groups (Average)" has a higher average than "(A-1) Issue-groups (Max)." Therefore, "(A-2) Issue-groups (Average)" can be seen as a risk-adverse method compared with "(A-1) Issue-groups (Max)." Also, "(C) Q-Factor" has almost the same optimality rate as "(A-1) Issue-groups (Max)."

Figure 10 and Figure 11 show the failure rate and optimality rate in the exponent distribution case, respectively. There are not large differences in the failure rates between the normal distribution case and the exponent distribution case. On the other hand, the optimality rates of "(A-1) Issue-groups (Max)" and "(A-2) Issue-groups (Average)" are higher compared with those in the normal distribution case. This is because the protocols by issue-groups proposed in this paper can achieve high optimality if the number of ignored interdependencies is low. In addition, there can be more independent issues or small interdependent issues in the exponent distribution case. Therefore, "(A-1) Issue-groups (Max)" and "(A-2) Issue-groups (Average)"

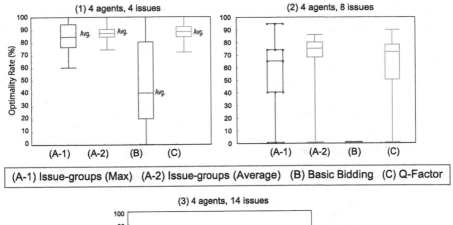

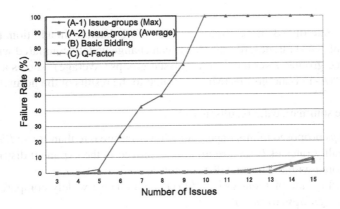

Fig. 9 Box-plots of the optimality rate (Normal distribution case)

Fig. 10 Failure rate (Exponent distribution case)

improve the optimality rate in the exponent distribution case. On the other hand, there is no significant difference in the optimality rates of "(B) Basic Bidding" and "(C) Q-Factor" between the normal distribution case and the exponent distribution case. Therefore, the structure of the interdependency graph doesn't greatly affect

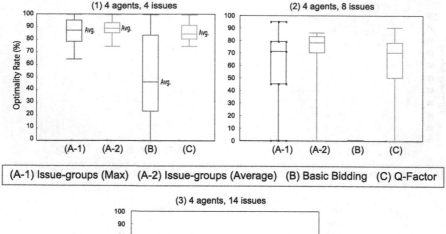

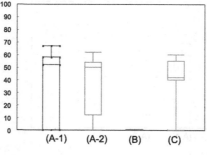

Fig. 11 Box-plots of the optimality rate (Exponent distribution case)

the optimality rate of "(B) Basic Bidding" and "(C) Q-Factor." In addition, the optimality rate of "(A-1) Issue-groups (Max)" can change widely compared with that of "(A-2) Issue-groups (Average)." "(A-2) Issue-groups (Average)" shows a higher average than "(A-1) Issue-groups (Max)," similar to the results of the normal distribution case.

Finally, we summarize the results as follows:

- *Issue-groups* proposed in this paper had a lower failure rate than *Basic Bidding*.
- The optimality rates of *Issue-groups* was improved in the exponent distribution case compared with the normal distribution case.
- The optimality rate of *Issue-groups(Max)* could change widely compared with that of *Issue-groups(Average)*.

6   Related Work

Recently, some works have regarded agents' privacy in the distributed constraint optimization problem (DCOP) field ([5], [13]). Even though the negotiation problem seems to involve a straightforward constraint optimization problem, we have been unable to exploit the high efficiency constraint optimizers. Such solvers attempt to find the solutions that maximize the weights of the satisfied constraints, but they do not account for the individual rationality of agents.

[7] proposed a bidding-based protocol for multiple interdependent issue negotiation. In this protocol, agents generate bids by sampling and searching for their utility functions, and the mediator finds the optimum combination of submitted bids from agents. [15, 16] proposed an auction-based protocol for nonlinear utility spaces generated using weighted constraints, and proposed a set of decision mechanisms for the bidding and deal identification steps of the protocol. They proposed the use of a quality factor to balance utility and deal probability in the negotiation process. This quality factor is used to bias bid generation and deal identification taking into account the agents' attitudes towards risk. The scalability on the number of issues is still problem in these works.

[12] explored a range of protocols based on mutation and selection on binary contracts. This paper does not describe what kind of utility function is used, nor does it present any experimental analyses, so it remains unclear whether this strategy enables sufficient exploration of utility space. [8] presented a protocol applied with near optimal results to medium-sized bilateral negotiations with binary dependencies. This work demonstrated both scalability and high optimality values for multilateral negotiations and higher order dependencies. [10] presented a protocol for multi-issue problems for bilateral negotiations. [18, 17] presented a multi-item and multi-issue negotiation protocol for bilateral negotiations in electronic commerce situations. [4] proposed a negotiation mechanism where the bargaining strategy is decomposed into a concession strategy and a Pareto-search strategy.[6] proposed an approach based on a weighted approximation technique to simplify the utility space. The resulting approximated utility function without dependencies can be handled by negotiation algorithms that can efficiently deal with independent multiple issues, and has a polynomial time complexity. Our protocol can find an optimal agreement point if agents don't have in common the expected negotiation outcome. [1] proposed bilateral multi-issue negotiations with time constraints. This method can find approximate equilibrium in polynomial time where the utility function is nonlinear. However, these papers focused on bilateral multi-issue negotiations. Our protocol focuses on multilateral negotiations. [11] proposed a method in which the mediator searches for a compromise direction based on an Equal Directional Derivative approach and computes a new tentative agreement in bilateral multi-issue negotiations. However, this method only focused on multilateral negotiation. [20] presents an axiomatic analysis of negotiation problems within task-oriented domains (TOD). In this paper, three classical bargaining solutions (Nash solution, Egalitarian solution, Kalai-Smorodinsky solution) coincide when they are applied to a TOD with mixed deals but diverge if their outcomes are restricted to pure deals.

7 Conclusion

We focused on negotiation with multiple interdependent issues in which agent utility functions are cone-shaped constraints that can represent more realistic and flexible constraint shapes in utility space. In this paper, we proposed the new negotiation protocol based on grouping the issues. In this protocol, agents generate the interdependency graph with analyzing all agents' constraints and the mediator identifies the issue-groups based on agents' interdependency graphs. Then, agents generate the bids for whole issues based on bidding-based negotiation protocol and divide the bids against issue-groups. Finally, the mediator identifies the final contract by finding all the combinations of bids based on the bidding-based protocol in each issue-group. We demonstrated that our proposed protocol has high scalability compared with existing works in the normal distribution case and the exponent distribution case.

Acknowledgements. This research is partially supported by PREST, JST.

References

1. Fatima, S.S., Wooldridge, M., Jennings, N. R.: An analysis of feasible solutions for multi-issue negotiation involving nonlinear utility functions. In: Proc. of the Eighth International Joint Conference on Autonomous Agents and Multi-agent Systems (AAMAS 2009), pp. 1041–1048 (2009)
2. Fujita, K., Ito, T., Klein, M.: A representative-based multi-round protocol for multi-issue negotiations. In: Proc. of the Seventh International Joint Conference on Autonomous Agents and Multi-agent Systems (AAMAS 2008), pp. 1573–1576 (2008)
3. Fujita, K., Ito, T., Klein, M.: A secure and fair negotiation protocol in highly complex utilityspace based on cone-constraints. In: Proc. of the 2009 International Joint Conference on Intelligent Agent Technology, IAT 2009 (2009); (short Paper)
4. Gerding, E., Somefun, D., Poutre, H.L.: Efficient methods for automated multi-issue negotiation: Negotiating over a two-part tariff. International Journal of Intelligent Systems 21, 99–119 (2006)
5. Greenstadt, R., Pearce, J., Tambe, M.: Analysis of privacy loss in distributed constraint optimization. In: Proc. of the 21th Association for the Advancement of Artificial Intelligence (AAAI 2006), pp. 647–653 (2006)
6. Hindriks, K.V., Jonker, C.M., Tykhonov, D.: Eliminating Interdependencies Between Issues for Multi-issue Negotiation. In: Klusch, M., Rovatsos, M., Payne, T.R. (eds.) CIA 2006. LNCS (LNAI), vol. 4149, pp. 301–316. Springer, Heidelberg (2006)
7. Ito, T., Hattori, H., Klein, M.: Multi-issue negotiation protocol for agents: Exploring nonlinear utility spaces. In: Proc. of the 20th International Joint Conference on Artificial Intelligence (IJCAI 2007), pp. 1347–1352 (2007)
8. Klein, M., Faratin, P., Sayama, H., Bar-Yam, Y.: Negotiating complex contracts. Group Decision and Negotiation 12(2), 58–73 (2003)
9. Kraus, S.: Strategic Negotiation in Multiagent Environments. Cambridge University Press (2001)
10. Lai, G., Li, C., Sycara, K.: A pareto optimal model for automated multi-attribute negotiations. In: Proc. of the Sixth International Joint Conference on Autonomous Agents and Multi-agent Systems (AAMAS 2007), pp. 1040–1042 (2007)
11. Li, M., Vo, Q.B., Kowalczyk, R.: Searching for fair joint gains in agent-based negotiation. In: Proc. of the Eighth International Joint Conference on Autonomous Agents and Multi-agent Systems (AAMAS 2009), pp. 1049–1056 (2009)

12. Lin R.J., Chou S.T.: Bilateral multi-issue negotiations in a dynamic environment. In: Proc. of the AAMAS Workshop on Agent Mediated Electronic Commerce (AMEC 2003) (2003)
13. Maheswaran, R.T., Pearce, J.P., Varakantham, P., Bowring, E.: Valuations of possible states (vps):a quantitative framework for analysis of privacy loss among collaborative personal assistant agents. In: Proc. of the Forth Inernational Joint Conference on Autonomous Agents and Multi-agent Systems (AAMAS 2005), pp. 1030–1037 (2005)
14. Malone, T.W., Klein, M.: Harnessing collective intelligence to address global climate change. Innovations Journal 2(3), 15–26 (2007)
15. Marsa-Maestre, I., Lopez-Carmona, M.A., Velasco, J.R., de la Hoz, E.: Effective bidding and deal identification for negotiations in highly nonlinear scenarios. In: Proc. of the Eighth International Joint Conference on Autonomous Agents and Multi-agent Systems (AAMAS 2009), pp. 1057–1064 (2009)
16. Marsa-Maestre, I., Lopez-Carmona, M.A., Velasco, J.R., Ito, T., Fujita, K., Klein, M.: Balancing utility and deal probability for negotiations in highly nonlinear utility spaces. In: Proc. of the Twenty-first International Joint Conference on Artificial Intelligence (IJCAI 2009), pp. 214–219 (2009)
17. Robu, V., Poutre, H.L.: Retrieving the structure of utility graphs used in multi-item negotiation through collaborative filtering of aggregate buyer preferences. In: Proc. of the 2nd International Workshop on Rational, Robust, and Secure Negotiations in Multi-Agent Systems, RRS 2006 (2006)
18. Robu, V., Somefun, D.J.A., Poutre, J.L.: Modeling complex multi-issue negotiations using utility graphs. In: Proc. of the 4th International Joint Conference on Autonomous Agents and Multi-Agent Systems (AAMAS 2005), pp. 280–287 (2005)
19. Russell, S.J., Norvig, P.: Artificial Intelligence: A Modern Approach. Prentice-Hall (2002)
20. Zhang, D.: Axiomatic characterization of task oriented negotiation. In: Proc. of the Twenty-first International Joint Conference on Artificial Intelligence (IJCAI 2009), pp. 367–372 (2009)

Automated Agents That Proficiently Negotiate with People: Can We Keep People Out of the Evaluation Loop*

Raz Lin, Yinon Oshrat, and Sarit Kraus

Abstract. Research on automated negotiators has flourished in recent years. Among the important issues considered is how these automated negotiators can proficiently negotiate with people. To validate this, many experimentations with people are required. Nonetheless, conducting experiments with people is timely and costly, making the evaluation of these automated negotiators a very difficult process. Moreover, each revision of the agent's strategies requires to gather an additional set of people for the experiments. In this paper we investigate the use of Peer Designed Agents (PDAs) – computer agents developed by human subjects – as a method for evaluating automated negotiators. We have examined the negotiation results and its dynamics in extensive simulations with more than 300 human negotiators and more than 50 PDAs in two distinct negotiation environments. Results show that computer agents perform better than PDAs in the same negotiation contexts in which they perform better than people, and that on average, they exhibit the same measure of generosity towards their negotiation partners. Thus, we found that using the method of peer designed negotiators embodies the promise of relieving some of the need for people when evaluating automated negotiators.

Raz Lin
Department of Computer Science, Bar-Ilan University, Ramat-Gan, Israel 52900
e-mail: linraz@cs.biu.ac.il

Sarit Kraus
Department of Computer Science, Bar-Ilan University, Ramat-Gan, Israel 52900 and
Institute for Advanced Computer Studies, University of Maryland,
College Park, MD 20742 USA
e-mail: sarit@cs.biu.ac.il

* This research is based upon work supported in part by the U.S. Army Research Laboratory and the U.S. Army Research Office under grant number W911NF-08-1-0144 and under NSF grant 0705587.

1 Introduction

An important aspect in research on automated negotiation is the design of proficient automated negotiators with people [1, 8, 9, 10, 11, 14, 18, 24]. We refer to these agents as EDNs (Expert Designed Negotiators). EDNs can be used with humans in the loop or without them. EDNs can be used in tandem with people to alleviate some of the efforts required of people during negotiations and also assist people that are less qualified in the negotiation process [4, 9, 24]. Additionally, there may be situations in which EDNs can even replace human negotiators. Another possibility is for people embarking on important negotiation tasks to use these agents as a training tool [3, 16], prior to actually performing the task. Thus, success in developing an automated agent with negotiation capabilities has great advantages and implications.

The design of automated agents that proficiently negotiate with people is a challenging task, as there are different environments and constraints that should be considered (for a recent survey of that describes studies that evaluate automatic agents that negotiate with people, see [13]). While the design issues of EDNs is important, we found little, if any, literature focusing on the important issue of the evaluation of EDNs designed to negotiate with people. Thus, in this paper we only focus on the evaluation process of these EDNs.

The evaluation and validation process of the automated negotiators is a vital part of the design process and allows demonstrating how successful the automated negotiators are. Yet, using people for experimentation purposes is timely and costly, making the evaluation process a very difficult task for researches. Designing agents that model the human behavior during negotiations only adds to the difficulties, due to the diverse behavior of people which makes it hard to capture it by a monolithic model. For example, people tend to make mistakes, and they are affected by cognitive, social and cultural factors, etc. [21]. Thus, it is commonly assumed that people cannot be substituted in the evaluation process of EDNs designed to negotiate with people.

The question which now arises is, even though people and agents behave differently, whether one can use agents to evaluate EDNs and reflect from the behavior of the EDNs with other agents to their behavior with people. Following this intuition, we turned to the strategy method [15, 20, 21] which is an experimental methodology which requires people to elicit their actions. The assumption behind this method is that people are able to effectively encapsulate their own strategies if they are properly motivated, monetarily or otherwise. This approach is well accepted in experimental economics and has also begun to be used within artificial intelligence research [2, 19]. The application of this methodology within the study of automated negotiator agents implies that peer designed agents (PDAs) that represent the negotiation preferences of diverse people can be developed.

In this paper we present an in-depth investigation into the question of whether PDAs can be used to keep people out of the evaluation loop and thus simplify the evaluation process required by designers of EDNs. As we will demonstrate in the rest of this paper, the use of PDAs has the potential of elevating some of the need for people in the evaluation of automated negotiators, yet people are still a mandatory factor in the final evaluation loop.

In this paper we provide results of extensive experiments involving more than 300 human subjects and 50 PDAs. The experiments involved negotiations of people that interact with other people, people that interact with PDAs and people that interact with EDNs. The experiments were conducted in two distinct negotiation environments that simulate real-world scenarios that require negotiators to reach agreement about the exchange of resources in order to complete their goals. In each experiment we investigate the behavior and dynamics of people and PDAs with respect to the EDNs in order to find out whether the behavior of PDAs is similar to that of people and whether they can be used as a substitute for people in the evaluation process of automated negotiators. The results shed light as to the prospect of using PDAs to better determine the proficiency of an automated negotiator when matched with people, as well as to compare the behavior of different EDNs.

This paper contributes to research on automated negotiations by tackling the important issue of the evaluation process of automated negotiators. We suggest using PDAs as an unbiased mechanism for evaluating automated negotiators, which can reflect on the behavior of people, as well as allowing fine-tuning and improving the strategy of the automated negotiators. We provide a general methodology for evaluating automated negotiators and different measures to compare the behavior of PDAs and people and show that it is important to understand the different behavioral aspects expressed by them to gain a better perspective on the prospects of the automated negotiator being proficient with people.

The rest of this paper is organized as follows. In Section 2 we review related work in the field of automated negotiators' evaluation. We provide an overview of the problem in Section 3. We continue and describe our methods for evaluating the negotiation behavior and dynamics in Section 4. In Section 5 we present the experiments we conducted, our methodology and our evaluation results. Finally, we provide a summary and discuss the results.

2 Related Work

Important differences exist between designing an automated agent that can successfully negotiate with a human counterpart and designing an automated agent to negotiate with other automated agents. As this paper does not focus on the design of such automated agents, we will not survey related work on this topic. A more detailed review on the design of automated agents capable of negotiating with people can be found, for example, in [13].

To date, in order to replace people in the evaluation loop, one cannot rely on specific automated agents. Instead, we examine the use of peer designed agent as a type of strategy method. Similarly to developing agents, using the strategy method requires subjects to specify their choices for all information sets of the game and not only the ones that occur during the course of a play of a game [15, 21, 20]. Despite the similarity, asking subjects to design and program agents is different from the strategy method. Developing agents requires subjects to implement much more complex strategies (e.g., using heuristics and learning algorithms), potentially, to make decisions in situations not originally considered.

The use of PDAs has been extensively studied within the context of the Trading Agent Competition for Supply Chain Management (TAC SCM) [36]. In TAC one needs to design a trading agent that participates in auctions for certain good. The use of PDA's within this domain demonstrates the benefits of a large set of PDAs for evaluation purposes of EDNs. Yet, in this context, both the PDAs and the EDNs were used for interacting with other computer agents, and not for interaction with people.

Grosz *et al.* [5] experimented with people designing agents for a game called Colored Trails. They observed that when people design agents, they do not always follow equilibrium strategies. Moreover, in their analysis they showed that people demonstrated more helpfulness, which led to higher scores, than their designed agents. Chalamish *et al.* [2] report on large-scale experiments in which people programmed agents which were shown to successfully capture their strategy in a set of simple games. They conclude that peer designed agents can be used instead of people in some cases. In another settings, Rosenfeld and Kraus [19] report on experiments done with PDAs designed for optimization problems. Based on the experiments they conclude that theories of bounded rationality can be used to better simulate people's behavior. However, the settings of Chalamish *et al.* [2] were relatively simple, while our settings have richer strategy space and are much more complicated. Hence, it is not straightforward that PDAs can be used to simulate the people's behavior and thus replace them for evaluation purposes.

3   Problem Description

We consider the problem of evaluating the proficiency of EDNs designed to negotiate with people. We consider a general environment of bilateral negotiation in which two agents, either automated negotiators or people, negotiate to reach an agreement on conflicting issues. We consider two distinct bilateral negotiation environments. The first involves a day-to-day scenario in which the parties negotiate to reach an agreement on conflicting goals, while the second involves playing a game. We describe both environments below.

The first negotiation environment involves a multi-attribute multi-issues negotiation environment (see Figure 1). In this environment, the negotiation can end either when (a) the negotiators reach a full agreement, (b) one of the agents opts out, thus forcing the termination of the negotiation with an opt-out outcome (OPT), or (c) a predefined deadline, dl, is reached, whereby, if a partial agreement is reached it is implemented or, if no agreement is reached, a status quo outcome (SQ) is implemented. It is assumed that the agents can take actions during the negotiation process until it terminates. Let I denote the set of issues in the negotiation, O_i the finite set of values for each $i \in I$ and O a finite set of values for all issues $(O_1 \times O_2 \times \ldots \times O_{|I|})$. We allow partial agreements, $\perp \in O_i$ for each $i \in I$. Therefore, an offer is denoted as a vector $\vec{o} \in O$. It is assumed that the agents can take actions during the negotiation process until it terminates. Let **Time** denote the set of time periods in the negotiation, that is **Time** $= \{0, 1, \ldots, dl\}$. Time also has an impact on the agents' utilities. Each agent is assigned a time cost which influences its utility as time passes. In each period $t \in$ **Time** of the negotiation, if the negotiation has not terminated earlier, each agent can propose a possible agreement, and the other agent can either accept the offer, reject it or opt out. Each agent can either propose an agreement which consists of all the issues in the negotiation, or a partial agreement. We use an extension of the model of alternating offers [17, p. 118-121], in which each agent can perform up to $M > 0$ interactions with its counterpart in each time period.

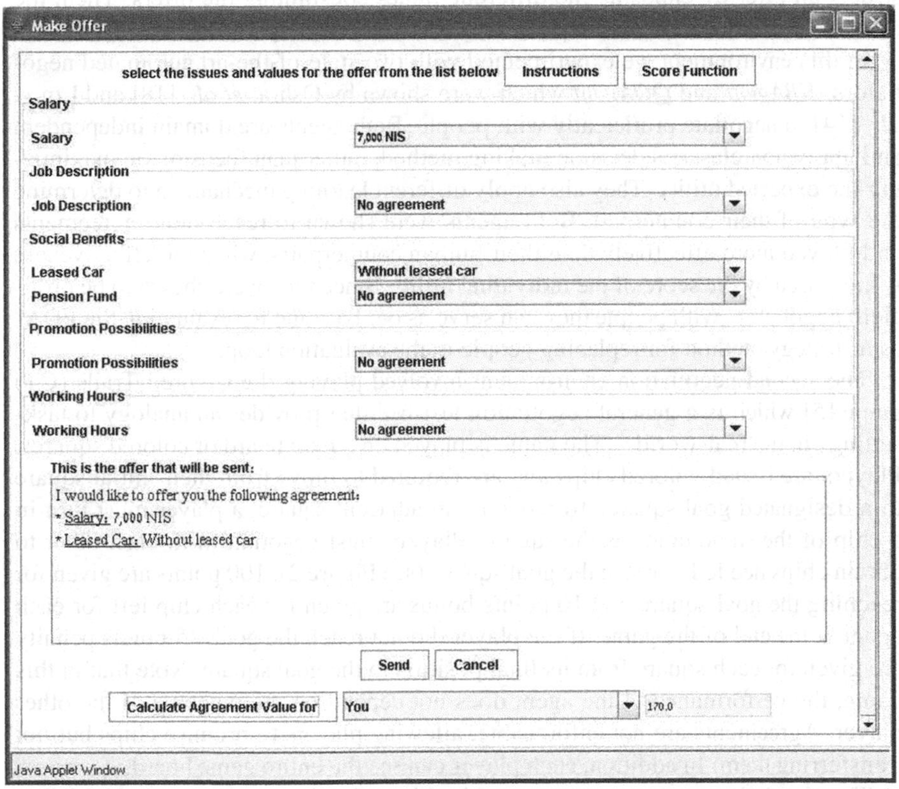

Fig. 1 Bilateral Negotiation: Generating offers screen

In order to make the settings more realistic, it also involved incomplete information concerning the opponent's preferences. We assume that there is a finite set of agent types. These types are associated with different additive utility functions (e.g., one type might have a long term orientation regarding the final agreement, while the other type might have a more constrained orientation). Formally, we denote the possible types of agents **Types** $= \{1, \ldots, k\}$. Given $l \in$ **Types**, $1 \leq l \leq k$, we refer to the utility of an agent of type l as u_l, and $u_l : \{(O \cup \{SQ\} \cup \{OPT\}) \times$ **Time**$\} \rightarrow \mathbb{R}$. Each agent is given its exact utility function. The negotiators are aware of the set of possible types of the opponent. However, the exact utility function of the rival is private information.

We developed a simulation environment which is adaptable such that any scenario and utility function, expressed as multi-issue attributes, can be used, with no additional changes in the configuration of the interface of the simulations or the automated agent. The agents (PDAs or EDNs) can play either role in the negotiation, while the human counterpart accesses the negotiation interface via a web address. The negotiation itself is conducted using a semi-formal language. Each agent constructs an offer by choosing the different values constituting the offers. Then, the offer is constructed and sent in plain English to its counterpart.

In this environment we experimented with two state-of-the-art automated negotiators, *KBAgent* and *QOAgent* which were shown by Oshrat *et al.* [18] and Lin *et al.* [14] to negotiate proficiently with people. Both agents are domain independent and apply non-classical decision making method, rather than focusing on maximizing the expected utility. They also apply different learning mechanism to determine the type of their counterpart. Both agents were shown to reach more agreements and played more effectively than their human counterparts, when the effectiveness is measured by the score of the individual utility. Since they were shown to be proficient negotiators with people they can serve as our baseline for evaluating the PDAs as a strategy method for replacing people in the evaluation loop.

The second negotiation environment involved playing the Colored Trails (CT) game [5] which is a general negotiation test-bed that provides an analogy to task-settings in the real-world [2]. The game is played on a nxm board of colored squares. Players are issued colored chips and are required to move from their initial square to a designated goal square. To move to an adjacent square, a player must turn in a chip of the same color as the square. Players must negotiate with each other to obtain chips needed to reach the goal square (see Figure 2). 100 points are given for reaching the goal square and 10 points bonus are given for each chip left for each agent at the end of the game. If the player did not reach the goal, 15 points penalty are given for each square from its final position to the goal square. Note that in this game, the performance of the agent does not depend on the outcome of the other player. Agreements are not enforceable, allowing players to promise chips but not transferring them. In addition, each player can see the entire game board.

The simulation environment we used in this setting is adaptable such that different variations of the game can be set. The size of the board, number and color of total chips and chips given to each player can be changed. The automated agents can play both sides in the game, while the human counterpart accesses the game via a web address. The game itself is split into turns, where each turn is divided to a

---

[2] Colored Trails is Free Software and can be downloaded at
http://www.eecs.harvard.edu/ai/ct

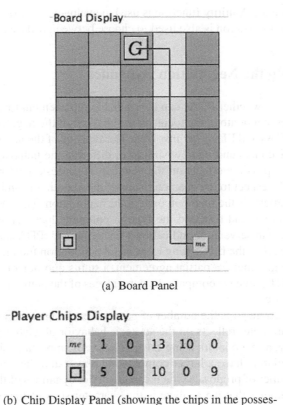

(a) Board Panel

(b) Chip Display Panel (showing the chips in the possession of both participants)

Fig. 2 Snapshots of Colored Trails GUI

negotiation phase, a commitment phase and a movement phase. In the negotiation phase the players can request or promise to send chips. Then, in the commitment phase, the players can send the actual chips or withdraw from the agreement. This might result in one agent sending the promised chips in return to be given other chips, while the other agent fails to deliver. In the movement phase, the players can choose to move to adjacent squares, given they have the required colored chips. The game terminates when either side reaches the goal square or if no player has moved in three consecutive turns.

An EDN was used in this environment as well. The agent, which is called the Personality Based (*PB*) agent, extends the agent reported by Talman *et al.* [23] to allow it to play proficiently with people. It combines a social utility function that represented the behavioral traits of other participants, as well as a rule-based mechanism that used the utility function to make decisions in the negotiation process. Its behavior is characterized as high reliability and medium generosity. The agent can change its personality level for cooperation and reliability and also model these traits of its opponent. A utility function is used to evaluate each possible action and proposal and randomization is also used to choose between different choices.

4 Evaluating the Negotiation Dynamics

In order to analyze whether PDAs can be a suitable replacement for people in evaluating automated negotiators, we focus on the analysis of the negotiation dynamics when people, PDAs and EDNs are involved. The analysis of the negotiation dynamics is a mean of understanding how similar or different the behavior of each population is and its possible effects on the negotiation process. Different parameters investigated with respect to the characteristics of the negotiators and its dynamics.

The first parameter is the final outcome of the negotiation. This includes the final utility score, average end turn and the type of outcome that was reached. The final utility scores can serve as an indication whether people, PDAs and EDNs reach similar outcomes. For the first setting of bilateral negotiation the final outcome can either be a full agreement, a partial agreement, a status quo outcome or an opt out option. For the CT game we compare the percentages of the games ending by reaching the goal square.

The second parameter is the number of proposal exchanged during a negotiation session. This parameter reflects on the strategic behavior of each negotiator. Comparing the different negotiators using this parameter is important, since PDAs and automated negotiators have a larger computation power and they may excess it to send a large number of proposals which can affect the dynamics of the negotiation.

Another parameter that we analyze is the characteristics of the proposals made by the negotiators. For the bilateral negotiation setting we characterize the proposals based on the step-wise analysis method ("negotiation moves") suggested by Hindriks et al. [7] (see Figure 3. In this analysis, each proposal is compared to its preceding proposal (made by the same negotiator) based on the utilities of the offers for both sides. Thus, an offer can be characterized as either:

1. Selfish (better for the proposer, worse for the other side),
2. Fortunate (better for the proposer and for the other side),
3. Unfortunate (worse for both sides),
4. Concession (worse for the proposer, better for the other party),
5. Silent (both utilities are changed only within a given small threshold), or
6. Nice oriented (utility for the proposer is only within a given small threshold, and higher for the other party).

Formally, let $u^t_{diff_A}$ and $u^t_{diff_B}$ denote the difference between the utilities of side A and B, respectively, at time t, that is: $u^t_{diff_A} = u_A(\vec{o}, t+1) - u_A(\vec{o}, t)$ (and similarly for side B). Let T_A and T_B denote a small threshold for sides A and B, respectively. In our settings we chose the threshold to be equal to the time discount value. A proposal made by agent A at time $t+1$ is characterized based on the algorithm listed in Listing 2.

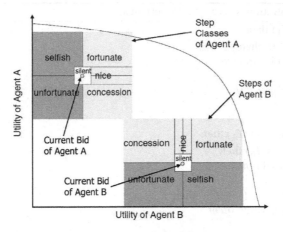

Fig. 3 Classification of negotiation moves. Figure taken from [6].

For the CT settings, the characteristics of the proposals is based on the type of the proposal. Three proposal types were available for the players, each represents the ratio between the number of chips the player is willing to send and in return the number of chips it requires:

- A *Give* proposal is a proposal in which the player proposes to send less chips than it will receive (e.g., send 1 chip in exchange of receiving 2 chips from the other party).
- A *Take* proposal is one in which the player proposes to give more than it receives.
- A *Reciprocal* proposal is one in which the two sides send and receive the same number of chips.

The type of the proposal the players choose to propose can reflect on the cooperativeness level of each player. The cooperativeness level or the reliability level can also be reflected by the extent of fulfilling the agreements.

5 Experiments

The experiments were conducted using the simulation environments mentioned in Section 3. We begin by describing the environments which were used in the different experiments and then continue to describe the experimental methodology and results.

5.1 The Negotiation Environments

For the bilateral negotiation environment two domains were used, which were described by Lin *et al.* [14]. The first domain is a Job Candidate domain, which is

Listing 2. Negotiation Moves Characteristics

```
 1: if (u^t_{diff_A} > T_A) then
 2:    if (u^t_{diff_B} > 0) then
 3:       Proposal is fortunate
 4:    else
 5:       Proposal is selfish
 6:    end if
 7: else
 8:    if (u^t_{diff_A} ≥ (−1) · T_A) then
 9:       if (u^t_{diff_B} > T_B) then
10:          Proposal is nice
11:       else
12:          if (u^t_{diff_B} ≥ (−1) · T_B) then
13:             Proposal is silent
14:          else
15:             if (u^t_{diff_A} > 0) then
16:                Proposal is selfish
17:             else
18:                Proposal is unfortunate
19:             end if
20:          end if
21:       end if
22:    else
23:       if (u^t_{diff_B} ≥ (−1) · T_B) then
24:          Proposal is concession
25:       else
26:          Proposal is unfortunate
27:       end if
28:    end if
29: end if
```

related to the subjects' experience, and thus they could better identify with it. In this domain, a negotiation takes place after a successful job interview between an employer and a job candidate. In the negotiation both the employer and the job candidate wish to formalize the hiring terms and conditions of the applicant. In this scenario, 5 different attributes are negotiable with a total of 1,296 possible agreements that exist.

The second domain involved reaching an agreement between England and Zimbabwe evolving from the World Health Organization's Framework Convention on Tobacco Control, the world's first public health treaty. The principal goal of the convention is "to protect present and future generations from the devastating health, social, environmental and economic consequences of tobacco consumption and exposure to tobacco smoke". In this domain, 4 different attributes are under negotiation, resulting with a total of 576 possible agreements.

In both domains if an agreement is not reached by the end of the allocated time a status quo agreement is implemented. In addition, time also has an impact and the sides might lose or gain as time progresses. In the Job candidate domain both sides lose as time progresses, while in the England-Zimbabwe domain, England gains while Zimbabwe loses as time progresses. Also, each side can choose to opt out of the negotiation at any time. As there is also incomplete information in each domain, we assume that there are three possible types of agents for each role. These types are associated with different additive utility functions. The different types are characterized as ones with short-term orientation regarding the final agreement, long-term and a compromising orientation.

In the CT game, a 7×5 board was used. Two types of games were used, where each differed by the ability of each player to reach the goal square with or without the assistance of the other player:

1. *Asymmetric* game, which was characterized by one of the player having 15 chips and being *dependant* of the other player and needed to exchange chips in order to reach the goal square, while the other player had 24 chips and was *independent* of its counterpart, and thus could reach the goal square without doing any exchange.
2. *Symmetric* game, which was characterized by the two players having 24 chips and being dependent and needing each other's chips to reach the goal square.

5.2 Experimental Methodology

We ran an extensive set of simulations, consisting of more than 300 human negotiators and more than 50 PDAs. The human negotiators were mostly computer science undergraduate and graduate students, while a few were former students who are currently working in the Hi-Tech industry. Each subject served only one specific role in the negotiations (e.g., in the bilateral negotiations either the employer role or the job candidate one, and in the CT game environment either the dependent player or the independent player). Prior to the experiments, the subjects were given oral instructions regarding the experiment and the domain. The subjects were instructed to play based on their score functions and to achieve the best possible agreement for them. The score function was private information and unknown for the other side.

The PDAs were automated negotiators designed by people. The students were given a task to implement an efficient automated agent for a given negotiation environment (different students designed PDAs for the different negotiation environments). The implementation was done in the same simulation environment as the negotiation itself. The students were provided skeleton classes, having all the necessary server-communication functionality, to help them implement their agents. This also allowed them to focus on the strategy and the behavior of the agent, and eliminate the need to implement the communication protocol or the negotiation protocol. In addition, it provided them with a simulation environment in which they could test their agents and their strategies. The students were able to first negotiate or play the CT game before submitting their PDAs.

5.3   Experimental Results

The main goal of the experiments was to analyze whether the strategy method of PDAs can be used to replace people in the evaluation process of EDNs designed to negotiate proficiently with people. In addition, we wanted to find whether this method can also be used to evaluate and compare different automated negotiators and obtain from it which will be a more proficient negotiator with people.

5.3.1   Evaluating EDNs When Matched with PDAs versus When Matched with People

In this section we analyze the final outcomes of the negotiation, mainly the final utility, the negotiation's duration and the success of it, to understand whether these parameters can be used, when matching EDNs with PDAs, as indicators to the results if the EDNs are later matched with people.

5.3.1.1   Analyzing Utility Outcomes

Table 1 summarizes the final utilities achieved by each side in each experiment for the Job candidate and England-Zimbabwe domains, while Table 2 summarizes the final utilities in the CT game environment. All results are statistically significant within the $p < 0.05$ range. To try and understand whether PDAs can be used in replacement of people to predict the performance of the EDNs we compare the results of the final utility values when the PDAs were involved, that is PDAs versus EDNs and PDAs versus PDAs, and when people were involved, that is, people versus EDNs and people versus people.

When the *KBAgent* was matched with PDAs it was able to achieve higher utility values than the average of the PDAs matched against themselves (lines (1),(2) in Table 1). This is also consistent with the *KBAgent*'s achievements when matched with people (lines (3),(4) in Table 1). Similar results are attained by the second EDN in this enviornment, the *QOAgent* (lines (2),(5) as compared to lines (4), (6) in Table 1).

A similar phenomenon was observed in the CT game. When the *PB* agent played in the symmetric settings and in the asymmetric game as the independent role, the final utilities achieved by it were higher than the average utilities of the PDAs. When it played the dependent role in the asymmetric game, its final utility was lower than the average utility of the PDAs (lines (1),(2) in Table 2). The same relation is revealed when comparing the *PB*'s utility when playing with people and the average utilities of people playing with one another (lines (3),(4) in Table 2).

In the bilateral negotiation domain we had two distinct EDNs (the *KBAgent* and the *QOAgent*). Thus, it is interesting to see whether the performance of them when matched with PDAs can be used as a prediction of whom will perform better when matched with people. The *KBAgent* was shown to perform better when matched with people than the *QOAgent* (lines (3),(6) in Table 1). In three out of the four sides in the two domains, this is also reflected when they are matched with the PDAs, with the *KBAgent* achieving higher utility scores than the *QOAgent* (lines (1),(5) in Table 1).

Table 1 Final utility results in the bilateral negotiation environment

	Job Can. Domain		Eng-Zim Domain	
	$u_{employer}$	$u_{job\ can}$	u_{eng}	u_{zim}
(1) *KBAgent* vs. PDAs	437.7	415.8	720.0	-14.5
(2) **PDAs vs. PDAs**	368.2	355.1	251.8	-83.7
(3) *KBAgent* vs. **People**	472.8	482.7	620.5	181.8
(4) **People vs. People**	423.8	328.9	314.4	-160.0
(5) *QOAgent* vs. **PDAs**	466.1	396.8	663.4	-36.5
(6) *QOAgent* vs. **People**	417.4	397.8	384.9	35.3

Table 2 Final utility results in the CT game environment

	Asymmetric game		Symmetric game
	$u_{independent}$	$u_{dependent}$	$u_{dependent}$
(1) *PB* vs. **PDAs**	180.53	35.00	131.36
(2) **PDAs vs. PDAs**	178.38	45.25	111.48
(3) *PB* vs. **People**	187.08	81.94	157.83
(4) **People vs. People**	181.45	97.26	130.67

Thus, the results support the hypothesis that the final utility values can serve as a good indication for evaluating the proficiency of the automated negotiator. Moreover, they can also be used to compare between different EDNs and reflect on their proficiency when matched with people.

5.3.1.2 Analyzing the Characteristics of the Negotiation Ending

Two more relevant questions are whether the end period of the negotiation and the type of agreements reached or whether the goal was reached when EDNs are matched with PDAs can also serve as an indication to the proficiency of the EDNs when matched with people. Tables 3 and 4 summarize the average end turn and the percentages of the negotiations terminated with full agreements, status quo (SQ) agreements, opting out (OPT) or partial agreements in the Job candidate and the England-Zimbabwe domains, respectively, while Table 5 summarizes the average end turn and the percentages of reaching the goal square in the CT game settings.

With regard to the duration of the negotiation, while in the Job candidate domain the negotiation lasted longer when the *KBAgent* was matched with people than when it was matched with PDAs, it lasted shorter time in the England-Zimbabwe domain (lines (5),(8) in Tables 3 and 4). The CT game also lasted shorter when the *PB* was matched with people than when it was matched with the PDAs (lines (3),(4) in

Table 3 Average end turn and percentages of reaching full agreements, status quo, opting out or partial agreements in the Job Candidate bilateral negotiation domain

	End Turn	Full	SQ	OPT	Partial
People vs. People					
(1)	6.68	91%	2%	4%	2%
PDAs vs. PDAs					
(2)	5.56	80%	4%	7%	9%
KBAgent vs. PDAs					
(3) $KBAgent_{Employer}$	6.14	86%	14%		
(4) $KBAgent_{Job\ can.}$	5.85	77%	8%	8%	8%
(5) Average	6	81%	11%	4%	4%
KBAgent vs. People					
(6) $KBAgent_{Employer}$	7.17	100%			
(7) $KBAgent_{Job\ can.}$	6.13	100%			
(8) Average	6.68	100%			

Table 4 Average end turn and percentages of reaching full agreements, status quo, opting out or partial agreements in the England-Zimbabwe bilateral negotiation domain

	End Turn	Full	SQ	OPT	Partial
People vs. People					
(1)	9.08	70%	8%	11%	11%
PDAs vs. PDAs					
(2)	8.30	69%	8%	12%	12%
KBAgent vs. PDAs					
(3) $KBAgent_{England}$	11.50	60%			40%
(4) $KBAgent_{Zimbabwe}$	10.00	50%	10%		40%
(5) Average	10.75	55%	5%		30%
KBAgent vs. People					
(6) $KBAgent_{England}$	10.06	94%			6%
(7) $KBAgent_{Zimbabwe}$	6.13	100%			
(8) Average	8.15	97%			3%

Table 5). Thus the negotiation duration when EDNs are matched with PDAs cannot be used as a good indication to the duration when the EDNs are matched with people.

With regard to the way the negotiation ended, when the *KBAgent* is matched with people more full agreements are reached than when it is matched with PDAs

Table 5 Average end turn and percentages of reaching the goal in the CT game settings. *Ind* stands for the independent player and *Dep* for the dependent player

	Asymmetric game				Symmetric game	
	End Turn		Goal Reached		End Turn	Goal Reached
	Ind	*Dep*	*Ind*	*Dep*		*Dep*
(1) **People vs. People**	4.1		97%	74%	5.8	60%
(2) **PDAs vs. PDAs**	4.5		98%	32%	4.8	47%
(3) *PB* **vs. PDAs**	6.1	5.9	100%	42%	4.6	73%
(4) *PB* **vs. People**	3.8	4.4	100%	61%	4.3	87%

(as shown in lines (5),(8) in Tables 3 and 4). Similarly, the *PB* agent reaches the goal square in more cases when matched with people than when matched with other PDAs (as shown in lines (3),(4) in Table 5).

5.3.2 Evaluating the Performance and Behavior of People Versus PDAs

It was demonstrated that PDAs behave differently than people (e.g., [5, 19]), yet our experiments indicated that playing against PDAs can reflect on the results when the EDNs are matched with people. Thus, it is important to understand whether the behavior of people is similar to that of PDAs when either matched with the same population or with the EDNs.

Our results show that most of the differences in the behavior of PDAs as compared to people lie in the negotiation dynamics. In the following subsections we investigate different parameters with respect to the negotiation dynamics.

5.3.2.1 Investigating the Final Outcome

With respect to the final outcome, when people were matched with people they were able to achieve higher utilities than when the PDAs were matched with other PDAs in one of the roles (the employer role in the Job candidate domain and the England role in the England-Zimbabwe domain) and worse in the other roles (lines (2),(4) in Table 1), while in the CT game they were better in both roles (lines (2),(4) in Table 2). Moreover, in most occasions the PDAs demonstrated lower percentages in reaching full agreements or reaching the goal square when matched against themselves than when people were matched against people (lines (1),(2) in Tables 3, 4 and 5).

5.3.2.2 Investigating the Negotiation Dynamics

To bolster our confidence from these results we examined the patterns of behavior demonstrated by people and the PDAs when matched with the EDNs. To make it concise, we only present the results on one of the domains and negotiation's sides, though the results are similar in the other domains and sides. Figure 4 compares the final utilities achieved by PDAs and people when matched with the EDNs in the job candidate domain when playing the role of the employer, while Figure 5 compares the times in which the negotiations terminated. Note, that we compare between the behavior of people and PDAs and not the EDNs behavior. The results demonstrate the similarity and trend between people and PDAs when matched with EDNs. For example, in Figure 4 we can observe that PDAs achieve somewhat higher utilities when matched with the *QOAgent* as compared to the *KBAgent*. The same trend is then observed when people are matched with both agents.

Another parameter we investigated is the number of proposals exchanged during the negotiation. Tables 6 and 7 summarize the total number of exchanges for the bilateral negotiation domains and the CT game settings, respectively. In most of the times, more exchanges are made when PDAs are involved than when only people are involved. This can be explained by the fact the automated agents have more computing power, making it easier for them to evaluate messages and send more messages. When the *KBAgent* was matched with people in three out of the four settings the average number of proposals it received from the other party was lower when received from people than when received from the PDAs (lines (3),(4) in Table 6). Similarly, in two out three cases the average number of proposals received from people was lower when matched with the *PB* agent than when the PDAs played against the *PB* agent (lines (3),(4) in Table 7). It is interesting to observe the exception when people exchange more proposals than agents. This happens in the bilateral negotiation domain when people play the job candidate role or the Zimbabwe role and in the CT game settings in the symmetrical game when both players are dependent of each other. It seems that the cause lies in the fact that in these roles the people play the "underdog" roles and have a greater incentive to propose more messages in the hope of one of them being accepted by the other side. The other side, in return, has the incentive to make the negotiation lasts longer or toughen its stands.

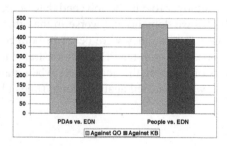

Fig. 4 Comparing overall performance between people and PDAs when matched with EDNs when playing the role of the employer in the job candidate domain

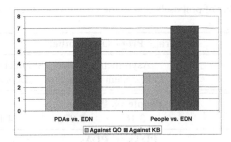

Fig. 5 Comparing end-turn between people and PDAs when matched with EDNs when playing the role of the employer in the job candidate domain

Table 6 Average number of proposals sent in the bilateral negotiation settings

	Average Proposals Number			
	Job Can. Domain		Eng-Zim Domain	
	Employer	Job can.	England	Zimbabwe
People vs. People				
(1)	7.06	6.87	8.00	9.13
PDAs vs. PDAs				
(2)	6.04	6.13	11.66	11.73
***KBAgent* vs. PDAs**				
(3) PDAs's proposals	10.85	7.14	12.80	27.90
***KBAgent* vs. People**				
(4) People's proposals	5.44	8.94	7.25	11.76

When comparing the PDAs' only negotiations to people's only negotiations slightly more exchanges are made by people in the Job candidate domain, yet higher exchanges are made by PDAs in the England-Zimbabwe domain (lines (1),(2) in Table 6). As for the CT game settings, while in the symmetric settings more proposals are made by people, in the asymmetric game settings more proposals are made by the PDAs when matched with other PDAs, as compared to proposals made by people (lines (1),(2) in Table 7). In both settings, it seems that more exchanges are made by the PDAs when the domain involves two sides, in which one has significantly more leverage than the other (in the England-Zimbabwe domain, England gains more as time progresses while Zimbabwe loses, and in the asymmetric settings the independent role can reach the goal without needing the dependent role). Thus, it seems to be the case that each side stands firmly and is inclined less to concede.

This leads us to the question whether the average behavior of people and the average behavior of PDAs differ. The results indeed demonstrate that there are

Table 7 Average number of proposals sent in the CT game settings

	Average Proposals Number		
	Asymmetric game		Symmetric game
	Independent	Dependent	Dependent
People vs. People			
(1)	3.16	2.87	4.15
PDAs vs. PDAs			
(2)	3.54	3.31	3.28
PB vs. PDAs			
(3) PDAs's proposals	2.68	2.95	2.09
PB vs. People			
(4) People's proposals	2.39	2.08	2.30

Table 8 Negotiation moves in the Job Candidate bilateral negotiation domain. Se = Selfish, F = Fortunate, U = Unfortunate, C = Concession, Si = Silent, N = Nice. Side A is the Employer and side B is the job candidate.

	Se	F	U	C	Si	N
People vs. People						
(1) People$_A$	10.92%	19.37%	13.73%	24.65%	29.23%	2.11%
(2) People$_B$	11.96%	17.03%	18.48%	19.93%	31.88%	0.72%
PDAs vs. PDAs						
(3) PDAs$_A$	6.70%	14.20%	21.30%	20.35%	32.13%	5.32%
(4) PDAs$_B$	9.04%	12.69%	15.45%	18.98%	37.85%	5.99%
KBAgent **vs. PDAs**						
(5) PDAs$_A$	16.15%	5.38%	13.08%	15.38%	13.08%	36.92%
(6) PDAs$_B$	19.32%	10.23%	9.09%	12.50%	27.27%	21.59%
KBAgent **vs. People**						
(7) People$_A$	12.50%	15.28%	9.72%	47.22%	9.72%	5.56%
(8) People$_B$	20.98%	16.78%	13.99%	34.27%	7.69%	6.29%

differences in the type of exchanges made by people when negotiating with EDNs and those made by the PDAs. Recall that we characterize the negotiation moves in the bilateral negotiation domains as selfish, fortunate, concession, unfortunate, concession, silent or nice (cf. Section 4), where cooperation is characterized by moves of type fortunate, concession, silent or nice. In the CT game settings we defined three types of proposals (*Give*, *Take* and *Reciprocal*, from which cooperation is characterized by *Take* exchanges.

Table 9 Negotiation moves in the England-Zimbabwe bilateral negotiation domain. Se = Selfish, F = Fortunate, U = Unfortunate, C = Concession, Si = Silent, N = Nice. Side *A* is England and side *B* is Zimbabwe.

	Se	F	U	C	Si	N
People vs. People						
(1) People$_A$	14.86%	10.54%	13.51%	43.51%	13.51%	4.05%
(2) People$_B$	30.47%	10.93%	9.77%	25.58%	17.67%	5.58%
PDAs vs. PDAs						
(3) PDAs$_A$	35.93%	14.10%	10.40%	26.19%	9.75%	3.64%
(4) PDAs$_B$	19.11%	3.36%	7.36%	37.77%	9.94%	22.47%
***KBAgent* vs. PDAs**						
(5) PDAs$_A$	19.49%	3.39%	11.86%	39.83%	18.64%	6.78%
(6) PDAs$_B$	23.60%	2.25%	7.49%	16.48%	31.46%	18.73%
***KBAgent* vs. People**						
(7) People$_A$	23.00%	12.00%	7.00%	49.00%	4.00%	5.00%
(8) People$_B$	31.69%	7.10%	7.10%	40.98%	11.48%	1.64%

The cooperation levels differ between people and PDAs in all settings we tested. Tables 8 and 9 summarize the negotiation moves for the Job candidate and the England-Zimbabwe domains, respectively. When matched with the EDNs, while the majority of the proposals made by people are concession oriented in both domains (lines (7),(8) in Tables 8 and 9), the majority of the proposals made by the PDAs are nice oriented and silent oriented in the Job candidate domain (lines (5),(6) in Table 8) and concession oriented and silent oriented in the England-Zimbabwe domain (lines (5),(6) in Table 9). In the Job candidate domain, people are also succeeding in making more fortunate offers and less silent offers than the PDAs when playing both roles against the same population (lines (1),(3) and (2),(4) in Table 8). It is also interesting to note that people are less selfish when playing the role of England and more selfish when playing the role of Zimbabwe, while the opposite is expressed in the behavior of the PDAs (lines (1),(3) and (2),(4) in Table 9). This might be derived from the leverage England role has over Zimbabwe and the fact that it gains as time progresses allows people to play more generously when playing England, while playing more rigidly when playing as Zimbabwe. A somewhat similar observation was also reported in experiments with the CT game run by Grosz *et al.* [5].

Figures 6 and 7 display the cooperation levels of people and PDAs in the Job candidate domain and England-Zimbabwe domain, respectively. We can observe similarity in the cooperation levels of both people and PDAs in the Job candidate domain (75.35% and 69.57% when people are matched with people, 77.78% and 65.03% when people are matched with the *KBAgent*, 72.01% and 75.51% when PDAs are matched with PDAs and 70.77% and 71.59% when PDAs are matched

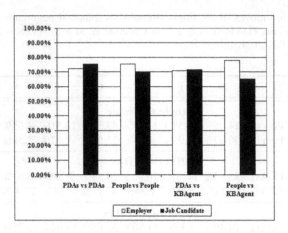

Fig. 6 Cooperation levels in the Job candidate domain

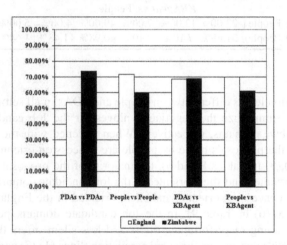

Fig. 7 Cooperation levels in the England-Zimbabwe domain

with the *KBAgent*), however the cooperation level is again inverted in the England-Zimbabwe domain. People are more cooperative when playing the England role (71.62% when matched with people and 70% when matched with the *KBAgent*) and less cooperative when playing the Zimbabwe role (59.77% and 61.2% when playing against people and against the *KBAgent*, respectively). Moreover, a similar trend is also noticed by the behavior of the PDAs. The cooperation levels when they are matched with other PDAs are 53.67% and 73.53% when playing the role of England and Zimbabwe, respectively, while playing against the *KBAgent* the cooperation levels are high and similar when playing both roles (68.64% and 68.91% when playing the roles of England and Zimbabwe, respectively).

Table 10 Negotiation moves in the CT game asymmetric settings. *Ind* stands for the independent player and *Dep* for the dependent player.

	Give	Reciprocal	Take	Fully Kept	Partially Kept	Unkept
People vs. People						
(1) *Ind*	14.3%	28.6%	57.1%	61%	10%	29%
(2) *Dep*	55.1%	19.1%	25.8%	32%	39%	29%
PDAs vs. PDAs						
(3) PDAs$_{Ind}$	15.4%	38.3%	46.4%	58%	2%	40%
(4) PDAs$_{Dep}$	21.5%	51.5%	67%	53%	2%	30%
PB vs. PDAs						
(5) PDAs$_{Ind}$	33.3%	39.2%	27.4%	50%	0%	50%
(6) PDAs $_{Dep}$	25.0%	51.8%	23.2%	50%	10%	41%
PB vs. People						
(7) People$_{Ind}$	20.9%	46.5%	32.6%	58%	11%	31%
(8) People$_{Dep}$	48.0%	36.0%	16.0%	19%	38%	44%

Table 11 Negotiation moves in the CT game symmetric settings. *Dep* stands for the dependent player

	Give	Reciprocal	Take	Fully Kept	Partially Kept	Unkept
People vs. People						
(1) *Dep*	26.8%	37.5%	35.6%	41%	21%	38%
PDAs vs. PDAs						
(2) PDAs$_{Dep}$	11.1%	62.9%	26.0%	58%	7%	35%
PB vs. PDAs						
(3) PDAs$_{Dep}$	15.2%	63.0%	21.7%	78%	0%	22%
PB vs. People						
(4) People$_{Dep}$	24.6%	47.8%	13.3%	54%	21%	24%

We also measured the cooperative levels, reflected by the types of exchanges, in the CT game settings, as summarized in Tables 10 and 11. Here as well, there is an apparent difference in the behavior of the PDAs and people. When playing with the *PB* agent in the asymmetric settings, the majority of proposals were *reciprocal* (46.51%) when people played the independent role and 48% of type *Give* when playing the dependent role. Yet, the majority of offers when the PDAs were

matched with the *PB* agent were 39.22% of type *Give* when they played the role
of the independent player and 51.79% *reciprocal* when playing the dependent role
(lines (5),(7) and (6),(8) in Table 10). In addition, in both cases the PDAs fully kept
50% of the agreements while people had fully kept 58% and 19% when playing the
independent and dependent roles, respectively. More full agreements were also fully
kept in the symmetric settings when the PDAs were involved as compared to when
people were involved (lines (1),(2) and (3),(4) in Table 11).

6   Conclusions

The importance of designing proficient automated negotiators to negotiate with people and evaluating them cannot be overstated. Yet, evaluating agents against people
is a tiresome task, due to the cost and time required. In this paper we presented
an extensive systematic experimentation to answer the question whether people can
be kept out of the evaluation loop when evaluating automated negotiators designed
specifically to negotiate proficiently with people. To do so, we evaluated several negotiation behavioral parameters in an extensive set of experiments with people and
with peer designed agents. In the bottom line, our results reveal that playing well
against peer designed agents can reflect on the proficiency of the automated negotiator when matched with people. Moreover, we showed that while PDAs results with
different negotiation outcomes than people, there is a common behavioral pattern
when they are both matched with EDNs.

Another interesting observation is that the EDNs strategy was mainly similar
when matched with people and when matched with PDAs. The *KBAgent* mostly sent
concession offers when playing both with PDAs and with people in the bilateral negotiation domains. The *PB* agent, when playing in the symmetric game settings sent
mainly *reciprocal* offers, when playing the independent role sent *Take* offers and
offered *Give* exchanges when it played the dependent role in the asymmetric game.
This is regardless of whether it played with people or PDAs; only the specific offers
had been adapted to the specific players. With respect to reliability, the *PB* agent
was highly reliable, yet its exact reliability level varied between 0.72 when playing the dependent role in the asymmetric game with people to 0.96 when playing
with PDAs in the symmetric case. The consistent strategies with some adaptation
the agent made for the specific player it was matched with probably contributed to
the similar outcomes with people and PDAs.

Surprisingly, the results also demonstrated that there is a difference in the behavior of PDAs and people, even though the PDAs were supposed to be designed in a
way that encapsulates people's strategy. Our results provide analysis of these differences by understanding the negotiation dynamics that occur when people and PDAs
are involved, and demonstrate the importance of an efficient opponent modeling
mechanism and the need for adapting to different attitudes towards the negotiation.

There are fundamental benefits of using PDAs instead of people. First, PDAs are
accessible 24/7 and can be used whenever needed. In addition, PDAs are not biased
and thus can be used several times to asses the EDN's behavior. Thus, they allow

the agent designer to revise and change her agent with the ability to evaluate each design and compare it to previous designs. Lastly, it allows different EDNs to be matched on the same set of PDAs and obtain an objective evaluation of the results.

While people cannot be kept completely out of the evaluation loop, we demonstrated the promise embodied in peer designed agents for evaluation purposes of automated negotiators. Thus, evaluating on peer designed agents could and should serve as a first extensive attempt to validate the agent's proficiency and strategy design before continuing on to evaluation with people.

As noted before, people behave differently than PDAs. Future work warrants careful investigation on the differences between the behavior of PDAs and people. This investigation might allow for a better understanding of people and the better design of automated agents specifically designed to negotiate with people.

Acknowledgements. We thank Ya'akov (Kobi) Gal for his helpful remarks and suggestions.

References

1. Byde, A., Yearworth, M., Chen, K.-Y., Bartolini, C.: AutONA: A system for automated multiple 1-1 negotiation. In: Proceedings of the 2003 IEEE International Conference on Electronic Commerc (CEC), pp. 59–67 (2003)
2. Chalamish, M., Sarne, D., Kraus, S.: Programming agents as a means of capturing self-strategy. In: Proceedings of the Seventh International Conference on Autonomous Agents and Multiagent Systems (AAMAS), pp. 1161–1168 (2008)
3. Fleming, M., Olsen, D., Stathes, H., Boteler, L., Grossberg, P., Pfeifer, J., Schiro, S., Banning, J., Skochelak, S.: Virtual reality skills training for health care professionals in alcohol screening and brief intervention. Journal of the American Board of Family Medicine 22(4), 387–398 (2009)
4. Gal, Y., Pfeffer, A., Marzo, F., Grosz, B.J.: Learning social preferences in games. In: Proceedinge of the Nineteenth AAAI Conference on Artificial Intelligence, pp. 226–231 (2004)
5. Grosz, B., Kraus, S., Talman, S., Stossel, B.: The influence of social dependencies on decision-making: Initial investigations with a new game. In: Proceedings of the Third International Conference on Autonomous Agents and Multiagent Systems (AAMAS), pp. 782–789 (2004)
6. Hindriks, K., Jonker, C., Tykhonov, D.: Analysis of negotiation dynamics. In: Klusch, M., Hindriks, K.V., Papazoglou, M.P., Sterling, L. (eds.) CIA 2007. LNCS (LNAI), vol. 4676, pp. 27–35. Springer, Heidelberg (2007)
7. Hindriks, K., Jonker, C., Tykhonov, D.: Negotiation dynamics: Analysis, concession tactics, and outcomes. In: Proceedings of the 2007 IEEE/WIC/ACM International Conference on Intelligent Agent Technology (IAT), pp. 427–433 (2007)
8. Jonker, C.M., Robu, V., Treur, J.: An agent architecture for multi-attribute negotiation using incomplete preference information. Autonomous Agents and Multi-Agent Systems 15(2), 221–252 (2007)
9. Katz, R., Kraus, S.: Efficient agents for cliff edge environments with a large set of decision options. In: Proceedings of the Fifth International Conference on Autonomous Agents and Multiagent Systems (AAMAS), pp. 697–704 (2006)

10. Kraus, S., Hoz-Weiss, P., Wilkenfeld, J., Andersen, D.R., Pate, A.: Resolving crises through automated bilateral negotiations. Artificial Intelligence 172(1), 1–18 (2008)
11. Kraus, S., Lehmann, D.: Designing and building a negotiating automated agent. Computational Intelligence 11(1), 132–171 (1995)
12. Lax, D.A., Sebenius, J.K.: Thinking coalitionally: party arithmetic, process opportunism, and strategic sequencing. In: Young, H.P. (ed.) Negotiation Analysis, pp. 153–193. The University of Michigan Press (1992)
13. Lin, R., Kraus, S.: Can automated agents proficiently negotiate with humans? Communications of the ACM 53(1), 78–88 (2010)
14. Lin, R., Kraus, S., Wilkenfeld, J., Barry, J.: Negotiating with bounded rational agents in environments with incomplete information using an automated agent. Artificial Intelligence 172(6-7), 823–851 (2008)
15. Offerman, T., Potters, J., Verbon, H.A.A.: Cooperation in an overlapping generations experiment. Games and Economic Behavior 36(2), 264–275 (2001)
16. Olsen, D.E.: Interview and interrogation training using a computer-simulated subject. In: Interservice/Industry Training, Simulation and Education Conference (1997)
17. Osborne, M.J., Rubinstein, A.: A Course In Game Theory. MIT Press, Cambridge (1994)
18. Oshrat, Y., Lin, R., Kraus, S.: Facing the challenge of human-agent negotiations via effective general opponent modeling. In: Proceedings of the Eighth International Conference on Autonomous Agents and Multiagent Systems (AAMAS), pp. 377–384 (2009)
19. Rosenfeld, A., Kraus, S.: Modeling agents through bounded rationality theories. In: Proceedings of the Tewenty-First International Joint Conference on Artificial Intelligence (IJCAI), pp. 264–271 (2009)
20. Selten, R., Abbink, K., Buchta, J., Sadrieh, A.: How to play (3x3)-games: A strategy method experiment. Games and Economic Behavior 45(1), 19–37 (2003)
21. Selten, R., Mitzkewitz, M., Uhlich, G.R.: Duopoly strategies programmed by experienced players. Econometrica 65(3), 517–556 (1997)
22. TAC Team: A trading agent competition. IEEE Internet Computing 5(2), 43–51 (2001)
23. Talman, S., Hadad, M., Gal, Y., Kraus, S.: Adapting to agents' personalities in negotiation. In: Proceedings of the Fourth International Conference on Autonomous Agents and Multiagent Systems (AAMAS), pp. 383–389 (2005)
24. Traum, D.R., Marsella, S.C., Gratch, J., Lee, J., Hartholt, A.: Multi-party, Multi-Issue, Multi-Strategy Negotiation for Multi-Modal Virtual Agents. In: Prendinger, H., Lester, J.C., Ishizuka, M. (eds.) IVA 2008. LNCS (LNAI), vol. 5208, pp. 117–130. Springer, Heidelberg (2008)

Matchmaking in Multi-attribute Auctions Using a Genetic Algorithm and a Particle Swarm Approach

Simone A. Ludwig and Thomas Schoene

Abstract. An electronic market platform usually requires buyers and sellers to exchange offers-to-buy and offers-to-sell. The goal of this exchange is to reach an agreement on the suitability of closing transactions between buyers and sellers. This paper investigates multi-attribute auctions, and in particular the matchmaking of multiple buyers and sellers based on five attributes. The proposed approaches are based on a Genetic Algorithm (GA) and a Particle Swarm Optimization (PSO) approach to match buyers with sellers based on five attributes as closely as possible. Our approaches are compared with an optimal assignment algorithm called the Munkres algorithm, as well as with a simple random approach. Measurements are performed to quantify the overall match score and the execution time. Both, the GA as well as the PSO approach show good performance, as even though not being optimal algorithms, they yield a high match score when matching the buyers with the sellers. Furthermore, both algorithms take less time to execute than the Munkres algorithm, and therefore, are very attractive for matchmaking in the electronic market place, especially in cases where large numbers of buyers and sellers need to be matched efficiently.

1 Introduction

The impact of E-commerce trading is rising rapidly due to the enhancement of the internet and the customer's need to comfortably search and buy products online.

Simone A. Ludwig
Department of Computer Science, North Dakota State University,
Fargo, North Dakota, U.S.A.
e-mail: simone.ludwig@ndsu.edu

Thomas Schoene
Department of Computer Science, University of Saskatchewan,
Saskatoon, Saskatchewan, Canada
e-mail: ths299@mail.usask.ca

T. Ito et al. (Eds.): New Trends in Agent-Based Complex Automated Negotiations, SCI 383, pp. 81–98.
springerlink.com   © Springer-Verlag Berlin Heidelberg 2012

E-commerce trading is more efficient than alternative methods because of active pricing mechanisms, up-to-date databases, and streamlined procurement processes.

An electronic market platform usually requires buyers and sellers to exchange offers-to-buy and offers-to-sell. The goal of this exchange is to reach an agreement on the suitability of a transaction between buyers and sellers. A transaction transfers one or more objects (e.g. a product, money, etc.) from one agent to another and vice versa. The transaction can be described by sets of properties such as the delivery date for the transaction, the color of the object, or the location of the agent. A property has a value domain with one or more values [1].

Two kinds of the negotiating strategies commonly used are distinguished by the relationship with markets [2]:

- Cooperative Negotiation: For cooperative negotiations, multiple items of trading attributes are negotiable (quality, quantity, etc.). Because the participants have their own preferences on different trading attributes, it is possible for the two parties to obtain satisfactory results out of the bargain.
- Competitive Negotiation: For this kind of negotiation, the objectives of both sides are conflicting. When one side gets more benefit out of a certain bargain, the other side will face some loss. This is a zero-sum game from the point of view of the game theory. Auction is an example of this kind of negotiation.

Hence, cooperative negotiation is a better bargaining method for two parties, as each can obtain a satisfactory result. Many trading attributes can be coordinated in such a bargain (quality, quantity, payment, etc.), and the participants can negotiate based on their preferences.

Automatic negotiation plays the most important role among processes in an E-marketplace as it seeks to maximize benefits for both sides. In advanced multi-agent systems, when a buyer and a seller are interested in trading with each other, both will be represented by agents who may hold opposite grounds initially, and then will start to negotiate based on available information in order to reach common ground. Two critical challenges are faced here. The first one is to provide a global platform in which efficient searching, publishing, and matching mechanisms can be enforced in order to minimize the load and make processes more efficient. The second challenge is to come up with autonomous processes that can capture essential human negotiation skills such as domain expertise, learning and inference. During the matching process, parties advertise offers-to-buy or offers-to-sell. These offers include consumer/provider properties and constraints [3]. Constraints expressed by one party represent the reservation value set on some aspect of a given transaction. The reservation value is the minimum value the party wants to achieve, and thus is similar to the reservation price in an auction [4].

Electronic negotiations are executed in the intention and agreement phase of an electronic market. A definition of matchmaking in electronic negotiations can be found in [5]. The steps in the intention phase are as follows:

- Offer and request specification: The agents have to specify offers and requests indicating their constraints and preferences towards the transaction object. This specification may also include the provision of signatures or the definition of timestamps. This specification can be executed instancing a candidate information object that has been designed for this specific marker.
- Offer and request submission: To submit an offer or a request, the agent can actively send the offer or request to a specific agent (middle agent) or notify the middle agent of the completion of the specification.
- Offer validation: When the middle agent receives the offer or request, the information object is checked for completeness and the compliance with certain rules.

The steps in the agreement phase are defined as follows:

- Offer and request matching: The aim of this phase is to find pairs of offers and requests that stratify potential counterparts for a transaction execution. This includes the identification of all offers that match a given request. In this phase, within the matchmaking framework, a ranking of all offers with respect to the current request is computed and returned as a ranked list to the requesting agent.
- Offer and request allocation: In this task the counterparts for a possible transaction is determined using the information from the matching and scoring phase. The duties of the single agents are determined and assigned. The final configuration has to be determined in this phase if the selected offers and requests still feature certain value ranges or options.
- Offer and request acceptance: This final stage confirms the acceptance of the terms and conditions, which have been determined. The agents have to accept the conditions in order to execute the transactions and complete the deal.

Electronic auctions, also referred to as internet auctions, are also widely studied [6,7,8]. Internet auctions are seen as an electronic mechanism, which provides a digital version of allocation mechanisms. Electronic auctions define an economic mechanism that is designed to solve competitive resource allocation problems. Generally speaking, an electronic auction is the implementation of auction protocols using electronic media. In electronic markets, many dimensions can be considered that are too complex to express in non-electronic negotiations. Bichler et al. [9] state that the item characteristics are an important factor for determining the type of the appropriate negotiation and matchmaking mechanism to be applied. There are several terms that provide a framework for the design of negotiation characteristic referring to concrete item characteristics:

- Multi-phased auctions: Several phases are carried out determining the auctions outcome.
- Multi-stage auctions: Similar to multi-phased auctions, several stages have to be passed before the auction terminates. In this case the order of the stages is relevant.
- Multi-unit auctions: Multi-unit auctions describe auctions in which several units of the same object are auctioned.

- Multi-item auctions: Multi-item auctions describe auctions in which several, possibly heterogeneous, items are auctioned.
- Multi-attribute auctions: As Bichler [10] defines multi-attribute auctions as auctions by which the overall score computation is not limited to bids on the mere price, but several aspects of a good can be bid on. Usually, a virtual currency is introduced to provide the overall score, which in turn is mapped to a price.
- Multi-dimensional auctions: Bichler and Werthner [11] see multi-dimensional auctions as an umbrella term for multi-unit, multi-item and multi-attribute auctions.

Matchmaking plays a crucial role within electronic auctions. Within each bidding procedure a winner has to be determined. In single-attribute auctions, where only the price can be bid on, the highest price wins. In this case, no sophisticated matchmaking mechanism has to be introduced. In multi-attribute actions however, where several attributes are bid on, mechanisms are needed to compute an overall score. In general, the more attributes and sub-attributes are provided, i.e. the more complex the bid structure is defined, the more complex matchmaking procedures have to be introduced.

This paper is organized as follows. The following section introduces related work that has been done in the past. Section 3 describes the approaches implemented. In Section 4, the experiment setup and the results are given. Section 5 concludes this paper with the findings and gives an account to future research.

2 Related Work

Many matchmaking models and frameworks have been introduced in the past and are introduced below.

SILKROAD [12] presents a matchmaking framework that is based on constraint satisfaction. The offers and requests can be supported in a subject-based structure. This enables a wide variety of application domains. However, the matchmaking mechanism is limited.

The INSPIRE system [13] provides communication support among offering parties and requesting parties to submit individual preferences. The matchmaking is performed by the parties that accept or reject an offer or request. The advantage of this system is the openness in negotiating the position. However, as system-based matchmaking is missing, the aim of the system is not to provide a complete matchmaking procedure, but to provide matchmaking support to the participating parties.

INSULA [14] provides a rule-base matchmaking unit. Several attributes can be supported in a domain specific way. These attributes are then matched using the constraints of the attributes of the counterpart applying the matching rules. This design limits the matchmaking complexity, but enables domain specific attributes.

The EMS framework by Stroebel and Stolze [1] contains a matchmaking unit that allows free definable offer and request structures. These structures are application dependent and adaptable. A disadvantage that limits the application domains of this framework is that the offer and request attributes are matched only based on the constraints and no discrete values are allowed.

The SHADE approach by Kuokka and Harada [15] defines one of the first generic free text matchmakers. This system has the advantage to provide distance functions from information retrieval that make it fairly flexible and domain independent. The main disadvantage of this system is that it does not provide mechanisms to match structure offer and requests.

The IMPACT matchmaker by Subrahmanian et al. [16] is based on a simple offer and request structure. It allows only verb-noun terms, consisting of two verbs and a noun as offer and request structures. On the other hand, hierarchies enable a powerful matchmaking that can also be applied in specific domains. However, it is limited due to its fixed offer and request structures.

LARKS [17] is one of the most powerful matchmaking approaches known so far. It provides several matchmaking stages responsible for different processes, which enables high matchmaking quality. However, the application domains are limited as the offer and request structure is static.

The GRAPPA framework [18] combines the benefits of a generic approach with the key advantages of domain specific solutions. The GRAPPA framework is explicitly defined to enable flexibility of generated offer and request structures as well as to be completely domain independent by supporting distance function and metric interfaces that allow easy integration of domain dependent and generic functions.

The aim of the research provided in this paper has a slightly different focus. First of all, it is envisioned that matchmaking support will increase in future, not only because electronic market places are seen to becoming more and more utilized, but also the number of participants in the marketplace (buyers and sellers) will rise gradually. Therefore, a robust, time-efficient and scalable assignment algorithm is needed to perform the task of matchmaking. One optimal algorithm, known as the Munkres algorithm, has a cubic time complexity and therefore, does not scale well with increasing numbers of buyers and sellers. Thus, approximate algorithms are necessary, which on one hand provide an optimized assignment, and on the other hand scale linearly with increasing numbers of buyers and sellers.

3 Matchmaking Approaches

Multi-attribute auctions not only use the price of the item, but a combination of different attributes of the deal, such as delivery time, terms of payment, product quality, etc. However, this requires a mechanism that takes multiple attributes of a deal into account when allocating it to a participant. The mechanism should automate multilateral auctions/negotiations on multiple attributes of a deal. Three matchmaking approaches are presented in the following subsections. The matchmaking function is introduced first. Then, the GA algorithm, PSO algorithm, Munkres algorithm and the Random approach are explained in detail.

3.1 Matchmaking Function

The problem of matchmaking in an electronic marketplace is having an effective algorithm that can match multiple buyers and sellers efficiently, while optimizing multiple attributes. The problem is twofold: firstly, multiple buyers requesting the same product should be satisfied, and secondly, the assignment process of the buyers and the sellers should be optimized. Please note that one buyer can only be matched with one seller.

Matchmaking is concerned with matching buyers with sellers based on a range of negotiation attributes. The five objective measures or negotiation attributes are: quality, quantity, price, delivery and payment. We assume that sellers have a fixed value s_i for each attribute i and the buyers have a fixed value b_i for each attribute i as well. The match value for each attribute i is calculated as follows based on the difference between b_i and s_i:

$$v_i = 1 - |b_i - s_i| \tag{1}$$

The match score m_i of one buyer-seller pair is the sum of all five match values multiplied by the weight value w_i for attribute i, divided by the number of negotiation attributes. The weight value w_i allows specifying preferences on the different attributes:

$$m_i = \frac{1}{n} \sum_{i=1}^{n} w_i v_i \tag{2}$$

The matchmaking algorithms implemented use Equations (1) and (2) to calculate the match score for each buyer-seller pair. However, the aim of this research is to match several buyers with several sellers as closely as possible. Therefore, the overall match score of p buyers and p sellers is defined as:

$$o = \frac{1}{p} \sum_{j=1}^{p} m_j \tag{3}$$

3.2 Genetic Algorithm

GA is a global optimization algorithm that models natural evolution [19]. In GA, individuals form a generation. An individual (similar to a particle in the PSO) corresponds to one match. The match is implemented as a vector, which is also referred to as a chromosome. Dimensions in the vector correspond to sellers, and values correspond to buyers. Thus, if the vector has value 3 at its 5th position (dimension), buyer 3 is matched with seller 5. Every number representing a buyer can only be present at one position in the vector, otherwise, the vector is a non-valid match.

At the beginning, the first population is randomly initialized. After that, the fitness of the individuals is evaluated using the fitness function (Equations (1)–(3)).

After the fitness is evaluated, individuals have to be selected for paring. The selection method used is tournament selection. Always two individuals are paired, resulting in an offspring of two new individuals. In the pairing phase, a random crossover mask is used, i.e. the positions (dimensions) for which crossover occurs are selected randomly. If crossover occurs at certain positions (dimensions), individuals that are mated exchange their values at that position and the resulting individuals are used as offspring. The crossover has to make sure that the offspring present a valid match. Therefore, if two values are exchanged, other positions in the two match vectors are usually effected as well. The offspring faces mutation with a certain low probability. After mutation, the fitness of the offspring is calculated. Then, either all individuals from the last generation compete against the whole offspring, or the offspring only compete with its corresponding parents. In this implementation, all individuals from the old generation compete with all individuals in the new generation. To achieve this all individuals are ordered by their fitness score, using a non-recursive advanced quicksort algorithm. After sorting, the lower half is truncated. After the new generation is selected, the GA will start over, and continue with parent selection and crossover again.

Configurable parameters in the implementation include number of iterations, tournament size (the size of the tournament used to select parents), crossover probability, effected positions (how many positions are set to crossover in the crossover mask), and mutation probability.

3.3   Particle Swarm Optimization Approach

PSO, as introduced in [20], is a swarm based global optimization algorithm. It models the behavior of bird swarms searching for an optimal food source. The movement of a single particle is influenced by its last movement, its knowledge, and the swarm's knowledge. In terms of a bird swarm this means, a bird's next movement is influenced by its current movement, the best food source it ever visited, and the best food source any bird in the swarm has ever visited.

PSO's basic equations are:

$$x(t+1) = x_i(t) + v_{ij}(t+1) \tag{4}$$

$$v_{ij}(t+1) = w(t)v_{ij}(t) + c_1r_{1j}(t)(xBest_{ij}(t) - x_{ij}(t))$$
$$+ c_2r_{2j}(t)(xGBest_j(t) - x_{ij}(t)) \tag{5}$$

where x represents a particle, i denotes the particle's number, j the dimension, t a point in time, and v is the particle's velocity. $xBest$ is the best location the particle ever visited (the particle's knowledge), and $xGBest$ is the best location any particle in the swarm ever visited (the swarm's knowledge). w is the inertia weight and used to weigh the last velocity, c_1 is a variable to weigh the particle's knowledge, and c_2 is a variable to weigh the swarm's knowledge. r_1 and r_2 are uniformly distributed random numbers between zero and one. PSO is usually used on real and not dis-crete problems. In order to solve the discrete assignment problem using the PSO approach, several operations and entities have to be defined. This implementation follows and adapts the implementation for solving the traveling salesman problem as described in [21]. First, a swarm of particles is required. A single particle repre-sents a match, i.e., every particle's position in the search space must correspond to a possible match. The match, that is the position, is implemented as a vector. Dimen-sions in the vector correspond to sellers, and values correspond to buyers. There-fore, if the vector has value 3 at its 5th position (dimension), buyer 3 is matched with seller 5. Every number representing a buyer has to be unique, otherwise, the vector represents a non-valid match.

Velocities are implemented as lists of changes that can be applied to a particle (its vector) and will move the particle to a new position (a new match). Changes are exchanges of values, i.e., an entity containing two values that have to be exchanged within a vector. This means that any occurrence of the first value is replaced by the second value, and any occurrence of the second is exchanged by the first value. Fur-ther, minus between two matches (particles), multiplication of a velocity with a real number, and the addition of velocities have to be defined. Minus is implemented as a function of particles. This function returns the velocity containing all changes that have to be applied to move from one particle to another in the search space. Multi-plication is implemented as a function of velocities. Multiplication randomly deletes single changes from the velocity vector, if the multiplied real number is smaller than one. If the real number is one, no changes are applied. For a real number larger than one, random changes are added to the velocity vector. Addition is also implemented as a function of velocities. When a velocity is added to another velocity, the two lists containing the changes will be concatenated.

The implemented PSO uses guaranteed convergence, which means that the best particle is guaranteed to search within a certain radius, implying that the global best particle will not get trapped in local optima.

Configurable parameters in the implementation include numbers of particles (size of the swarm), number of iterations, c_1 (the weighting of the local knowledge), c_2 (the weighting of the global knowledge), w (the weighting of the last velocity), radius (defines the radius in which the global best particles searches randomly), global best particle swarm optimization (determines whether global best particle swarm or local best particle swarm optimization is used), and neighborhood size (defines the neighborhood size for local best particle swarm optimization).

3.4 Munkres Algorithm

The Hungarian algorithm is a combinatorial optimization algorithm that solves the assignment problem in polynomial time. It was developed by Harold Kuhn in 1955 [22,23], who gave the name "Hungarian method" because the algorithm was largely based on the earlier works of two Hungarian mathematicians: Denes Koenig and Jeno Egervary.

In 1957 James Munkres reviewed the algorithm and observed that it is (strongly) polynomial. Since then, the algorithm has been known also as Kuhn-Munkres algorithm or Munkres assignment algorithm [24,25]. The time complexity of the original algorithm was $O(n^4)$, however, Edmonds and Karp, and independently Tomizawa noticed that it can be modified to achieve an $O(n^3)$ running time.

The Munkres algorithm is used to serve as a benchmark for the negotiation matchmaking as it is an optimal algorithm. However, one drawback that the Munkres algorithm has is the time complexity of $O(n^3)$ as previously mentioned, and hence is not very time efficient. The assignment problem as formally defined by Munkres [24]:

"Let r_{ij} be a performance ratings for a man M_i for job J_i. A set of elements of a matrix are said to be independent if no two of them lie in the same line ("line" applies both to a row and a column of a matrix). One wishes to choose a set of n independent elements of the matrix (r_{ij}) so that the sum of the element is minimum."

Similarly, the problem of matchmaking can be defined as an $n \times m$ buyer-seller matrix, representing the match scores of each buyer with every other seller. The match score matrix is the matrix where each element of the matrix represents the match score for an individual buyer-seller pair. The Munkres algorithm works on this matrix, to assign the buyer requests to the sellers, as to achieve an overall maximum total match score. Please note that one buyer can only be matched with one seller.

An implementation developed by Nedas in Java (freely available at [26]) was used and slightly adopted to serve as a benchmark for this matchmaking investigation, as it provides the base match score of the optimal assignment of buyer and seller pairs.

3.5 Random Approach

The random approach, as the name indicates, randomly selects and assigns buyer-seller pairs. Sellers and buyers are held in separate vectors and one buyer and one seller is randomly selected and matched up until both vectors are empty and all sellers and buyers are uniquely assigned.

4 Experiments and Results

All four algorithms as introduced in the previous section were implemented using Java. Experiments were designed to measure the overall match score and the execution time of all approaches. The GA and PSO algorithms were furthermore analyzed with regards to the population size (GA algorithm), as well as with regards to the particle sizes used (PSO algorithm). All measurement points shown are average results taken from 30 runs, in order to guarantee statistical equal distribution. The data sets for the buyers and sellers were randomly generated.

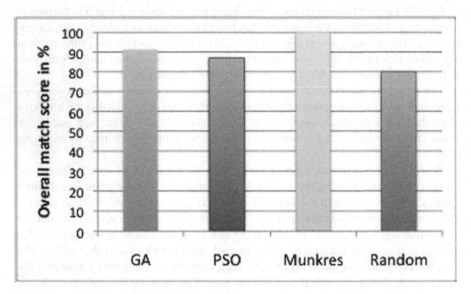

Fig. 1 Overall match score of all approaches

The following parameters are used throughout the experiments if not stated otherwise: The number of buyer-seller pairs is set to 500; for the GA algorithm the population size is set to 1000, the crossover and mutation probabilities are set to 0.6 and 0.05 respectively, the tournament size is 10%, and elitism is set; for the PSO algorithm the iteration is set to 100, the number of particles is set to 10, the weight is set to 0.001, the local and global constants are both set to 0.5, and guaranteed convergence is enabled.

Figure 1 shows the overall match score of all approaches. The optimal matchmaking of 100% is achieved by the Munkres algorithm, followed by the GA algorithm with an average match score of 91.34 ± 4.37, followed by the PSO algorithm with an average match score of $87.03 \pm 4.62\%$, and $79.82 \pm 0.63\%$ achieved by the random approach.

Figure 2 shows the execution time in seconds of all approaches. The Munkres algorithm has the longest running time with 212.97 seconds, followed by the PSO algorithm with 54.29 seconds. The GA algorithm is fairly fast with a run time of 35.59 seconds, whereas the fastest algorithm is, as expected, the random approach with 5.93 milliseconds (not visible on Figure 2 because its execution time is too small).

Figure 3 shows the overall match score for the increasing numbers of buyer-seller pairs. As expected, it can be seen that the GA as well as the PSO algorithm have higher match score values (GA is outperforming the PSO algorithm) than the random approach, and it can be stated that the match scores are slightly higher for smaller numbers of buyer-seller pairs.

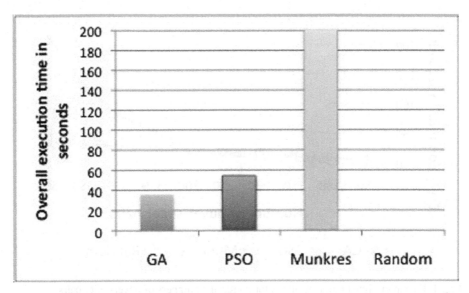

Fig. 2 Overall execution time of all approaches

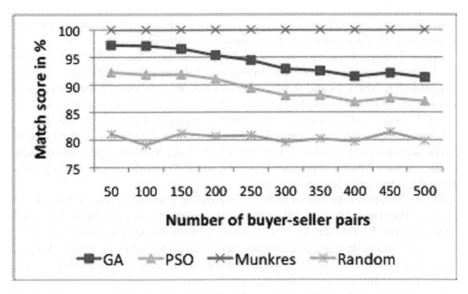

Fig. 3 Match score of all algorithms for increasing numbers of buyer-seller pairs

Figure 4 shows the execution time of the algorithms for the increasing numbers of buyer-seller pairs. As mentioned before, the execution time of the Munkres algorithm is cubic as can clearly be observed. The GA and PSO algorithm as well

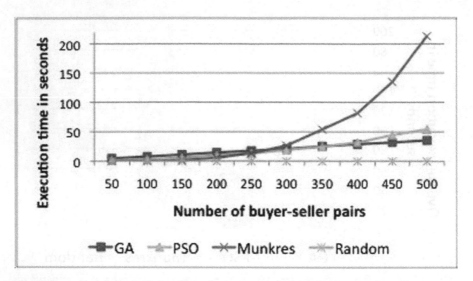

Fig. 4 Execution time of all algorithms for increasing numbers of buyer-seller pairs

as the Random approach are less time-consuming; whereby the Random approach outperforms all algorithms showing the smallest execution time.

Figure 5 shows the overall match score for an increase in the number of iterations. As the increase in the number of iterations has only an effect on the GA and PSO algorithm, the Munkres and Random approaches are only plotted for comparison reasons. It can be seen that the overall match score of the GA algorithm as well as for the PSO algorithm is increasing with the increase in the number of iterations.

Figure 6 shows the execution time for an increase in the number of iterations. It can be seen that the execution time of the GA, PSO and Random approach are much smaller than the Munkres approach. Both GA and PSO algorithm scale linearly with increasing numbers of iterations.

Figure 7 displays the match score for increasing population size of the GA algorithm. Population sizes of 250 up to 2500 were investigated and the match score varies between 83.71% and 85.77%, as observed with the previous measurements.

Figure 8 shows the execution time for increasing population size of the GA algorithm. Population sizes of 250 up to 2500 were investigated and the execution time shows a linear increase with larger population sizes.

Figure 9 shows the match score of the PSO algorithm for increasing numbers of particles. The match score varies between 82.62% and 84.33% as previously observed.

Figure 10 shows the execution time of the PSO algorithm for increasing numbers of particles. It can be seen that the execution time increases linearly with increasing numbers of particles. An execution time of 72.69 seconds is measured for the particle size of 500.

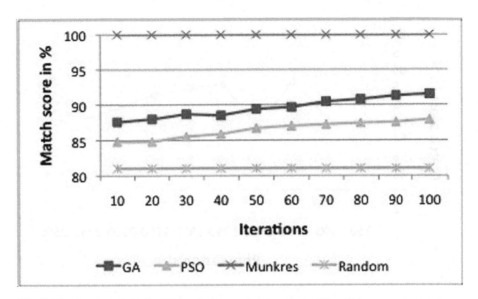

Fig. 5 Match score of all algorithms for increasing numbers of iterations

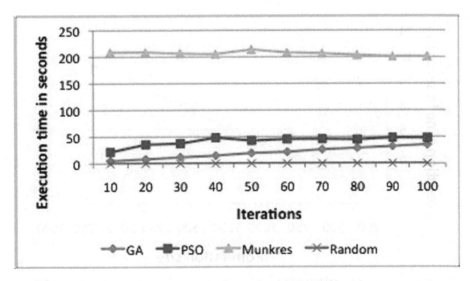

Fig. 6 Execution time of all algorithms for increasing numbers of iterations

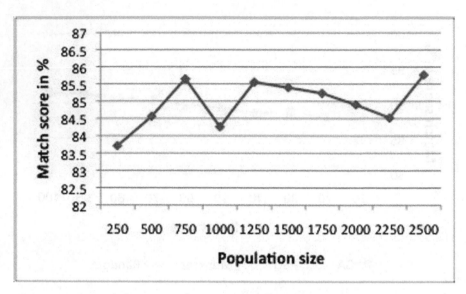

Fig. 7 Match score of GA algorithm for increasing numbers of population size

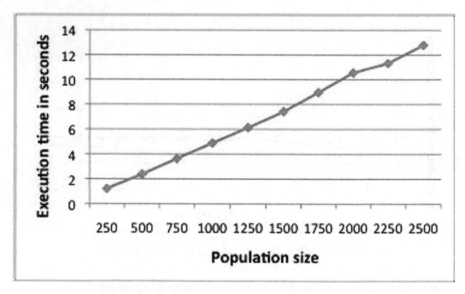

Fig. 8 Execution time of GA algorithm for increasing numbers of population size

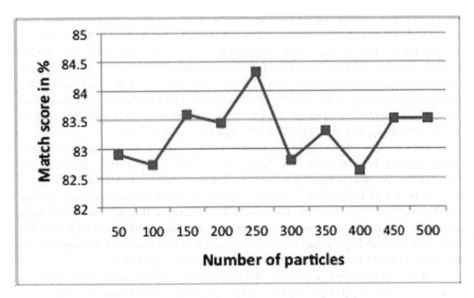

Fig. 9 Match score of PSO algorithm for increasing numbers of particles

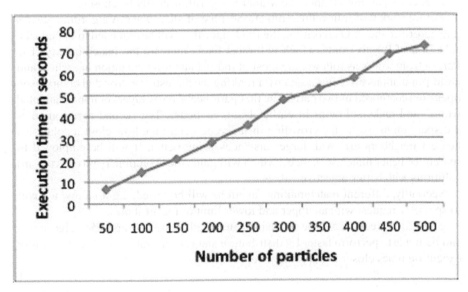

Fig. 10 Execution time of PSO algorithm for increasing numbers of particles

5   Conclusion

This paper investigated four approaches for the matchmaking of multi-attribute auctions, and in particular the matchmaking of multiple buyers and sellers based on five attributes. The scenario was envisioned, that in future the number of multi-attribute auctions will rise, and therefore, an efficient algorithm is necessary to provide this matchmaking functionality. The Munkres algorithm provides an optimal assignment, however, it has a cubic computational time complexity, and thus, does not scale very well. Therefore, other approximate approaches were investigated such as a GA-based approach as well as a PSO approach was chosen.

Overall, the two approximate algorithms performed fairly well, reducing the execution time by 17% with the GA algorithm and 25% with the PSO algorithm for 500 buyer-seller pairs, thereby achieving match scores of 91.1% and 87.0% respectively. The GA algorithm in particular achieved higher match scores as well as having shorter execution times. It seems that the theory following the evolutionary principles works very good for the scenario of matchmaking. Even though the PSO approach achieves higher match scores than the Random approach, the application of swarm intelligence is outscored by the evolutionary approach (GA).

As a recommendation, given that the demand for matchmaking services is rising, approximate algorithms are necessary as in most auction scenarios time is of essence. Therefore, the GA and PSO algorithm should be chosen, in particular the GA algorithm. However, if the time is not critical and if the match quality of buyers and sellers is paramount, then the Munkres algorithm should be chosen.

Future work will follow three directions. First of all, as the GA algorithm always outperformed the PSO algorithm, the PSO algorithm will be extended to include the selection of Pareto fronts, which will most likely increase the match quality. The Pareto front works as follows: the most fit individuals from the union of archive and child populations are determined by a ranking mechanism (or crowded comparison operator) composed of two parts. The first part 'peels' away layers of non-dominated fronts, and ranks solutions in earlier fronts as better. The second part computes a dispersion measure, the crowding distance, to determine how close a solution's nearest neighbors are, with larger distances being better. It will be employed to search for better match sequences, which will guide the evolutionary process toward solutions with better objective values.

Secondly, different matchmaking functions will be tested, e.g. to allow a buyer to specify a request with an upper and lower limit of the attributes.

The third direction will be to investigate whether the GA and PSO algorithms can be made to perform faster by distribution and parallelization in order to achieve execution times closer to the Random approach.

References

1. Ströbel, M., Stolze, M.: A Matchmaking Component for the Discovery of Agreement and Negotiation Spaces in Electronic Markets. Group Decision and Negotiation 11, 165–181 (2002)

2. Guttman, R.H., Maes, P.: Cooperative vs. Competitive Multi-Agent Negotiations in Retail Electronic Commerce. In: Klusch, M., Weiss, G. (eds.) CIA 1998. LNCS (LNAI), vol. 1435, p. 135. Springer, Heidelberg (1998)
3. Hoffner, Y., Schade, A.: Co-operation, Contracts, Contractual Matchmaking and Binding. In: Proceedings EDOC 1998, pp. 75–86. IEEE, Piscataway (1998)
4. Ströbel, M.: Communication design for electronic negotiations on the basis of XML schema. In: Proceedings of the tenth international conference on World Wide Web, Hong Kong, May 01-05, pp. 9–20 (2001)
5. Stroebel, M., Weinhardt, C.: The Montreal Taxonomy for Electronic Negotiations. Group Decision and Negotiation 12(2), 143–164 (2003)
6. Kumar, M., Feldman, S.: Internet auctions. In: Proceedings of the 3rd USENIX Workshop on Electronic Commerce, Boston, USA, pp. 49–60 (1998)
7. Koppius, O.: Electronic multidimensional auctions: Trading mechanisms and applications. In: Homburg, V., Janssen, M., Wolters, M. (eds.) Edispuut Workshop 1998: Electronic Commerce — Crossing Boundaries. The Netherlands, Rotterdam (1998)
8. Koppius, O., Kumar, M., van Heck, E.: Electronic multidimensional auctions and the role of information feedback. In: Hansen, H.R., Bichler, M., Mahrer, H. (eds.) Proceedings of the 8th European Conference on Information Systems (ECIS 2000), Vienna, Austria, pp. 461–468. Association for Information Systems (2000)
9. Bichler, M., Kaukal, M., Segev, A.: Multi-attribute auctions for electronic procurement. In: Proceedings of the 1st IBM IAC Workshop on Internet Based Negotiation Technologies, Yorktown Heights, USA, pp. 154–165 (1999)
10. Bichler, M.: Trading financial derivatives on the web - an approach towards automating negotiations on OTC markets. Information Systems Frontiers 1(4), 401–414 (2000)
11. Bichler, M., Werthner, H.: Classification framework for multidimensional, multi-unit procurement auctions. In: Proceedings of the DEXA Workshop on Negotiations in Electronic Markets, Greenwich, U.K, pp. 1003–1009 (2000)
12. Stroebel, M.: Design of roles and protocols for electronic negotiations. Electronic Commerce Research Journal, Special Issue on Market Design 1(3), 335–353 (2001)
13. Lo, G., Kersten, G.: Negotiation in electronic commerce: Integrating negotiation support and software agent technologies. In: Proceedings of the 29th Atlantic Schools of Business Conference (1999)
14. Benyoucef, M., Alj, H., Vezeau, M., Keller, R.K.: Combined negotiations in e-commerce: Concepts and architecture. Electronic Commerce Research Journal - Special issue on Theory and Application of Electronic Market Design 1(3), 277–299 (2001)
15. Kuokka, D., Harada, L.: Integrating information via matchmaking. Journal of Intel ligent Information Systems 6(2-3), 261–279 (1996)
16. Subrahmanian, V.S., Bonatti, P., Dix, J., Eiter, T., Kraus, S., Ozcan, F., Ross, R.: Heterogenous Agent Systems. MIT Press (2000); ISBN: 0262194368
17. Sycara, K., Widoff, S., Klusch, M., Lu, J.: Larks: Dynamic matchmaking among heterogeneous software agents in cyberspace. Autonomous Agents and Multi-Agent Systems 5, 173–203 (2002)
18. Veit, D., Müller, J.P., Schneider, M., Fiehn, B.: Matchmaking for autonomous agents in electronic marketplaces. In: Proceedings of the Fifth International Conference on Autonomous Agents AGENTS 2001, Montreal, Quebec, Canada, pp. 65–66. ACM, New York (2001)
19. Holland, J.H.: Adaptation in Natural and Artificial Systems. University of Michigan Press, Ann Arbor (1975)

20. Kennedy, J., Eberhart, R.C.: Particle swarm optimization. In: Proceedings of IEEE International Conference on Neural Networks (1995)
21. Clerc, M.: Discrete particle swarm optimization – illustrated by the traveling salesmen problem. In: New Optimization Techniques in Engineering. Springer, Heidelburg (2004)
22. Kuhn, H.W.: The Hungarian method for the assignment problem. Naval Research Logistics 52(1) (2005)
23. Kuhn, H.W.: The hungarian method for solving the assignment problem. Naval Research Logistics 2, 83 (1955)
24. Munkres, J.: Algorithms for the Assignment and Transportation Problems. Journal of the Society for Industrial and Applied Mathematics 5, 32 (1957)
25. Bourgeois, F., Lassalle, J.C.: An extension of the munkres algorithm for the assignment problem to rectangular matrices. Commun. ACM 14(12) (1971)
26. Nedas, K.: Munkres' (Hungarian) Algorithm, Java implementation (2008), http://konstantinosnedas.com/dev/soft/munkres.htm (last retrieved on March 2009)

A Coalition Structure-Based Decision Method in B2B E-Commerce Model with Multiple-Items

Satoshi Takahashi and Tokuro Matsuo

Abstract. This paper proposes a new B2B electronic commerce model by using bidding information in double auctions. In B2B electronic commerce, buyers try to purchase in multiple items at the same time, since a buyer develops something products by using purchased items. Also suppliers have an incentive of making coalitions, since buyers want to purchase in multiple items. A mechanism designer has to consider an optimal mechanism which calculates an optimal matching between buyers and suppliers. But to find an optimal matching is very hard, since a mechanism calculates all combinations between buyers and suppliers. Consequently, we propose a calculation method which has two steps, first a mechanism determines winners of buyers' side, then, determines coalitions and winners of suppliers by using the result of buyers' side. This paper also discusses the improved method with dynamical mechanism design by using the bidding information. The auction protocol trees are expressed by all possible results of auctions. The result of each auction is recorded and stored with bidding data and conditions for subsequent auctions. Advantages of this paper are that each developer can procure the components to develop a certain item and tasks are allocated to suppliers effectively. The previous result of auction data can be available to shorten the period of winner determinations.

1 Introduction

In recent years, electronic commerce has been developing as multiple forms of trading and is researched and analyzed by many researchers. Regarding B2C (Business

Satoshi Takahashi
1-1-1, Tennoudai, Tsukuba, Ibaraki, Japan, Graduate School of System & Information
Engineering, University of Tsukuba
e-mail: takahashi2007@e-activity.org

Tokuro Matsuo
4-3-16, Johnan, Yonezawa, Yamagata, Japan, Graduate School of Science & Engineering,
Yamagata University
e-mail: matsuo@yz.yamagata-u.ac.jp

T. Ito et al. (Eds.): New Trends in Agent-Based Complex Automated Negotiations, SCI 383, pp. 99–109.
springerlink.com   © Springer-Verlag Berlin Heidelberg 2012

to Consumer) and C2C (Consumer to Consumer) trading models like electronic auctions, many researchers work on them in researches of multi-agent system and mechanism design [4][3][7]. In existing researches, they assume incomplete information on their trading mechanism and did not design auction protocols with bidding value information. However, in actual e-marketplace, if we are able to use the information, it is possible to design further effective trading schemes and to make high-speed algorithms. Also, in B2B (Business to Business), there are many auction-based trading models; auctions where a company procures resources/items are typical. In procurement auctions, the procurement parties generally try to acquire multiple items simultaneously. On the other hands, many suppliers provide same sort of items in the market. When they negotiate with each other to trade, multiple suppliers have an incentive to make a coalition if procurement parties bid complementary. However, coalition formation needs negotiations by suppliers and has demerit regarding the negotiation monetary costs and time costs. To solve the problem, we propose a new dynamical coalition formation method for suppliers by using result of auctions of procurement parties. The procurement auction is a combinatorial auction with multiple bidders and suppliers. In such auctions, it is difficult to analyze the properties and features and is not available since the computational costs of tasks/items allocation are very large. Thus, we propose, in this paper, a method to improve the mechanism dynamically by reusing bidding trends information.

The rest of the paper is organized as follows. Section 2 clarifies the problem establishment for our work. In Section 3, we give some definitions and assumptions. We also describe the proposed model. In Section 4, we describe the coalition formation method for suppliers. Then, in Section 5, an auction protocol with bidding information. We provide our concluding remarks in Section 6.

2 Problem Setting

We analyze our proposed model using procurement dealings of manufacturers. A product in which a manufacturer provides to end-users in an actual marketplace is configured from some parts. Manufacturers procure their parts for creating a new added value as the product. For example, in auto industry, when a manufacturer makes a auto, he use tens of thousands parts, and he purchases these parts from each supplier. There are attractive values of these parts not for end-users but for the manufacturer. The value of these parts influences the value of the product. Therefore, it is important for manufacturers not only to reduce the purchasing cost, but also to optimize their costs.

The parts which compose the product is complementally goods for manufacturers, since if manufacturers can purchase whole of parts they can make and provide the product stably. Hence the suppliers have an incentive that they cooperate with each other, they sell their parts as bundle for developing newly profit. Also, there is an advantage in which manufacturers do not procure from two or more suppliers.

In this paper, we define the dealing for end-users as primary trading, and the B2B dealing of parts suppliers and manufacturers as secondary trading. On secondary trading, manufacturers have to purchase optimal parts bundle, since there are two or more parts suppliers. Our proposed B2B trading is included in the class of multi-sided and multi-items dealing in which several sellers and purchasers exist. There are some double auction models as the class of multi-sided trading[8][6][11][1]. The double auction is the dealing in which buyers and sellers bid for the same kind of good, and the mechanism makes several pair of the buyer and the seller. We are not able to use this auction protocol to our proposed model, since in double auction model it is not possible to deal different kind of goods. Therefore, we use combinatorial auction model to secondary trading scheme. In this scheme, suppliers and manufacturers bid individually. Suppliers' auction is single item auction, and manufacturers' auction is combinatorial auction. We use a manufacturers auction's result to decide to coalition structures for suppliers' side, and use a bidding result of suppliers' auction to make coalitions. In our research, we discuss a case that suppliers bid individual value.

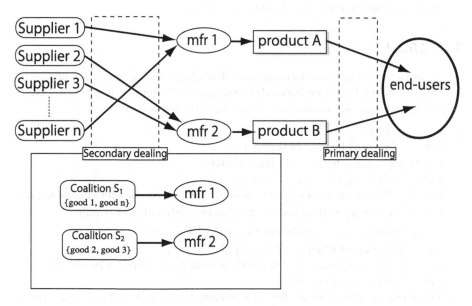

Fig. 1 Business Model

3 Preliminaries

We show the secondary trading model. Figure 1 shows relationship primary trading and secondary trading. Manufacturers try to optimize procurement plan for to reduce costs in business process. When we analyze concretely in market supply process, manufacturer 1 and 2 procure necessary parts from some parts n suppliers. Manufacturer 1 and 2 make product A, B by using procured parts. And they provide their products to end-users. In secondary trading, suppliers who provide some parts make pseudo-coalitions. It seems that a manufacturer dealing the parts with a pseudo-coalition. In primary trading, each manufacturer is providing the end user with service that assembles procured-parts.

It is effective to use auction mechanism when manufacturers procure in secondary trading. In optimal auction mechanism, it is known to satisfy the following properties under some conditions[5][10].

Strategyproofness. An auction satisfies strategyproofness if bidding true evaluations is dominant strategy.

Pareto efficiency/optimal. We say an auction protocol is Pareto optimal when the sum of all participants' utilities (including that of the auctioneer), i.e., the social surplus, is maximized in a dominant strategy equilibrium.

Individual rationality. When all participants' utilities are not negative, the designed mechanism exhibits individual rationality.

3.1 Model

We describe several terms and assumptions. Participators of the auction is suppliers and manufacturers. Each suppliers bids nonnegative value for providable items. Also each manufacturers bids nonnegative value for desirable items.

Definition of term
- Let $N = \{1,...,n\}$ be a set of suppliers.
- Let $M = \{1,...,m\}$ be a set of manufacturers.
- Let $G = \{a_1,...,a_k\}$ be a set of items(parts).
- Let $\mathscr{B} \subseteq 2^G$ be a subset family of G, and $B_j \subseteq \mathscr{B}$ be a bundle set of manufactures j. If the auction treat kth items, the number of combination of bundle is $2^{|k|} - 1$. Hence a bundle set of manufacturer j is $B_j = \{B_j^1,...,B_j^{2^{|k|}-1}\}$.
- Let $c_i(a_1)$ be a cost when supplier i supplies item a_k.
- Let $v_j(B_j^l)$ be a evaluate value in which manufacturer j bids for bundle B_j^l.
- Let p_j be a payoff of manufacture j.
- We define an allocation set for each manufacturers as $X = \{x_1,...,x_m | \forall a,b \in M, x_a, x_b \in X, x_a \cap x_b = \phi\}$. In this regard, for all $j \in M$, $x_j \in B_j$.

Assumption 1(quasi-linear utility). An utility u_j of manufacture j is defined by difference between manufacture j's payoff p_j and his evaluate value v_j as $u_j = v_j - p_j, \forall j \in M$. Such an utility is called a quasi-linear utility, we assume quasi-linear utility. Also, we define supplier i's utility as $u_i = \hat{p}_j - c_i, \forall i \in N, \forall j \in M$. In this regard, $\hat{p}_j$ is payoff to supplier $i(\hat{p}_j \leq p_j)$.

Assumption 2(complementarily bids). We assume manufacturers' evaluation values satisfy a following condition. If there exists $B_j^l, B_j^{\hat{l}}(B_j^l \cap B_j^{\hat{l}} = \phi)$ for all $j \in M$:

$$v_j(B_j^l) + v_j(B_j^{\hat{l}}) \leq v_j(B_j^l \cup B_j^{\hat{l}})$$

Assumption 3(completeness). There is no shill bidding, since all participators in this auction finish registration before auction.

Assumption 4(no risk). There is no risk to dealing items. Item's quality does not different for same kind of it, and there is no non-performance of contract.

In this paper, we assume that each supplier can provide only one item.

3.2 Allocation Mechanism

Our trading model uses GVA(Generalized Vickrey Auction)mechanism. GVA mechanism satisfies optimal auction's properties when there is no shill bidding[3]. We calculate allocation of each manufacturers and their payoff by using GVA mechanism. If $X^\star$ is an optimal allocation, then $X^\star$ is calculated as follows.

$$X^\star = argmax_{X=\{x_1,...,x_m\}} \sum_{j \in M} v_j(x_j) \tag{1}$$

We define an allocation without manufacturer j as follows.

$$X_{-j}^\star = argmax_{X \setminus x_j} \sum_{M \setminus j} v_m(x_m) \tag{2}$$

We calculate manufacturer j's payoff p_j.

$$p_j = \sum_{m \neq j, x_m^\star \in X_{-j}^\star} v_m(x_m^\star) - \sum_{m \neq j, x_m^\star \in X^\star} v_m(x_m^\star) \tag{3}$$

We define a winner set by GVA mechanism as $M' = \{j | x_j^\star \neq \phi\}$.

4 Suppliers Coalition

We get information which is an optimal allocation $X^\star$ and payoff $p_j, \forall j \in M'$. Next, we make a coalition of suppliers by using the manufacturer's auction result. The manufacturer's auction result includes an allocation set $X^\star$ and a payoff set $P = \{p_j \mid j \in M'\}$. These two sets are supplier's coalition structures and outcomes of each coalition. We show a concept of coalition decision in figure 2. In this figure, the auction result is given by $X^\star = \{\{a,b,c\}_1, \{d,e\}_2, \{f,g,h,i\}_3\}$, $P = \{150_1, 70_2, 200_3\}$. Then, suppliers are able to make three coalitions which depends on the auction result $X^\star$. It seems that the $X^\star$-induced coalitions are a best coalition set by the auction property. Also, we reduce a calculation cost, since we use

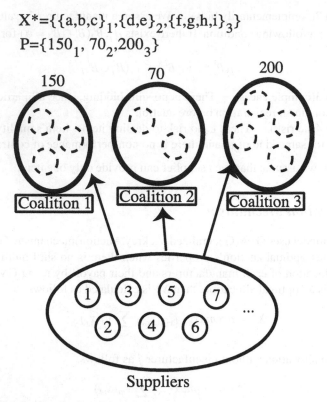

$$X^*=\{\{a,b,c\}_1,\{d,e\}_2,\{f,g,h,i\}_3\}$$
$$P=\{150_1, 70_2, 200_3\}$$

Fig. 2 Auction result reuse concept

decided coalition structure information beforehand. If we do not use the auction information, we do not make a best coalition set, since the manufacturer's information is incomplete.

When we use an auction information, we do not hope to show the information to suppliers. Because the suppliers have an incentive of bid-rigging by knowing the information. Hence, the information is used by only an auctioneer or a mechanism.

Let $S = \{S_{x_1},...,S_{x_m}\}$ be a coalition set based on manufacturers' auction. Our mechanism do the suppliers' auction for making coalition. In suppliers' auction, we analyze only individual bidding. Individual bidding is that each supplier's bid does not demand each other. In this condition, each coalition's total cost is defined as summation of each supplier's cost. Hence total cost of coalition S_{x_j} is defined as follows.

$$C(S_{x_j}) = \sum_{i \in S_{x_j}, a_k \in x_j} c_i(a_k)$$

We make an optimal coalition set by following formula.

$$S^\star = argmax_{S=\{S_{x_1},...,S_{x_m}\}} \sum_{j \in M} \{p_j - C(S_{x_j})\} \tag{4}$$

We define an optimal coalition set without supplier i as $S^\star_{-i} = \{S^\star_{x_1,-i}, ..., S^\star_{x_m,-i}\}$.

$$S^\star_{-i} = argmax \sum_{j \in M} \{p_j - C(S_{x_j,-i})\}$$

Also each supplier's outcome o_i when he participates in coalition S_{x_j} is calculated as follows.

$$o_i = c_i(a_k) + \{C(S^\star_{x_j,-i}) - C(S^\star_{x_j})\} \tag{5}$$

Suppliers can get additionally surplus, since we assume complementally bidding.

$$p_j \geq \sum_{i \in S^\star_{x_j}} o_i$$

Let $Sur = p_j - \sum_{i \in S^\star_{x_j}} o_i$ be a coalition $S^\star_{x_j}$'s surplus. Each supplier who participates in coalition $S^\star_{x_j}$ reallocates it. In this paper we assume a reallocation method as equitable distribution.

4.1 Protocol

We describe secondary trading's protocol.

Step1.   Each supplier declares own providable item and its nonnegative cost. Also each manufacturer declares own desirable bundles and its nonnegative evaluation values. Manufacturers' evaluation values satisfies complementally.

Step2.   We calculate an optimal allocation and its payoff based on GVA mechanism.

Step3.   We decide supplier's coalition structure based on the manufacturers' auction result. And we calculate a supplier's optimal coalition set.

Step4.   We calculate each supplier's outcome, and surplus.

We show an example. Given a manufacturers' bid set as table 1, and a suppliers' bid set as table 2. The result of manufacturers' auction is (mfr 1, $\{a,b,c\}$). Manufacture 1's payoff is $p_3 = \$35$. Hence we decide optimal coalition with structure $\{a,b,c\}$ in suppliers' auction. In this case, coalition $S_{x^\star_3}$ is $S_{x^\star_3} = \{sup. 3, sup. 5, sup. 6\}$, and its total cost is $C(S_{x^\star_3}) = 24$. We calculate each supplier's outcome as follows.

$$o_3 = c_3(\{a\}) + \{C(S^\star_{x_3,-3}) - C(S^\star_{x_3})\} \tag{6}$$
$$= 7 + (27 - 24) = 10$$
$$o_5 = c_5(\{b\}) + \{C(S^\star_{x_3,-5}) - C(S^\star_{x_3})\} \tag{7}$$
$$= 7 + (25 - 24) = 8$$
$$o_7 = c_7(\{a\}) + \{C(S^\star_{x_3,-7}) - C(S^\star_{x_3})\} \tag{8}$$
$$= 10 + (26 - 24) = 12$$

As a result, the surplus of coalition $S_{x^\star_3}$ is $\$5$. Therefore each supplier gets $\$5/3$.

Table 1 Bidding table of manufacturers

Mfr	{a}	{b}	{c}	{a,b}	{b,c}	{a,c}	{a,b,c}
mfr 1	5	7	3	20	15	10	30
mfr 2	7	5	0	20	0	0	20
mfr 3	6	4	9	14	18	25	35

Table 2 Bidding table of suppliers

Suppliers	{a}	{b}	{c}
supplier 1	10	0	0
supplier 2	0	8	0
supplier 3	7	0	0
supplier 4	0	0	12
supplier 5	0	7	0
supplier 6	0	0	10

5   Dynamic Protocol Developing Based on Bidding Information

In this section, we consider a dynamic protocol developing method by using the
auction information. Also we treat a simple combinatorial auction model. A general
combinatorial auction calculates an item allocation and winners, which it tries to
maximize social surplus by using all bids' values. The more items and participators
are increasing, the more huge calculation cost is spent in an allocation problem.
Also, if we try to find an optimal solution, we have to do a full search.

If we solved the problems, we earn two advantages. First, we can adapt combinatorial auction mechanism to real-time Internet auction. Second, we can prevent
illegal tenders by the pattern of bidding.

In an auction which assumes incomplete information, a mechanism is able to
know only bundles and its evaluations. The mechanism is not able to do high speed
calculation in this condition. Therefore our new method is to declare own bidding
patterns. The bidding pattern is, for example, complementally bids, substitute bids
and normal bids. Each pattern has an accent which is about bidding. For example, a
complementally bid is that a agent interests to only to get multi object at one auction.
If a bidding pattern is complementally bid, we search only bundles and reduce the
searching area. Also, in a substitute bid, we can ignore bundles bidding.

We introduce an new method in which a mechanism develops a protocol dynamically for high speed calculation.

Protocol's steps used by dynamic protocol are stored in database. A mechanism
uses their stored steps for to develop an better protocol. When a mechanism develops
it, a mechanism selects best steps from database by using the auction information.
In the concrete, a mechanism adapts following sub-protocols to dynamic protocol
based on whole of bidders' bids patterns.

(1) A case of one or more complementally bidders. A mechanism compares the most large size bundles in complementally bidders. A agent who bids the most expensive value is to be semi-winner(complementally bidder). Other agents are loser. By way of exception, let B be a bundle and $\mathscr{B}$ be a set of bundles, if $\mathscr{B} \setminus B \neq \phi$ and $\mathscr{B} \setminus B \cap B = \phi$ is justified, then we can find B' such that $B \cap B' = \phi$ and $B' \in \mathscr{B} \setminus B$. Hence we decide the semi-winner(complementally bidder) from $\mathscr{B} \setminus B$.

(2) A case of one or more substitute bidders. A mechanism focuses on single object bids in substitute bidders, calculates a combination that social surplus becomes the maximum. Also the mechanism decides the semi-winners(substitute bidder).

(3) Final determination. A mechanism determines the final winners from step 1's semi-winner, step 2's semi-winner and standard bidders.

(4) Payoff determination. In the case of that the semi-winner(complementally bidder) was final winner, we consider two situations. If the semi-winner(complementally bidder)'s allocation is equal the number of total items, the payoff of the final winner is a maximum social surplus which is decided from the semi-winner(substitute bidders) and standard bidders. Else if there exists two or more semi-winners(complementally bidders), a mechanism focuses on bundle bids, and calculates payoffs based on GVA method.

In the case of that the semi-winner(substitute bidder) was final winner, a mechanism focuses on single object bids in substitute bidders, calculates a payoff based on GVA method.

Otherwise, a mechanism calculate a payoff by using evaluation without complementally bidders based on GVA method.

We can reduce a calculation cost by using dynamic protocol developing. The auction process including upgrade protocol is shown by figure 3 A mechanism selects some sub-protocols from the database, and develops an upgrade protocol by using the bidding patterns. In upgrade protocol, a mechanism decides the semi-winners by using the patterns, and does the re-auction. On the special case, a mechanism reduces some sub-protocols. In the concrete, if all bidders are complementally bidders, a mechanism can reduce the sub-protocols without considering substitute and standard bidders.

6   Discussion

In this section, we discuss our protocol's strategyproofness. According to Norm et al.[9], it is known that there does not exist approximately mechanisms which satisfy fully strategyproofness. Hence, our mechanism does not satisfy fully strategyproofness, since it is approximately allocation method. However, Kothariet al.[2] discusses approximately strategyproofness. They give an upper bound of increasing utility by using approximation degree when agents do all strategy operations. And they propose approximately strategyproofness by using upper bound. We give a definition of approximately strategyproofness as follows.

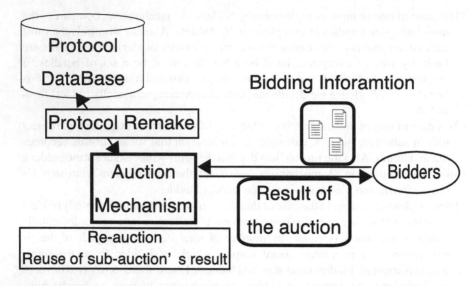

Fig. 3 Modified auction process

Definition 1(approximately strategyproofness). A mechanism is $\varepsilon-$approximately
strategyproofness if and only if agents' utilities increase at most ε by all
strategies.

We assume our mechanism has $\varepsilon-$approximately allocation method($\varepsilon > 0$). Let
V be a maximum social surplus by using our mechanism. Our mechanism satisfy
$(\varepsilon/1+\varepsilon)V-$approximately strategyproofness. Also, it is known that true bidding is
dominant strategy in approximately strategyproof mechanism

7 Conclusion

We propose a trading protocol for a class of multi-sided combinatorial auction. We
use gradual GVA mechanism which operates GVA to both manufacturers' auction
and suppliers' auction. Gradual GVA mechanism uses manufacturers' auction result
to decide the supplier's coalition structures.

Also we introduce a new method which is dynamic protocol developing based
on information reuse. In dynamic protocol developing, we reuse the sub-protocol in
the database. If a mechanism develops an upgrade protocol dynamically, the winner
determination problem is solved faster.

Our future work is, first, to develop an algorithm which guarantees an approximately allocation. Second, we verify the stability of the protocol by running the
simulation of bidding pattern. And we consider a new mechanism for our trading
model, since GVA mechanism has a very large computation cost.

References

1. Friedman, D., Rust, J.: The Double Auction Market. Addison-Wesley Publishing Company (1993)
2. Kothari, A., Parkes, D.C., Suri, S.: Approximately-strategyproof and tractable multi-unit auction. Decision Support System 39, 105–121 (2005)
3. Krishna, V.: Auction Theory. Academic Press (2002)
4. Leyton-Brown, K., Shoham, Y., Tennenholtz, M.: Bidding clubs: institutionalized collusion in auctions. In: Proceedings of the 2nd ACM Conference on Electronic Commerce, pp. 253–259 (2000)
5. Matsuo, T.: A reassuring mechanism design for traders in electronic group buying. Applied Artificial Intelligence 23, 1–15 (2009)
6. McAfee, P.R.: A dominant strategy double auction. Journal of Economic Theory 56, 434–450 (1992)
7. Milgrom, P.: Putting Auction Theory to Work. Cambridge University Press (2004)
8. Myerson, R.B., Satterthwaite, M.A.: Efficient mechanism for bilateral trading. Journal of Economic Theory 29(2), 265–281 (1983)
9. Nisan, N., Rosen, A.: Computationally feasible vcg mechanism. In: Proc. on 2nd ACM Conference on Electronic Commerce EC 2000, pp. 242–252 (2000)
10. Yokoo, M., Sakurai, Y., Matsubara, S.: The effect of false-name bids in combinatorial auctions: new fraud in internet auctions. Games and Economic Behavior 46, 174–188 (2004)
11. Yokoo, M., Sakurai, Y., Matsubara, S.: Robust double auction protocol against fales-name bids. Decision Support System 39, 241–252 (2005)

Part II
Automated Negotiating Agents Competition

The First Automated Negotiating Agents Competition (ANAC 2010)

Tim Baarslag, Koen Hindriks, Catholijn Jonker, Sarit Kraus, and Raz Lin

Abstract. Motivated by the challenges of bilateral negotiations between people and automated agents we organized the first automated negotiating agents competition (ANAC 2010). The purpose of the competition is to facilitate the research in the area bilateral multi-issue closed negotiation. The competition was based on the GENIUS environment, which is a **G**eneral **E**nvironment for **N**egotiation with **I**ntelligent multi-purpose **U**sage **S**imulation. The first competition was held in conjunction with the Ninth International Conference on Autonomous Agents and Multi-agent Systems (AAMAS-10) and was comprised of seven teams. This paper presents an overview of the competition, as well as general and contrasting approaches towards negotiation strategies that were adopted by the participants of the competition. Based on analysis in post–tournament experiments, the paper also attempts to provide some insights with regard to effective approaches towards the design of negotiation strategies.

1 Introduction

Negotiation is an important process to form alliances and to reach trade agreements. Research in the field of negotiation originates from various disciplines including

Tim Baarslag · Koen Hindriks · Catholijn Jonker
Man Machine Interaction Group
Delft University of Technology
e-mail: {T.Baarslag,K.V.Hindriks,C.M.Jonker}@tudelft.nl

Sarit Kraus · Raz Lin
Computer Science Department
Bar-Ilan University
e-mail: {linraz,sarit}@cs.biu.ac.il

Sarit Kraus
Institute for Advanced Computer Studies
University of Maryland

T. Ito et al. (Eds.): New Trends in Agent-Based Complex Automated Negotiations, SCI 383, pp. 113–135.
springerlink.com   © Springer-Verlag Berlin Heidelberg 2012

economics, social science, game theory and artificial intelligence (e.g., [2, 19, 28]). Automated agents can be used side by side the human negotiator embarking on an important negotiation task. They can alleviate some of the efforts required of people during negotiations and also assist people that are less qualified in the negotiation process. There may even be situations in which automated negotiators can replace the human negotiators. Another possibility is for people to use these agents as a training tool, prior to actually performing the task. Thus, success in developing an automated agent with negotiation capabilities has great advantages and implications.

In order to help focus research on proficiently negotiating automated agents, we have organized the first automated negotiating agents competition (ANAC). The principal goals of the ANAC competition are as follows:

- Encouraging the design of agents that can proficiently negotiate in a variety of circumstances,
- Objectively evaluating different bargaining strategies,
- Exploring different learning and adaptation strategies and opponent models, and
- Collecting state-of-the-art negotiating agents, negotiation domains, and preference profiles, and making them available and accessible for the negotiation research community.

A number of successful negotiation strategies already exist in literature [8, 9, 14, 15, 24]. However, the results of the different implementations are difficult to compare, as various setups are used for experiments in ad hoc negotiation environments [13, 22]. An additional goal of ANAC is to build a community in which work on negotiating agents can be compared by standardized negotiation benchmarks to evaluate the performance of both new and existing agents.

In designing proficient negotiating agents, standard game-theoretic approaches cannot be directly applied. Game theory models assume complete information settings and perfect rationality [29]. However, human behavior is diverse and cannot be captured by a monolithic model. Humans tend to make mistakes, and they are affected by cognitive, social and cultural factors [3, 7, 21, 26]. A means of overcoming these limitations is to use heuristic approaches to design negotiating agents. When negotiating agents are designed using a heuristic method, we need an extensive evaluation, typically through simulations and empirical analysis.

We have recently introduced an environment that allowed us to evaluate agents in a negotiation competition such as ANAC: GENIUS [22], a General Environment for Negotiation with Intelligent multi-purpose Usage Simulation. GENIUS helps facilitating the *design* and *evaluation* of automated negotiators' strategies. It allows easy development and integration of existing negotiating agents, and can be used to simulate individual negotiation sessions, as well as tournaments between negotiating agents in various negotiation scenarios. The design of general automated agents that can negotiate proficiently is a challenging task, as the designer must consider different possible environments and constraints. GENIUS can assist in this task, by allowing the specification of different negotiation domains and preference profiles by means of a graphical user interface. It can be used to train human negotiators by means of negotiations against automated agents or other people. Furthermore, it can be used to teach the design of generic automated negotiating agents.

With GENIUS in place, we organized ANAC with the aim of coordinating the research into automated agent design and proficient negotiation strategies for bilateral multi-issue closed negotiation, similar to what the Trading Agent Competition (TAC) achieved for the trading agent problem [36].

We believe ANAC is an important and useful addition to existing negotiation competitions, which are either aimed at human negotiations or have a different focus, as we explain in Section 2.8.

The remainder of this paper is organized as follows. Section 2 provides an overview over the design choices for ANAC, including the model of negotiation, tournament platform and evaluation criteria. In Section 3, we present the setup of ANAC 2010 followed by Section 4 that layouts the results of competition. In Section 5 we discuss proposed outline for the future ANAC competitions, and finally, Section 6 outlines our conclusions and our plans for future competitions.

2   General Design of ANAC

One of the goals of ANAC is to encourage the design of agents that can negotiate in a variety of circumstances. This means the agents should be able to negotiate against any type of opponent within arbitrary domains. Such an open environment lacks a central mechanism for controlling the agents' behavior, and the agents may encounter human decision-makers who make mistakes and whose behavior is diverse, cannot be captured by a monolithic model, is affected by cognitive, social and cultural factors, etc. [3, 21]. Examples of such environments include online markets, patient care-delivery systems, virtual reality and simulation systems used for training (e.g., the Trading Agent Competition (TAC) [36]). The use of open environments is important as the automated agent needs to be able to interact with different types of opponents, who have different characteristics, e.g. people that origin from different countries and cultures. Automated negotiation agents capable of negotiating proficiently with people thus must deal with the fact that people are diverse in their behavior and each individual might negotiate in a different manner.

The design of the competition was focused on the development of negotiating strategies, rather than other aspects of the negotiation process (though not less important aspects) such as preference elicitation, argumentation or mediation. The setup of ANAC was designed to make a balance between several concerns, including:

- Strategic challenge: the game should present difficult negotiation domains in a real–world setting with real–time deadlines.
- Multiplicity of issues on different domains, with a priori unknown opponent preferences.
- Realism: realistic domains with varying opponent preferences.
- Clarity of rules, negotiation protocols, and agent implementation details.

We specify the general choices that were made for ANAC with regard to negotiation model and the tournament setup.

2.1 Negotiation Model

In order to define the setup of the negotiation competition, we first introduce the model of negotiation that we use. In this competition, we only consider *bilateral* negotiations, i.e. a negotiation between two parties. The parties negotiate over *issues*, and every issue has an associated range of alternatives or *values*. A negotiation outcome consists of a mapping of every issue to a value, and the set Ω of all possible outcomes is called the negotiation *domain*. The domain is common knowledge to the negotiating parties and stays fixed during a single negotiation session.

We further assume that both parties have certain preferences prescribed by a *preference profile* over Ω. These preferences can be modeled by means of a utility function U that maps a possible outcome $\omega \in \Omega$ to a real-valued number in the range $[0, 1]$. In contrast to the domain, the preference profile of the players is private information.

Finally, the interaction between negotiating parties is regulated by a *negotiation protocol* that defines the rules of how and when proposals can be exchanged. We use the alternating-offers protocol for bilateral negotiation as proposed in [32], in which the negotiating parties exchange offers in turns. The alternating-offers protocol conforms with our criterion to have simplicity of rules. It is widely studied and used in literature, both in game-theoretic and heuristic settings of negotiation (a non-exhaustive list includes [10, 19, 20, 28, 29]).

2.2 Tournament Platform

As a tournament platform to run and analyze the negotiations, we use the GENIUS environment (General Environment for Negotiation with Intelligent multi-purpose Usage Simulation) [22]. GENIUS is a research tool for automated multi–issue negotiation, that facilitates the design and evaluation of automated negotiators' strategies. It also provides an easily accessible framework to develop negotiating agents via a public API. This setup makes it straightforward to implement an agent and to focus on the development of strategies that work in a general environment.

GENIUS incorporates several mechanisms that aim to support the design of a general automated negotiator. The first mechanism is an analytical toolbox, which provides a variety of tools to analyze the performance of agents, the outcome of the negotiation and its dynamics. The second mechanism is a repository of domains and utility functions. Lastly, it also comprises repositories of automated negotiators. In addition, GENIUS enables the evaluation of different strategies used by automated agents that were designed using the tool. This is an important contribution as it allows researchers to empirically and *objectively* compare their agents with others in different domains and settings.

2.3   Domains and Preference Profiles

The specifications of the domain and preferences, such as the constitution and valuation of issues, can be of great influence on the negotiation outcome. We assume that all agents have complete information about the domain, but the preference profile of the players is private information. Thus, if a strategy attempts to tailor its offers to the needs of the opponent, it is required to model the opponent. As the amount of information exchanged during the negotiation is limited in a closed negotiation, the size of the domain has a big impact on the learning capabilities of the agents.

For example, for ANAC 2010, we used a domain named Itex–Cypress [18], in which a buyer and a seller of bicycle components negotiate about issues such as the price of the components and delivery times. There are few possible values per issue, creating a domain of only 180 potential offers. Such a small domain simplifies the task of getting a good picture of the opponent's preferences by studying its proposal behavior.

Due to the sensitivity to the domain specifics, negotiation strategies have to be assessed on negotiation domains of various sizes and of various complexity [13]. Therefore, we selected several domains for ANAC, with different characteristics.

Negotiation strategies can also depend on whether preferences of the negotiating parties are opposed or not. The notion of weak and strong opposition can be formally defined [17]. Strong opposition is typical of competitive domains, when a gain for one party can be achieved only at a loss for the other party. Conversely, weak opposition means that both parties achieve either losses or gains simultaneously.

Negotiation strategies may depend on the opposition of the preferences. For example, in the case of Itex–Cypress the opposition is strong as it concerns a manufacturer negotiating with a consumer. In such a case the parties have naturally opposing requirements. Hence, the selection of preference profiles should also take into account that the preference profiles have a good variety of opposition.

As stated in the negotiation model, we assume that the negotiating parties have a certain preference profile which can be modeled by a utility function U that maps a possible outcome to a real-valued number in $[0, 1]$.

There are various ways to represent such a utility function (cf. [16]). For ANAC, we have chosen domains without (preferential) dependencies between issues, i.e.: the contribution of every issue to the utility is linear and does not depend on the values of other issues. An advantage of independence between issues is that algorithms that search for a proposal with a particular utility can be implemented in a computationally efficient way. It also makes it easier for negotiation strategies to efficiently model the opponent's preferences, as it reduces the amount of information that is to be learned by a preference learning technique.

When the tradeoffs between issues are (preferentially) independent, then an additive scoring system is appropriate [30]. Therefore, we assume that utility functions are additive [16, 30, 31]. That is, in a domain Ω with an outcome $\omega = (\omega_1, \ldots, \omega_m)$, we assume the utility function has the following form:

$$U(\omega) = \sum_{i=1}^{m} w_i \cdot u_i(\omega_i),$$

where the w_i are normalized weights (i.e. $\sum w_i = 1$) and $u_i(\omega_i)$ is an evaluation function with range $[0, 1]$ for every individual issue x_i.

2.4 Protocol and Deadline

To add to the realism of the protocol, we can supplement it with a deadline and discount factors. We impose a real–time deadline on the negotiation process for both theoretical and practical reasons. The pragmatic reason is that without a deadline, the negotiation might go on forever, especially without any discount factors. Secondly, with unlimited time an agent may simply try a huge amount of proposals to learn the opponent's preferences. Another reason for introducing a real-time deadline in the alternating offers protocol is the various challenges it poses to the competitors, as described in Section 2.7. We believe that a negotiation model with a real-time deadline comes closer to realistic negotiation environment.

2.5 Scoring

We now move on to a formal description of the utility and final scoring functions. Let $\mathscr{D}$ be our set of domains. For every domain $D \in \mathscr{D}$ two preference profiles exist, $\mathscr{P}^D = \{P_1^D, P_2^D\}$. Let $\mathscr{A}$ be the set of competing agents, with $|\mathscr{A}| = n$. Every agent competes against all other agents on all domains (see Section 2.3), alternating between the two preference profiles defined on that domain.

Suppose agent A negotiates with B on domain $D \in \mathscr{D}$, where A has the first preference profile P_1^D and B uses P_2^D. If they reach a certain outcome ω, in which A receives the associated utility $U(\omega)$, then we denote this utility with

$$U_{A \to B}^D.$$

Our evaluation metric is defined as follows. Every agent A plays against all agents $B \in \mathscr{A}$, with the exception that A will not play itself. The score for A is averaged over all trials, playing with both preference profiles P_1 and P_2 (e.g., on the Itex–Cypress domain, A will play both as Itex and as Cypress against all others). That is, for each profile $P \in \mathscr{P}^D$ an average utility $\bar{u}_D(A, P)$ is calculated for each agent:

$$\bar{u}_D(A, P_1) = \frac{1}{n-1} \sum_{B \in \mathscr{A} \setminus \{A\}} U_{A \to B}^D,$$

and

$$\bar{u}_D(A, P_2) = \frac{1}{n-1} \sum_{B \in \mathscr{A} \setminus \{A\}} U_{B \to A}^D.$$

The average utility is then re-normalized using the maximum and minimum utility achieved by all other agents for that profile:

$$\tilde{u}_D(A, P) = \frac{\bar{u}_D(A, P) - \min\limits_{B \in \mathscr{A} \setminus \{A\}} \bar{u}_D(B, P)}{\max\limits_{B \in \mathscr{A} \setminus \{A\}} \bar{u}_D(B, P) - \min\limits_{B \in \mathscr{A} \setminus \{A\}} \bar{u}_D(B, P)}.$$

This gives a score per profile, which is averaged over the two profiles in the domain to give an overall normalized domain-score. The domain-score is then averaged over all trials and the final score $s(A)$ of an agent is determined as an average of the domain-scores:

$$s(A) = \frac{1}{|\mathscr{D}|} \sum_{D \in \mathscr{D}} \left(\frac{1}{2} \tilde{u}_D(A, P_1) + \frac{1}{2} \tilde{u}_D(A, P_2) \right).$$

2.6 ANAC Rules

To enter the competition the teams had to register and upload their agents by a given deadline. The time frame for the submission allowed teams to upload their agents for compliance checks prior to the official deadline.

The tournament itself consists of several rounds. If more than 16 agents are submitted, a qualifying round is held to select the best 16 agents. If an insufficient number of agents is submitted we preserved the right to either run one final round, or to add agents from the repository available at GENIUS at the moment in which we run the qualifying rounds. During the qualifying rounds, tournaments will be played with four agents each, the winners of those tournaments go to the next qualifying round until there are 16 agents in the competition. The top 16 agents proceed to the quarter finals. During the quarter finals rounds, four tournaments are played with four agents each, the top two agents then proceed to the semi-finals. During the semi-finals, two tournaments are played with four agents each, the best two agents proceed to the finals. The finals consist of one tournament with the best four agents of the semi-finals. The best agent wins the competition. In case two agents share the best score, the average results of the two agents when playing against each other will be used to determine the winner.

The domains and preference profiles used during the competition are not known in advance and were designed by the organizing team. The size of the domains can be up to 10 issues. The structure of a negotiation domain is fixed before the tournament and cannot be changed during negotiation.

An agent's success is measured using the evaluation metric (see Section 2.5) in all negotiations of the tournament for which it is scheduled.

Agents can be disqualified for violating the spirit of fair play. In particular, the following behaviors were strictly prohibited:

- Design an agent in such a way that it benefits some specific other agent.
- Communication with the agent during the competition.
- Alter the agent during the competition.
- Denial-of-service attacks.
- Agents employ API operations for the purpose of occupying or loading the game servers.

2.7 Challenges

Our aim in designing ANAC was to provide a strategic challenge on multiple accounts. According to our goals set forward in Section 1, we selected real-life domains containing multiple issues and preference profiles that are unknown to the opponent. Our setup guarantees the participating teams have to deal with the following challenges:

Incomplete information

Suppose an agent wants to model the utility function of the opponent. Because of incompleteness of information about the opponent, it has to learn the opponent's preferences during the negotiation by studying the proposal behavior. Our protocol only allows proposals to be exchanged, so the communication between the agents is very limited. This prevents agents to share information about their preferences, other than by their proposal behavior. Consequently, if agents want to use an opponent model to make effective proposals, they have to make use of a sophisticated learning technique.

Domain complexity

Analyzing the domain for beneficial outcomes is essential when implementing an efficient proposal strategy. Even when an agent has gathered information about its opponent, it still has to be able to find win–win situations, for example by computing the Pareto frontier.

Real-time deadline

Dealing with all of the above while taking into account a real–time deadline. Agents should be more willing to concede near the deadline, as a break-off yields zero utility for both agents. A real–time deadline also makes it necessary to employ a strategy to decide when to accept an offer. Deciding when to accept involves some prediction whether or not a significantly better opportunity might occur in the future.

Some parts of ANAC are less demanding: the utility functions are (linearly) additive functions, so there are no dependencies between issues. This means the utility of an offer can be effectively computed and conversely, a proposal of a certain utility can be easily generated. Moreover, additive utility functions make it fairly easy to enumerate proposals from best to worst. Secondly, agents are completely aware of their own preferences and the corresponding utility values.

2.8 Related Competitions

Some parts of the Trading Agent Competition (TAC) and the Agent Reputation Trust (ART) Competition also relate to automated agents negotiation. We provide a short description of both competitions and outline the differences with ANAC. For 2010, TAC is divided into three games [12, 27, 34, 36]:

TAC SCM

TAC Supply Chain Management was designed to simulate a dynamic supply chain environment. Agents have to compete to secure customer orders and components required for production. In order to do so, the agents have to plan and coordinate their activities across the supply chain. Participants face the complexities of supply chains, which admits a variety of bidding and negotiation strategies.

TAC/AA

In the TAC Ad Auction, game entrants design and implement bidding strategies for advertisers in a simulated sponsoring environment. The agents have to bid against each other to get an ad placement that is related to certain keyword combinations in a web search tool. The search engine simulates clicks and sale conversions, yielding revenues for the advertiser. The advertiser strategies have to decide which keywords to bid on, and what prices to offer. Therefore, the strategies have to optimize their data analysis and bidding tactics to maximize their profit.

CAT

The CAT Competition or TAC Market Design is a reverse of the normal TAC game: as an entrant you define the rules for matching buyers and sellers, while the trading agents are created by the organizers of the competition. Entrants have to compete against each other to build a robust market mechanism that attracts buyers and sellers. The market needs to make profit by setting appropriate commission fees and at the same time attract enough traders, while also adapting to changing environmental conditions.

Some elements of TAC have similar challenges as posed by ANAC, especially the games of TAC SCM and TAC/AA. The games can get very complex and the domains of the games are specifically chosen to model a certain scenario of a trading agent problem. On the other hand, the entrants of ANAC have to consider very general negotiation domains when they design their agents. This makes it very easy to implement a simple agent that can participate in ANAC.

We believe that the ANAC is already very challenging despite its seemingly simple setup. The current design of ANAC poses a lot of challenges, such as a real timeline and the incomplete information of the opponent's preferences (see also Section 2.7). Moreover, a transparent setup makes it easier to allow insight into the implications of design choices.

The Agent Reputation Trust (ART) Competition [6, 11] is also a negotiating agent competition with a testbed that allows the comparison of different strategies. The ART competition simulates a business environment for software agents that use the reputation concept to buy advices about paintings. Each agent in the game is a service provider responsible for selling its opinions when requested. The agent can exchange information with other agents to improve the quality of their appraisals. The challenge is to perceive when an agent can be trusted and to establish a trustworthy reputation.

Compared to ANAC, the focus of ART is more on trust: the goal is to perceive which agents can be trusted in a negotiation process and what reputation should be attributed to each agent.

3 Design of ANAC 2010

ANAC 2010 was held at the Ninth International Conference on Autonomous Agents and Multiagent Systems (AAMAS-10) in Toronto, Canada, with presentations of the participating teams and a closing discussion (see Section 5). AAMAS is a well-suited platform to host the competition, as it is the premier scientific conference for research on autonomous agents and multiagent systems, which includes researchers on automated negotiation. It brings together an international community of researchers that are well-suited to tackle the automated agents negotiation challenges posed by ANAC.

3.1 Teams

ANAC 2010 had seven participating teams from five different universities, as listed in Table 1.

Table 1 The participating teams of ANAC 2010

IAMhaggler	University of Southampton
IAMcrazyHaggler	University of Southampton
Agent K	Nagoya Institute of Technology
Nozomi	Nagoya Institute of Technology
FSEGA	Babes Bolyai University
Agent Smith	TU Delft
Yushu	University of Massachusetts Amherst

3.2 Domains

We approached the design of ANAC to comply with the goals that were described in Section 1. Because ANAC is aimed towards multi-issue negotiations under uncertainty in open environments, we selected the following domains and profiles after the participating agents had been submitted. We aimed for a good spread of the relevant parameters, such as the number of issues, the number of possible proposals and the opposition of the domain (see Table 2).

Itex–Cypress

Our first scenario, taken from [18], is a buyer–seller business negotiation for one commodity. It involves representatives of two companies: Itex Manufacturing, a producer of bicycle components and Cypress Cycles, a builder of bicycles. There are four issues that both sides have to discuss: the price of the components, delivery times, payment arrangements and terms for the return of possibly defective parts. An example outcome would be:

($3.98, 45 days, payment upon delivery, 5% spoilage allowed).

The opposition is strong in this domain, as the manufacturer and consumer have naturally opposing needs and requirements. Altogether, there are 180 potential offers that contain all combinations of values for the four issues.

Zimbabwe–England

The second domain taken from [22, 23] involves a case where England and Zimbabwe are negotiating to reach an agreement in response to the world's first public health treaty: the World Health Organization's Framework Convention on Tobacco Control. The leaders of both countries must reach an agreement on five issues:

1. Funding amount
The total amount to be deposited into a fund to aid countries that are economically dependent on tobacco production. This issue has a negative impact on the budget of England and a positive effect on the economy of Zimbabwe. The possible values are no agreement, $10, $50 or $100 billion. Thus, this issue has a total of four possible values.

2. Other aid programs
The impact on other aid programs. If other aid programs are reduced, then this will create economic difficulties for Zimbabwe. Possible values are:

a. No reduction;
b. Reduction equal to half of the fund;
c. Reduction equal to the whole size of the fund;
d. No agreement.

Thus, a total of four possible values are allowed for this issue.

3.&4. Trade barriers

Trade issues for both countries. Zimbabwe and England can use trade barriers such as tariffs (taxes on imports) or they can abstain from restrictive trade barriers to increase imports from the other party.

There is a trade-off in revenue of these policies: tariffs increases short-time revenue, but can lead to higher consumers prices. Decreasing import is good for local industries but it can decrease costumer welfare due to the increase in costumer costs. There are actually two issues here: the trade barriers that either side decides to use. Zimbabwe's possible values are divided between

a. Reducing tariffs on imports;
b. Increasing tariffs on imports;
c. No agreement.

While England can choose between

a. Reducing imports;
b. Increasing imports;
c. No agreement.

Thus, a total of three possible values are allowed for each of the two issues.

5. Creation of a forum

A forum can be created to explore other arrangements for health-issues. Zimbabwe would like to establish such a fund, to be able to apply to other global health agreements in the future, while this would be costly for England. The four possible values are:

a. Creation of a fund;
b. Creation of a committee that will discuss the creation of a fund;
c. Creation of a committee that will develop an agenda for future discussions;
d. No agreement.

Consequently, the domain has a total of $4^3 \cdot 3^2 = 576$ possible agreements. England and Zimbabwe have contradictory preferences for the first two issues, but the other issues have options that are jointly preferred by both sides, making it a domain of medium opposition.

Travel

Our final domain has two persons negotiating to go on holiday to a location. From a small travel recommendation system we obtained multiple real–life profiles of travelers. They can each list their preferences on properties of a holiday destination:

1. Atmosphere.
2. Amusement.
3. Culinary.
4. Shopping.

5. Culture.
6. Sport.
7. Environment.

There are seven issues to discuss, all with a fairly large amount of choices. This leads to a big offers space of 188,160 possibilities. A sample negotiation outcome reads:

(*Hospitable, Nightlife and entertainment, International cuisine, Small boutiques, Art galleries, Outdoor activities, Parks and gardens*).

The opposition is weak in this domain, because traveling friends may have very compatible interests. Still the challenge is to find this optimal outcome in such a big search space.

Table 2 The domains used in ANAC 2010

	Itex–Cypress	**Zimbabwe–England**	**Travel**
Number of issues	4	5	7
Size	180	576	188,160
Opposition	Strong	Medium	Weak

3.3 Deadline

We impose a real–time deadline on the negotiation process for reasons stated in Section 2.4. In ANAC 2010, the agents are bound to three minutes *each* to deliberate. This allowed every agent to utilize the same CPU time, but it forces agents to keep track of the time that the opponent has left. This feature may undergo a small change in the next ANAC (see Section 5).

While the domains of the first ANAC competition did not include any discount factors, we do plan to add this feature to the next ANAC competition to be held in 2011 (see the discussion in Section 5).

4 Tournament Results

We describe the normalized domain scores of every agent in ANAC 2010 in Table 3. The normalized domain score is obtained by averaging the score against the other agents on multiple trials. All agents use both of the profiles that are linked to a domain (see Section 2.5 for more details on the scoring). The final score is listed in the last column, thus making *Agent K* the winner of ANAC 2010.

4.1 Overall Scoring

For every domain, due to the normalization of the scores, the lowest possible score is 0 and the highest is 1. The fact that the maximum and minimum score are not

Table 3 Normalized scores of every agent per domain

Rank	Agent	Score per domain			Avg
		Itex-Cyp	Eng-Zimb	Travel	
1	Agent K	0.901	0.712	0.685	0.766
2	Yushu	0.662	1.0	0.250	0.637
3	Nozomi	0.929	0.351	0.516	0.599
4	IAMhaggler	0.668	0.551	0.500	0.573
5	FSEGA	0.722	0.406	0	0.376
6	IAMcrazyHaggler	0.097	0.397	0.431	0.308
7	Agent Smith	0.069	0.053	0	0.041

always achieved, can be explained by non-deterministic behavior of the agents: the top-ranking agent on one domain does not always obtain the maximum score on every trial.

Agent K has won by a big margin, but it only managed to dominate on the Travel domain. On both Itex–Cypress and England–Zimbabwe, it earned second place after *Nozomi* and *Yushu*, respectively. Its consistent high scoring made *Agent K* the winner of ANAC. Only *IAMhaggler* managed to mirror this consistent scoring on all three domains.

4.2 Negotiation Strategies

We present a discussion of the strategies used by the participating agents. We compare the strategies by highlighting both common and contrasting approaches taken in the general strategic design. We are concerned with the following aspects of proposal strategies:

Proposal behavior

For every agent, we give a brief overview of the basic decisions that comprise the agents' inner proposal loop. We also describe the criteria for accepting an offer. Either of the two can be decided in a deterministic or non-deterministic manner.

Learning

In order to reach an advantageous negotiation agreement, it is beneficial to have as much information about the preference profile of an opponent as possible. If an agent can take into consideration the opponent's interests and learn during their interactions, then their utility might increase [37]. Because of the closed negotiation setting of ANAC, the negotiating parties exchange only proposals, but they do not share any information about their preferences. To overcome this problem, a negotiating agent may try to obtain a model of the preference profile of its opponent by means of learning.

For the participating agents, we are concerned how their strategies model the opponent.

Timing aspects

There are substantial risks associated with delaying the submission of a proposal at the end of the negotiation. These risks arise from unpredictable delays and can cause proposals to be received when the game is already over. Agents can try to estimate the length of their negotiation cycles to cope with these risks. The agents can then concede in the final phase of the negotiation, or place their proposals in some calculated amount of time before the end. We examine whether the agents make any predictions on how many time is left and how they use this information.

Table 4 gives an overview of the strategies of all agents. In the "Time dependent" column, we address whether the proposal strategies keep track of the time that is left and change their proposals accordingly. The next column specifies what kind of learning method the agents use to generate the next offer. When agents decide to accept an offer, all take the offer's utility in account (U), but some of them also consider the remaining time (T).

Finally, most of the agents are non-deterministic. For example, *Agent K* may decide on a certain proposal target. But if it previously received even better offers *B*, then it will counteroffer a random offer taken from *B*. Otherwise, it will also select a random proposal; in this case it will choose any offer that satisfies its proposal target. Most agents have this same mechanism: when they are indifferent between certain offers, they will choose randomly.

Table 4 Strategies of the agents participated in ANAC

	Time dependent	Learning method	Acceptance Criteria	Deterministic
Agent K	Yes	All proposals	T/U	No
Yushu	Yes	Best proposals	T/U	No
Nozomi	No	Compromises	T/U	No
IAMhaggler	Yes	Bayesian	U	No
FSEGA	Yes	Bayesian	U	Yes
IAMcrazyHaggler	No	None	U	No
Agent Smith	Yes	Weights	T/U	Yes

We continue to report on the individual strategies of the ANAC agents, starting with the winner.

4.2.1 Agent K

The proposal mechanism of *Agent K* [33] works as follows: based on the previous proposals of the opponent and the time that is left, it sets a so-called *proposal target* (initially set to 1). If it already received an offer that matches at least the utility of the proposal target, it will offer this proposal to improve the chances of acceptance.

Otherwise, it searches for random proposals that are at at least as good as the proposal target. If no such proposals are found, the proposal target is slightly lowered.

The agent has a sophisticated mechanism to accept an offer. It uses the mean and variance of the utility of all received offers, and then tries to determine the best offer it might receive in the future and sets its proposal target accordingly. It then accepts or rejects the offer, based on the probability that a better offer might be proposed. For more information and technical details on Agent K, see [33].

4.2.2 Yushu

Yushu [1] is a fairly simple agent that makes use of a target utility to make its next offer. As a learning mechanism, it uses the ten best proposals made by the opponent, called *suggested proposals*. It also makes an estimate of how many rounds are still left for the negotiation. Combining this information, *Yushu* obtains the target utility. It also keeps track of the acceptability-rate: the minimum utility it is willing to accept. To set the acceptability-rate, *Yushu* first finds the best possible utility that can be obtained in the domain, and accepts no less than 96% of it. When the number of estimated future rounds becomes short, this percentage is lowered to 92%.

The agent can only accept a proposal when the offered utility is above the target utility or when the utility reaches the acceptability-rate. Provided that either of the two is the case it accepts, when there are less than eight rounds left. When there is more time, it will accept only if it cannot find a suggested proposal with a better utility. If a better suggested proposal is available, it will offer that instead.

4.2.3 Nozomi

The proposal strategy of *Nozomi* [33] starts with an offer of maximum utility. It defines the gap between two parties as the differences in utility of their last offers. Depending on the gap and time that is left, it then chooses to make a certain proposal type, such as making a compromise, or staying put. *Nozomi* keeps track of the compromises made, but the agent does not model the utility function of the opponent.

The agent splits the negotiation into four intervals around 50%, 80% and 90% of the negotiation time. Based on previous offers, the gap between the two parties, and the time that is left in the negotiation, it will choose whether to accept an offer or reject it.

4.2.4 IAM(crazy)Haggler

IAMhaggler and *IAMcrazyHaggler* (cf. [5]) are both implementations of a framework called *SouthamptonAgent*, thus creating a lot of similarity between the two agents. The *SouthamptonAgent* provides standard methods for handling offers, proposing offers and keeping track of time. The framework is the only one that also keeps track of the time that the opponent uses.

IAMcrazyHaggler is a very simple take-it-or-leave-it strategy: it will make random proposals with a utility that is above a constant threshold, set to 0.9 (without discount factors it is set to 0.95). The proposal is done without regard to time or opponent moves.

IAMHaggler, on the other hand, is a fully fledged negotiation strategy, which incorporates a model of the opponent using Bayesian learning. It starts with a proposal of maximum utility and successively sets a target utility based on multiple factors, such as: the utility offered by the opponent, the time left for both agents, and the perceived opponent's profile, such as hardheadedness. Upon receiving an offer, it analyzes the previous proposals of the opponent and adapts the hypotheses on the opponent's utility function. With this opponent model, it tries to find trade-offs that satisfy the target utility.

Let u be the utility of the last opponent's offer. Both agents accept an offer depending on u, namely when either of the following three conditions is met:

1. When u is at least 98% of the utility of its own previous offer.
2. When u is at least 98% of a *maximum aspiration* constant. The default value is 0.9, but if there are discount factors it is set to 0.85 for *IAMcrazyHaggler* to make it reach an agreement sooner.
3. When u is at least 98% of the utility of its own upcoming offer.

Note that the three conditions only depend on the utility of the offer and not on the available time.

4.2.5   FSEGA

Similar to *Nozomi*, the *FSEGA* strategy [25] splits the negotiation into three intervals of time and applies different sub-strategies to each interval:

1. The first interval consists of the starting 85% of the negotiation time and is mainly used to acquire the opponent's profile from the counter-offers.
2. In the next 10%, the proposal strategy still does not concede, but relaxes some conditions for selecting the next proposal to improve the chances that the opponent accepts. The agent makes only small concessions and still tries to learn the opponent's profile.
3. In the final 5%, *FSEGA* considers the time restrictions and employs a concession-based strategy to select the next offer up to its reservation value.

In the first phase of the negotiation, the accept mechanism will admit any opponent offer that is 3% better than the utility of *FSEGA*'s last proposal. It will also always accept the best possible proposal. Otherwise, it selects a new proposal, but if the previous opponent's offer is better than the upcoming proposal it will accept it instead. After interval 1, it will also accept when it cannot find a better proposal for the opponent.

4.2.6 Agent Smith

Agent Smith [35] constructs an opponent model that represents the importance and preference for all values of each issue. The agent starts by making a first proposal of maximum utility and subsequently concedes slowly towards the opponent.

The agent accepts an offer given the following circumstances. The agents' threshold for acceptance slowly decreases over time. In the last 10 seconds of the negotiation session, *Agent Smith* will propose the best proposal that the opponent already proposed (even when the offer is very bad for itself). Since it previously proposed it, it is likely for a rational opponent to accept this proposal. However, an error was made in the implementation, resulting in the fact that the agent already shows this behavior after two minutes instead of three. This explains the poor performance of the agent in the competition.

4.3 Timing Aspects

All agents of ANAC 2010, except for *IAMcrazyHaggler*, make concessions when the deadline approaches. Because a break-off yields zero utility for both agents, an agent that waits until the end of the negotiation takes a substantial risk. The other agent may not know that the deadline is approaching and may not concede fast enough. In addition, either the acceptance of a proposal or the (acceptable) counter-offer may be received when the game is already over. In the same manner, a real-time deadline also makes it necessary to employ a mechanism for deciding when to accept an offer.

We study the inclination of the agents of ANAC 2010 to exhibit either risk averse or risk seeking behavior regarding the timing of their proposals. In order to get a good picture of the risk management of the agents, we consider the number of break-offs that occur for every agent. Table 5 lists for each agent the percentage of negotiations that result in a break-off. All break-offs occur due to the deadline being reached or an occasional agent crash on a big domain.

Table 5 Percentage of all failed negotiations of every agent per domain

Agent	Break-off percentage per domain			
	Itex-Cyp	Eng-Zimb	Travel	Avg
Agent K	22%	6%	63%	30%
Yushu	36%	0%	90%	42%
Nozomi	25%	17%	75%	39%
IAMhaggler	11%	0%	63%	25%
FSEGA	22%	0%	100%	41%
IAMcrazyHaggler	72%	23%	83%	59%
Agent Smith	0%	0%	98%	33%

The number of break-offs in the Travel domain stands out compared to the other domains. Recall that this is the biggest domain of ANAC 2010, with 188,160 possible proposals. Most of the agents had a lot of problems dealing with this domain. In a large domain it takes too much time to enumerate all proposals or to work with an elaborate opponent model. For example the *FSEGA* agent was unable to finish a single negotiation. Only *Agent K, Nozomi* and *IAM(crazy)Haggler* were able to effectively negotiate with each other on this domain.

With respect to the number of break-offs, *IAMHaggler* performs very well on all domains, while *IAMcrazyHaggler* ranks as the worst of all agents. This is to be expected, as its proposal generating mechanism does not take into account the time or the opponent (see Section 4.2.4 for an overview of its strategy). There is an interesting trade-off here: when *IAMcrazyHaggler* manages to reach an agreement, it always scores a utility of at least 0.9, but most of the time it scores 0 because the opponent will not budge.

The exact opposite of *IAMcrazyHaggler* is the strategy of *Agent Smith*. Because of an implementation error, *Agent Smith* accepts any proposal after two minutes, instead of three minutes. This explains why it did not have any break-offs on Itex–Cypress and England–Zimbabwe. The reason for the break-offs on the Travel domain is due to crashing of its opponent model. The importance of the timing aspects is underlined by the performance of *Agent Smith*: a small timing error resulted in very poor scoring on all three domains.

5   Design of Future ANAC

After ANAC 2010 was held at AAMAS-10, the participating teams had a closing discussion. This discussion yielded valuable suggestions for improving the design of future ANAC competitions. The consensus among participants was that the basic structure of the game should be retained. In the discussion between the participating teams and interested parties, we decided to leave further complicating factors out and not introduce too many innovations for the next year. This includes issue interdependencies, a richer negotiation protocol, different evaluation criteria, repeating scenarios (i.e.: multiple negotiation sessions), self–play and changes to the real–time deadline setup.

For the next ANAC in 2011 we decided that the teams participated in the first ANAC agree on the modifications to the rules and thus it was jointly agreed that the following modifications should be made into effect:

- Domains with discount factors should be included in the tournament.
- Changes should be made to the deadline setup and the selection criteria of the domains, that is, how to select a wide variety of domains without bias.

We detail the specific changes below.

5.0.1 Domains

ANAC 2011 will have domains that have discount factors. Without discount factors the current negotiation setup offers no incentive to accept an offer, except for right before the deadline. Waiting until the end of the round is an optimal strategy, except in the rare case that the opponent makes a mistake that might be retracted in the following round. Because of the lack of discount factors, almost every negotiation between the agents took the entire negotiation time of three minutes each to reach an agreement. Adding discount factors should provide more interesting negotiations with faster deals. The future ANAC setup could also be made more challenging by adding domains that contain continuous issues, such as real–valued price issues.

5.0.2 Issue Predictability

When learning the opponent's preference profile, a learning technique usually makes assumptions about the structure of the domain and preference profile (e.g., [4, 9, 38]). Negotiation strategies can try to exploit the internal structure of the issues in order to improve their proficiency. For example, a learning technique benefits from the information that a certain issue is *predictable*. Informally, an issue is called predictable when the global properties of its evaluation function is known. To illustrate, let us consider the discrete issue "Amount of funding" from the Zimbabwe–England domain (cf. Section 3.2). Its values are: *no agreement, $10 billion, $50 billion*, or *$100 billion*. Even when we do not know which party we are dealing with, we can be confident that the utility of a particular value is either increasing or decreasing in the amount of funding. A price issue like this is typically predictable, where more is either better or worse for a negotiating party. Other issues, e.g. color, can be less predictable and therefore learning the preferences of color is more difficult. In order to improve the efficiency of the learning algorithms, we intend to eventually introduce (un)predictability labels.

5.0.3 Deadline

The real–time deadline of ANAC is considered a nice challenge to the competitors. The agents had three minutes each to deliberate. This means agents have to keep track of both their own time and the time the opponent has left. For a future ANAC setup, we may choose a simpler protocol where both agents have a shared time window of three minutes.

6 Summary and Conclusion

This paper describes the first automated negotiating agents competition. Based on the process, the submissions and the closing session of the competition we believe that our aim has been accomplished. Recall that we set out for this competition in order to steer the research in the area bilateral multi-issue closed negotiation. The competition has achieved just that. Seven teams have participated in the first competition and we hope that many more will participate in the following competitions.

One of the successes of ANAC lies in the development of state-of-the-art negotiation strategies that co–evolve every year. This incarnation of ANAC already yielded seven new strategies and we hope that next year will bring even more sophisticated negotiation strategies. ANAC also has an impact on the development of GENIUS. We have released a new, public build of GENIUS[1] containing all relevant aspects of ANAC. In particular, this includes all domains, preference profiles and agents that were used in the competition. This will make the complete setup of ANAC available to the negotiation research community.

Not only have we learnt from the strategy concepts introduced in ANAC, we have also gained understanding in the correct setup of a negotiation competition. The joint discussion with the teams gives great insights into the organizing side of the competition.

To summarize, the agents developed for ANAC are the first step towards creating autonomous bargaining agents for real negotiation problems. We plan to organize the second ANAC in conjunction with the next AAMAS conference in 2011.

Acknowledgements. We would like to thank all participating teams of the University of Southampton, Nagoya Institute of Technology, Babes Bolyai and TU Delft and their representatives at AAMAS. We thank the ACAN and AAMAS organizers for enabling us to host ANAC. We are indebted to Dmytro Tykhonov his work on GENIUS and for helping organize this event. Furthermore, we would like to thank the sponsors of Bar-Ilan University and TU Delft.

References

1. An, B., Lesser, V.: Yushu: a heuristic-based agent for automated negotiating competition. In: Ito, T., et al. (eds.) New Trends in Agent-Based Complex Automated Negotiations. SCI, vol. 383, pp. 145–149. Springer, Heidelberg (2011)
2. Aumann, R.J., Hart, S. (eds.): Handbook of Game Theory with Economic Applications, March 1992. Handbook of Game Theory with Economic Applications, vol. 1. Elsevier, Amsterdam (1992)
3. Max, H.: Bazerman and Margaret A. Neale. Negotiator rationality and negotiator cognition: The interactive roles of prescriptive and descriptive research. In: Young, H.P. (ed.) Negotiation Analysis, pp. 109–130. The University of Michigan Press (1992)
4. Coehoorn, R.M., Jennings, N.R.: Learning an opponent's preferences to make effective multi-issue negotiation tradeoffs (2004)
5. Gerding, E.H., Williams, C.R., Robu, V., Jennings, N.R.: Iamhaggler: A negotiation agent for complex environments. In: Ito, T., et al. (eds.) New Trends in Agent-Based Complex Automated Negotiations. SCI, vol. 383, pp. 151–158. Springer, Heidelberg (2011)

[1] http://mmi.tudelft.nl/negotiation/index.php/Genius

6. Diniz da Costa, A., de Lucena, C.J.P., Torres da Silva, V., Azevedo, S.C., Soares, F.A.: Art Competition: Agent Designs to Handle Negotiation Challenges. In: Falcone, R., Barber, S.K., Sabater-Mir, J., Singh, M.P. (eds.) Trust 2008. LNCS (LNAI), vol. 5396, pp. 244–272. Springer, Heidelberg (2008)
7. Erev, I., Roth, A.: Predicting how people play games: Reinforcement learning in experimental games with unique, mixed strategy equilibrium. American Economic Review 88(4), 848–881 (1998)
8. Faratin, P., Sierra, C., Jennings, N.R.: Negotiation decision functions for autonomous agents. Int. Journal of Robotics and Autonomous Systems 24(3-4), 159–182 (1998)
9. Faratin, P., Sierra, C., Jennings, N.R.: Using similarity criteria to make negotiation trade-offs. Journal of Artificial Intelligence 142(2), 205–237 (2003)
10. Shaheen, S.: Fatima, Michael Wooldridge, and Nicholas R. Jennings. Multi-issue negotiation under time constraints. In: AAMAS 2002: Proceedings of the First International Joint Conference on Autonomous Agents and Multiagent Systems, pp. 143–150. ACM Press, New York (2002)
11. Fullam, K.K., Klos, T.B., Muller, G., Sabater, J., Schlosser, A., Barber, K.S., Rosenschein, J.S., Vercouter, L., Voss, M.: A specification of the agent reputation and trust (art) testbed: experimentation and competition for trust in agent societies. In: The 4th International Joint Conference on Autonomous Agents and Multi Agent Systems (AAMAS), pp. 512–518. ACM Press (2005)
12. Greenwald, A., Stone, P.: Autonomous bidding agents in the trading agent competition. IEEE Internet Computing 5(2), 52–60 (2001)
13. Hindriks, K., Tykhonov, D.: Towards a quality assessment method for learning preference profiles in negotiation (2008)
14. Ito, T., Hattori, H., Klein, M.: Multi-issue negotiation protocol for agents: Exploring nonlinear utility spaces (2007)
15. Jonker, C.M., Robu, V., Treur, J.: An agent architecture for multi-attribute negotiation using incomplete preference information. Journal of Autonomous Agents and Multi-Agent Systems 15(2), 221–252 (2007)
16. Keeney, R.L., Raiffa, H. (eds.): Decisions with Mutliple Objectives. Cambridge University Press (1976)
17. Kersten, G.E., Noronha, S.J.: Rational agents, contract curves, and inefficient compromises report. In: Working papers, International Institute for Applied Systems Analysis (1997)
18. Kersten, G.E., Zhang, G.: Mining inspire data for the determinants of successful internet negotiations. InterNeg Research Papers INR 04/01 Central European Journal of Operational Research (2003)
19. Kraus, S.: Strategic Negotiation in Multiagent Environments. MIT Press (October 2001)
20. Kraus, S., Wilkenfeld, J., Zlotkin, G.: Multiagent negotiation under time constraints. Artificial Intelligence 75(2), 297–345 (1995)
21. Lax, D.A., Sebenius, J.K.: Thinking coalitionally: party arithmetic, process opportunism, and strategic sequencing. In: Young, H.P. (ed.) Negotiation Analysis, pp. 153–193. The University of Michigan Press (1992)
22. Lin, R., Kraus, S., Tykhonov, D., Hindriks, K., Jonker, C.M.: Supporting the design of general automated negotiators. In: Proceedings of the Second International Workshop on Agent-Based Complex Automated Negotiations, ACAN 2009 (2009)
23. Lin, R., Kraus, S., Wilkenfeld, J., Barry, J.: Negotiating with bounded rational agents in environments with incomplete information using an automated agent. Artificial Intelligence 172(6-7), 823–851 (2008)

24. Lin, R., Oshrat, Y., Kraus, S.: Investigating the benefits of automated negotiations in enhancing people's negotiation skills. In: AAMAS 2009: Proceedings of the 8th International Conference on Autonomous Agents and Multiagent Systems, pp. 345–352 (2009)
25. Litan, C.M., Serban, L.D., Silaghi, G.C.: Agentfsega - time constrained reasoning model for bilateral multi-issue negotiations. In: Ito, T., et al. (eds.) New Trends in Agent-Based Complex Automated Negotiations. SCI, vol. 383, pp. 159–165. Springer, Heidelberg (2011)
26. McKelvey, R.D., Palfrey, T.R.: An experimental study of the centipede game. Econometrica 60(4), 803–836 (1992)
27. Niu, J., Cai, K., Parsons, S., McBurney, P., Gerding, E.: What the 2007 tac market design game tells us about effective auction mechanisms. Journal of Autonomous Agents and Multi-Agent Systems (2010)
28. Rubinstein, A., Osborne, M.J., Rubinstein, A.: Bargaining and Markets Economic Theory, Econometrics, and Mathematical Economics. Academic Press (April 1990)
29. Osborne, M.J., Rubinstein, A. (eds.): A Course in Game Theory. MIT Press (1994)
30. Raiffa, H.: The Art and Science of Negotiation. Harvard University Press (1982)
31. Raiffa, H., Richardson, J., Metcalfe, D.: Negotiation Analysis: The Science and Art of Collaborative Decision Making. Harvard University Press (2003)
32. Rubinstein, A.: Perfect equilibrium in a bargaining model. Econometrica 50(1), 97–109 (1982)
33. Fujita, K., Kawaguchi, S., Ito, T.: Compromising strategy based on estimated maximum utility for automated negotiating agents. In: Ito, T., et al. (eds.) New Trends in Agent-Based Complex Automated Negotiations. SCI, vol. 383, pp. 137–144. Springer, Heidelberg (2011)
34. Stone, P., Greenwald, A.: The first international trading agent competition: Autonomous bidding agents. Electronic Commerce Research 5(2), 229–265 (2005)
35. van Galen Last, N.: Opponent model estimation in bilateral multi-issue negotiation. In: Ito, T., et al. (eds.) New Trends in Agent-Based Complex Automated Negotiations. SCI, vol. 383, pp. 167–174. Springer, Heidelberg (2011)
36. Wellman, M.P., Wurman, P.R., O'Malley, K., Bangera, R., de Lin, S., Reeves, D., Walsh, W.E.: Designing the market game for a trading agent competition. IEEE Internet Computing 5(2), 43–51 (2001)
37. Zeng, D., Sycara, K.: Benefits of learning in negotiation (1997)
38. Zeng, D., Sycara, K.: Bayesian learning in negotiation. International Journal of Human Computer Systems 48, 125–141 (1998)

AgentK: Compromising Strategy Based on Estimated Maximum Utility for Automated Negotiating Agents

Shogo Kawaguchi, Katsuhide Fujita, and Takayuki Ito

Abstract. The Automated Negotiation Agents Competition (ANAC-10) was held and our agent won the tournament. Our agent estimates the alternatives the opponent will offer based on the history of the opponent's offers. In addition, our agent tries to compromise to the estimated maximum utility of the opponent by the end of the negotiation. Also, we modify the basic strategy to not exceed the limit of compromise when the opponent is uncooperative. We can adjust the speed of compromise depending on the negotiation. We introduce the new $ratio(t)$ to control our agent's actions at the final phase. With this improvement, our agents try to reach agreement when the opponent's proposal is closer to our estimated maximum values. The main reason our agent outperforms the others is its ability to reach a last-minute agreement as often as possible.

1 Introduction

The Automated Negotiation Agents Competition (ANAC2010) was held in International Joint Conference on Autonomous Agents and Multi-Agent Systems (AAMAS2010)[1]. In fact, our agent(*AgentK*) won the tournament of ANAC2010!

In this paper, we describe the details of our agent (Agent-K). Our agent estimates the alternatives the opponent will offer based on the history of the opponent's offers.

Shogo Kawaguchi
Department of Computer Science, Nagoya Institute of Technology, Nagoya, Aichi, Japan
e-mail: kawaguchi@itolab.mta.nitech.ac.jp,

Katsuhide Fujita
Department of Computer Science and Engineering, Nagoya Institute of Technology, Nagoya, Aichi, Japan
e-mail: fujita@itolab.mta.nitech.ac.jp

Takayuki Ito
School of Techno-Business Administration, Nagoya Institute of Technology, Nagoya, Aichi, Japan
e-mail: ito.takayuki@nitech.ac.jp

T. Ito et al. (Eds.): New Trends in Agent-Based Complex Automated Negotiations, SCI 383, pp. 137–144.
springerlink.com   © Springer-Verlag Berlin Heidelberg 2012

In addition, our agent tries to compromise to the estimated maximum utility of the opponent by the end of the negotiation. By compromising to the estimated maximum utility of the opponent, our agent reaches agreement with high possibility.

However, our basic strategy has a problem in that our agent can't make effective responses when an opponent takes a hard stance. We modify the basic strategy to not exceed the limit of compromise when the opponent is uncooperative. In addition, we can adjust the speed of compromise depending on the negotiation. The proposal of bids and acceptance of the opponent's bids at the final phase are most important. We introduce the new *ratio(t)* to control our agent's actions at the final phase. By this improvement, our agents try to reach agreement when the opponent's proposal is closer to our estimated maximum values.

The remainder of the paper is organized as follows. First, we describe our agent's basic strategy. Second, we describe the way of control of compromising. Third, we describe the selection of coefficient and some strategies at final phase. Finally, we draw a conclusion.

2 Basic Strategy

Our agent estimates the alternatives the opponent will offer in the future based on the history of the opponent's offers. In particular, we estimate it using the values mapping the opponent's bids to our own utility function. The agent works at compromising to the estimated optimal agreement point.

Concretely, our behavior is decided based on the following expressions((1), (2)).

$$emax(t) = \mu(t) + (1 - \mu(t))d(t) \tag{1}$$

$$target(t) = 1 - (1 - emax(t))t^{\alpha} \tag{2}$$

$emax(t)$ means the estimated maximum utility of bid the opponent will propose in the future. $emax(t)$ is calculated by $\mu(t)$ (the average of the opponent's offers in our utility space), $d(t)$ (the width of the opponent's offers in our utility space) when the timeline is t. $d(t)$ is calculated based on the deviation. We can see how favorable the opponent's offer is based on the deviation ($d(t)$) and the average ($\mu(t)$).

If we assume that the opponent's offer is generated based on uniform distribution $[\alpha, \alpha + d(t)]$, the deviation is calculated as follows.

$$\sigma^2(t) = \frac{1}{n}\sum_{i=0}^{n} x_i^2 - \mu^2 = \frac{d^2(t)}{12} \tag{3}$$

Therefore, $d(t)$ is defined as follows.

$$d(t) = \sqrt{12}\sigma(t) \tag{4}$$

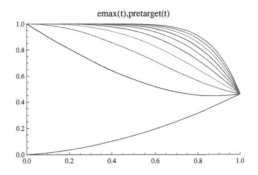

Fig. 1 Example of $target(t)$ when $emax(t)$ is $\mu(t) = \frac{1}{10}t$ $d(t) = \frac{2}{5}t^2$

We consider the averages as the weights for the following reason. When the average of the opponent's action is located at the center of the domain of the utility, $emax(t)$ is the average plus half of the width of the opponent's offers. However, it is possible to move only in the high direction when the average of the utility value is low, and the action can be expanded only in the low direction when the average is high. Therefore, an accurate estimation is made by introducing the weights.

$target(t)$ is a measure of proposing a bid when time is t, and α is a coefficient for adjusting the speed of compromise. We have no advantages for finishing the negotiation in a short time in this competition rule. Therefore, it is effective to search the opponent's utility information by repeating the proposal to each other as long as time allows. However, our utility value is required to be as high as possible. Our bids are the higher utility for the opponent at the first stage, and approach asymptotically to $emax(t)$ as the number of negotiation rounds increases.

Figure 1 is an example of $target(t)$ when α is changed from 1 to 9. $emax(t)$ is $\mu(t) = \frac{1}{10}t, \quad d(t) = \frac{2}{5}t^2$.

3   Control of Compromise

Our basic strategy has a problem in that our agent can't make effective responses when the opponent takes a hard stance. This is because our agent compromises excessively in such a situation.

Figure 2 is an example of $target(t)$ in changing α. $emax(t)$ is $\mu(t) = 0$, $d(t) = \frac{1}{10}$ from 1 to 9. When we don't employ the control of compromise, our agent makes a low quality agreement as the left side of Fig. 2 shows.

Then, the speed of compromise becomes slow if the degree of compromise is too much larger than the width of the opponent's offers. The degree of compromise is estimated by the following expressions (5). The degree of lowest compromise at time t is $g(t)$.

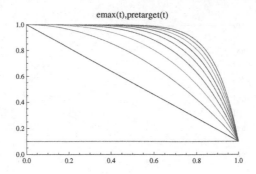

Fig. 2 Example of *target*(*t*) when *emax*(*t*) is set by $\mu(t) = 0$, $d(t) = \frac{1}{10}$

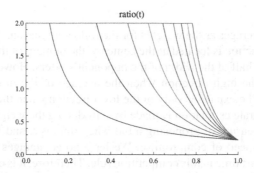

Fig. 3 Example of *ratio*(*t*) when *emax*(*t*) is $\mu(t) = 0$, $d(t) = \frac{1}{10}$

$$ratio(t) = \begin{cases} \frac{d(t)+g(t)}{1-target(t)} & \text{if } \frac{d(t)+g(t)}{1-target(t)} < 2 \\ 2 & \text{otherwise} \end{cases} \qquad (5)$$

Figure 3 shows the changes of *ratio*(*t*) when *emax*(*t*) is $\mu(t) = 0$, $d(t) = \frac{1}{10}$.

When our agent compromises too much compared with the opponent, *ratio*(*t*) becomes closer to 0 as figure 3 showing. We can control excessive compromises by regarding *ratio*(*t*) as a weight of *emax*(*t*). *target*(*t*) is defined as expression (3).

$$target(t) = ratio(t) * (1 - (1 - emax(t))t^{\alpha}) + (1 - ratio(t)) \qquad (6)$$

When the opponent is cooperative, our agent can decompose to the opponent's offers quickly by adjusting the degree of compromise. In other words, our agent doesn't exceed the limit of compromise when the opponent is uncooperative.

Figure 2 shows the changes of *target*(*t*) with *ratio*(*t*). As Fig. 2 shows, we can find that an excessive decomposing is controlled by *ratio*(*t*).

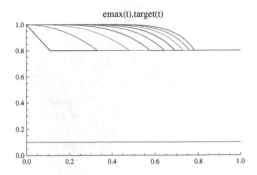

Fig. 4 *target*(*t*) that introduced *ratio*

4 Selection of Coefficient

We show the function of selecting the coefficient α. Our agent's offers swing at the early stage of the negotiation and stay fixed at the final phase by selecting the coefficient α. We introduce the random nature into selecting the coefficient (α). However, *target*(*t*) should not swing in judging whether to accept the opponent's offers. Therefore, we generate two coefficient functions as the situation demands. We define $\alpha(t)$, $\beta(t)$, and τ as the following expressions.

$$\alpha(t) = 1 + \tau + 10\mu(t) - 2\tau\mu(t) \tag{7}$$

$$\beta(t) = \alpha + random[0,1] * \tau - \frac{\tau}{2} \tag{8}$$

Expression (7) is a coefficient for judging whether to accept the opponent's offer. Expression (8) is a coefficient for proposing our bids. $random[0,1]$ in expression (8) generates a random value from 0 to 1.

5 Decisions on Our Agent's Bids and Evaluations of Opponent's Offers

First, we show the method of selecting the bids from our utility space. Our agent searches for alternatives whose utility is *target*(*t*) by changing the starting points randomly by iteratively deepening the depth-first search method. Next, we show the decision of whether to accept the opponent's offer. Our agent judges whether to accept it based on *target*(*t*) and the average of the opponent's offers. Expression (9) defines the probability of acceptance.

$$P = \frac{t^5}{5} + (Offer - emax(t)) + (Offer - target(t)) \tag{9}$$

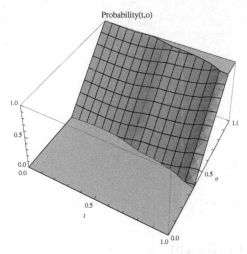

Fig. 5 Acceptance probability space

Acceptance probability P is calculated using t, $Offer$, $target(t)$ and the estimated maximum value $emax(t)$. $Offer$ is the utility of the opponent's bid in our utility space. In $target(t)$, $\alpha(t)$ is defined as expression (7).

Figure 5 shows the acceptance probability space when $emax(t)$ is $\mu(t) = \frac{1}{10}t$, $d(t) = \frac{2}{5}t^2$. The horizontal axis is time t and the vertical axis is a utility value o of the opponent's offer.

6 Corrections at the Final Phase of Negotiation

In this section, we show the correction at the final phase of the negotiation. An excessive compromise can be controlled by introducing $ratio(t)$. However, our agent stops compromising even if the opponent proposes a great offer that is close to $target(t)$ in the final phase. We modify this problem in the final phase by introducing the corrections.

$$\mu(t) = \frac{4}{5}t \quad d(t) = 0 \tag{10}$$

Figure 6 is an example of $target(t)$ when $emax(t)$ is defined as expression (10). The following expressions are given as a correction when the opponent proposes a bid whose utility is close to that of our target.

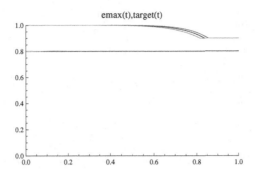

Fig. 6 Example of $target(t)$ when $emax(t)$ is set by expression 10

$$\gamma(t) = -300t + 400 \tag{11}$$

$$\delta(t) = target(t) - emax(t) \tag{12}$$

$$\varepsilon(t) = \begin{cases} \frac{1}{\delta(t)^2} & \text{if } \frac{1}{\delta(t)^2} < \gamma(t) \\ \gamma(t) & \text{otherwise} \end{cases} \tag{13}$$

$$approach(t) = \frac{\delta(t) * \varepsilon(t)}{\gamma(t)} \tag{14}$$

$$fintarget(t) = \begin{cases} target(t) - approach(t) & target(t) > emax(t) \\ emax(t) & \text{otherwise} \end{cases} \tag{15}$$

$\gamma(t)$ is a function of adjusting the correction in the final phase. $\delta(t)$ means the difference between a targeted value $target(t)$ and a estimated maximum value $emax(t)$. $\varepsilon(t)$ is inversely proportional to the square of $\delta(t)$, and limited by $\gamma(t)$. $approach(t)$ is a function of deciding correction based on $\gamma(t), \delta(t)$, and $\varepsilon(t)$. When $target(t)$ with $beta(t)$ is larger than $emax(t)$, our agent fixes the final objection $target(t)$, which is corrected by $approach(t)$. $target(t)$ is replaced with $emax(t)$ when it is smaller than $emax(t)$. We regard the expression (15) as a final standard function.

This compromise doesn't stop as Fig. 6 shows by introducing the correction in the final phase.

7 The Reasons Our Agent (*AgentK*) Wins in ANAC2010

In this section, we analyze the reasons our agent outperforms others in ANAC2010. As considering our strategies, we believe the following reasons.

- We try to improve the possibility of making agreements by adjusting the opponent's actions. We didn't introducing the EndNegotiation action because we can't get the utility without making agreement between the opponent.

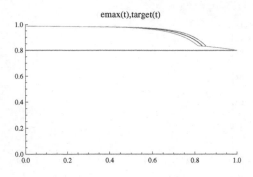

Fig. 7 Example of applying correction function to figure 6

- Our agent tried to compromise positively when the opponent is cooperative. On the other hand, our agent didn't compromise without exceeding the limit when the opponent is uncooperative.
- Our agent used all the time allowed for searching the effective solutions between our agent and the opponent.

8 Conslusion

In this paper, we described our agent, (*AgentK*), which won the ANAC-10. Our agent estimated the alternatives the opponent would offer based on the history of the opponent's offers. Our agent tried to compromise to the estimated maximum utility of the opponent by the end of the negotiation. We modified the basic strategy to not exceed the limit of compromise when the opponent is uncooperative. We also analyzed the reasons our agent outperformed the others.

Reference

1. Baarslag, T., Hindriks, K., Jonker, C., Kraus, S., Lin, R.: The First Automated Negotiating Agents Competition (ANAC 2010). In: Ito, T., et al. (eds.) New Trends in Agent-Based Complex Automated Negotiations. SCI, vol. 383, pp. 113–135. Springer, Heidelberg (2010)

Yushu: A Heuristic-Based Agent for Automated Negotiating Competition

Bo An and Victor Lesser

Abstract. This article analyzes important issues regarding the design of a successful negotiation agent for ANAC and presents the design of **Yushu**, one of the top-scoring agents in the first Automated Negotiating Agents Competition (ANAC). **Yushu** uses simple heuristics to implement a conservative concession strategy based on a dynamically computed measure of competitiveness and number of negotiations rounds left before the deadline.

1 Introduction

The first Automated Negotiating Agents Competition (ANAC) challenged its entrants to design an automated negotiation agent that was capable of competing with other automated negotiation agents in bilateral multi-issue negotiation in a variety of negotiation domains [3]. During negotiation, agents exchange offers following the alternating-offers bargaining protocol and the negotiation terminates when they make an agreement or their negotiation deadline approaches. The negotiation agent's objective is to reach negotiation agreements to maximize its utility. It is very challenging to design a successful negotiation agent. In the ANAC game, each agent has no knowledge about other agents' preferences, which makes it is impossible to drive agents' optimal (or equilibrium) strategies by directly applying game-theoretic approaches. Instead, heuristic based approaches have to be used to design negotiating agents. However, the lack of information about other agents' strategies makes it difficult to design efficient heuristics given that an agent's optimal strategy depends on the strategies of its opponents in such competitive bargaining games. Furthermore, time constraints, different utility functions, and different negotiation domains make it difficult to construct an integrated framework in which all these factors are optimized concurrently.

Bo An · Victor Lesser
Department of Computer Science
University of Massachusetts, Amherst, USA
e-mail: {ban, lesser}@cs.umass.edu

T. Ito et al. (Eds.): New Trends in Agent-Based Complex Automated Negotiations, SCI 383, pp. 145–149.
springerlink.com   © Springer-Verlag Berlin Heidelberg 2012

This article analyzes important issues regarding designing a successful negotiation agent for ANAC. In addition, this article introduces **Yushu**, one of the top-scoring agents in the first Automated Negotiating Agents Competition. The reminder of this article is organized as follows. Section 2 discusses the issues need to be taken into account while designing a negotiation agent. Section 3 presents the design of the agent **Yushu**. Section 4 concludes this article and outlines future research directions.

2 Designing Issues

This section discusses some important issues related to designing a successful negotiation agents for ANAC.

2.1 The Tougher, the Better?

Consider the following scenario: My opponent is very conservative and it never makes (large) concessions. Do I need to make concessions? The answer is yes. If there are only two agents in the competition, you can easily win (at least not lose) the competition by making no agreements since both agents get a utility of 0. If there are $n > 2$ agents participating in the competition, you need to consider your competition with agents other than your current opponent. Intuitively, a very conservative agent cannot win a competition since it may fail to make an agreement with some agents. If you get 0.5 utility by making some concessions to the conservative opponent, it is better than getting 0 utility by making no agreement. However, if most agents are willing to make (large) concessions, a very conservative agent will win the competition since it will get a high utility when it negotiates with a conciliatory agent.

One key feature of ANAC is that an agent's utility of an agreement is independent of the agreement making time. Therefore, to maximize its utility, an agent may not be willing making large concession at the beginning and prefers to wait for its opponent to make concessions.

2.2 Domain Competitiveness

In the tournament, each agent will compete with other agents in different domains. For some domains, each agent can achieve a high utility for an agreement. In contrast, in some other domains, an agent has a very low utility at an agreement. Assume that two negotiation agents **a** and **b** are negotiating for an agreement in a domain. Let D be the space of all possible offers. Let $u_{\mathbf{a}}(x)$ ($u_{\mathbf{b}}(x)$, respectively) be **a**'s (**b**'s, respectively) utility of an offer $x \in D$. The *competitiveness* $c(D)$ of the domain D can be measured by

$$c(D) = \max_{x \in D} \quad \min\left(u_{\mathbf{a}}(x), u_{\mathbf{b}}(x)\right)$$

A high value of $c(D)$ implies that the two agents can make an agreement which leads to a high utility to both agents. The higher the value of $c(D)$, the more competitive the domain D. It is easy to see that agents have to make more concessions in competitive domains. In contrast, in less competitive domains, agents can make an agreement with small concessions. Knowing the competitiveness of a domain helps with an agent's decision making. However, each agent has no knowledge about the competitiveness of a domain in advance and it is appealing to design some mechanisms to measure it.

2.3   Learning Is Difficult

A key feature of ANAC is that each agent only knows its preference and has no knowledge of its opponent's utility function. The lack of knowledge about other agents's preferences makes it difficult for agents to resolve their conflicts. Learning the preferences of other agents is very difficult in uncooperative game theory and it is also true in bargaining theory. While the literature provides a number of learning approaches for negotiation agents, those approaches often make unrealistic assumptions about agents' rationality or strategies, e.g., an agent will follow a specified strategy which is known to its opponent. The main difficult of learning in automated negotiation is that the offers from one agent is the output of a function (i.e., strategy) of the agent's private information. Without knowing an agent's strategy, it is impossible to learn the agent's private information. One simple argument is that different teams used totally different strategies in ANAC.

2.4   Do You Want to Be Learned

Another interesting question is whether an agent wants to learned by the other agent during negotiation. The answer is yes which is mainly due to the fact that both agents have the same deadline. The better the two agents know each other, the more likely that they can make an agreement which is pareto optimal. Note that you can always control which offer to make and whether to accept an offer or not, no matter whether your opponent knows your preference.

3   The Strategy of Yushu

Agent **Yushu** is a simple negotiation agent based on a number of heuristics. The features of **Yushu** include:

- **No learning**: **Yushu** does not learn the private information of its opponent given the difficulty of learning in selfish negotiation.
- **Making sufficiently minimal concessions**: **Yushu** always wants to make an agreement with its opponent, which implies that it makes concessions during

negotiation. The core concept of **Yushu**'s strategy is making sufficiently minimal concessions. By making a sufficient concession, **Yushu** increases its probability of reaching an agreement. At the same time, **Yushu** tries to make minimally sufficient concessions in order to achieve the highest possible utility while maintaining a minimum probability of reaching an agreement. Specifically, **Yushu** adopts a time dependent strategy as in [1, 2, 4]. The reserve utility (i.e., the lowest utility an agent can accept) is based on domain competitiveness and the utilities of the opponent's offers. **Yushu** adopts a conservative concession making strategy and will only make large concessions when a limited number of negotiation rounds are left.

- **Measure domain competitiveness**: **Yushu** uses a simple heuristic to measure a domain's competitiveness: if its opponent's offers always give **Yushu** a low utility, the domain is more competitive.
- **Measure the time for each negotiation round**: **Yushu** will make larger concessions when the negotiation deadline approaches. Instead of considering how much time left, **Yushu** measures how many negotiation rounds left based on the average time for each negotiation round. The number of negotiation rounds left provides a better estimation of remaining negotiation time since some opponents may respond very slowly but some others may make offers very fast.
- **Randomization**: While making an offer, **Yushu** first computes its reserve utility μ following the time-dependent strategy. Then it generates all offers whose utilities fall in the range of $[\mu - \varepsilon, \mu + \varepsilon]$ and it randomly chooses one offer to send to its opponent. By this kind of exploration (or randomness), **Yushu**'s opponent has a better understanding of **Yushu**'s acceptable offers and the chance of making a pareto optimal agreement increases.

4   Conclusion

This article analyzes important issues regarding the design of a successful negotiation agent for ANAC and presents the design of **Yushu** which uses a heuristic-based strategy. Future work will focus on designing a strategy for negotiation with 1) different deadlines and 2) discount factors. With different negotiation deadlines, two negotiation agents may have different attitudes toward how to make concessions with time's elapse. With discount factors, each agent needs to make tradeoffs between 1) making less concessions to exploit its opponent's concession making and 2) making more concessions in order to make an agreement earlier to avoid utility loss due to the discount factor.

Reference

1. An, B., Lesser, V., Sim, K.M.: Strategic agents for multi-resource negotiation. Journal of Autonomous Agents and Multi-Agent Systems (to appear)
2. An, B., Sim, K.M., Tang, L., Li, S., Cheng, D.: A continuous time negotiation mechanism for software agents. IEEE Trans. on Systems, Man and Cybernetics, Part B 36(6), 1261–1272 (2006)

3. Baarslag, T., Hindriks, K., Jonker, C., Kraus, S., Lin, R.: The First Automated Negotiating Agents Competition (ANAC 2010). In: Ito, T., et al. (eds.) New Trends in Agent-Based Complex Automated Negotiations. SCI, vol. 383, pp. 113–135. Springer, Heidelberg (2010)
4. Kraus, S.: Strategic Negotiation in Multiagent Environments. MIT Press, Cambridge (2001)

IAMhaggler: A Negotiation Agent for Complex Environments

Colin R. Williams, Valentin Robu, Enrico H. Gerding, and Nicholas R. Jennings

Abstract. We describe the strategy used by our agent, IAMhaggler, which finished in third place in the 2010 Automated Negotiating Agent Competition. It uses a concession strategy to determine the utility level at which to make offers. This concession strategy uses a principled approach which considers the offers made by the opponent. It then uses a Pareto-search algorithm combined with Bayesian learning in order to generate a multi-issue offer with a specific utility as given by its concession strategy.

1 Introduction

We present the negotiation strategy, called IAMhaggler, which we developed and entered into the first Automated Negotiating Agent Competition [1]. The competition is a tournament between a set of agents which each perform bilateral negotiation using the alternating offers protocol. The negotiation environment consists of multiple issues (which may take continuous or discrete values), where there is uncertainty about the opponent's preferences. The environment also uses a real-time discounting factor and deadline.

Our negotiation strategy consists of two parts. First, it uses a principled approach to choose a concession rate based on the concession of the opponent, the real-time discounting factor and the negotiation deadline. Then, using the utility level chosen according to this concession strategy, it generates a multi-issue offer at that utility level, whilst attempting to maximise the likelihood that the opponent will form an agreement by accepting the offer.

Our strategy makes two main contributions to the literature on multi-issue negotiation with uncertainty about the opponent's preferences. Specifically, we propose

Colin R. Williams · Valentin Robu · Enrico H. Gerding · Nicholas R. Jennings
School of Electronics and Computer Science, University of Southampton,
University Road, Southampton, SO17 1BJ
e-mail: {crw104,vr2,eg,nrj}@ecs.soton.ac.uk

T. Ito et al. (Eds.): New Trends in Agent-Based Complex Automated Negotiations, SCI 383, pp. 151–158.
springerlink.com   © Springer-Verlag Berlin Heidelberg 2012

a novel approach to determining our agent's concession rate by taking into account
the real-time discounting factor and deadline in order to choose a concession rate
which is a best response to that of the opponent (see Section 2). Furthermore, we
combine two existing approaches (Pareto-search and Bayesian learning) which represent the state of the art in continuous and discrete domains respectively, to generate
a multi-issue offer in domains containing both discrete and continuous issues (see
Section 3).

2 Setting the Concession Rate

Our concession strategy is composed of two parts. The first tries to estimate the
opponent's future concession based on observations of the opponent's offers. This
estimate is then used to set an appropriate rate of concession. We discuss both parts
in turn in Sections 2.1 and 2.2.

2.1 Learning the Opponent's Concession

Our agent's approach is to set its concession as a best response to the opponent's
concession throughout the negotiation. Therefore, the first stage of our agent's concession strategy is to build an estimate of the opponent's future concession. To this
end, our agent observes the offers that the opponent makes, recording the time that
the offer was made, along with the utility of that offer according to our own agent's
utility function. We assume that the opponent is concessive over time, and therefore,
for any offer which gives us a lower utility than that of a previous offer from the opponent, we instead record the highest utility that we have observed so far. A further
motivation for this behaviour is that we make the assumption that our opponent is
likely to accept any offer that it has previously made. Therefore, if we were to propose the best offer that we have seen so far again, we assume that the opponent will
accept. We believe this to be a reasonable assumption and one which holds for our
agent. This gives us a function over time of the opponent's concession in terms of
our own utility. At this stage we ignore the effect of the discounting factor on our
utility.

Now, in order to predict the opponent's concession at any future point in time
during the game, we apply non-linear regression to the opponent's concession curve,
and in doing so assume that the observed points will roughly fit to a curve of the
form:

$$U_o(t) = U_0 + e^a t^b, \tag{1}$$

where $U_o(t)$ is the utility of the offer made by the opponent at time t and U_0 is the
utility of the opponent's first offer. The constants a and b are the parameters which
we find using a least mean squares curve fitting algorithm.

Once the constants a and b have been found, we use Equation 1 to estimate the
utility of the opponent's offers at any time during the negotiation session. After
every offer from the opponent, we update our estimation, by repeating the process
described in this section.

2.2 Choosing the Concession Rate

Once we have an approximation of our opponent's concession in terms of our utility, we are able to use this information to set our own rate of concession.

Our approach to this is to use standard time dependent concession, as described by [2]. Specifically, we use the polynomial time concession function:

$$U_p(t) = U_0 - (U_{\min} - U_0)\frac{t}{t_{\max}}^{1/\beta},\tag{2}$$

where U_0 is the initial utility, $U_{\min}$ is the reservation utility (which the agent will not concede beyond), $t_{\max}$ is the negotiation deadline and β is the parameter that affects the rate of concession. The β value can be partitioned into three types: tough ($\beta < 1$), linear ($\beta = 1$) or weak ($\beta > 1$). Our approach to finding an appropriate value for β proceeds as follows, and we give a graphical example of this approach in Fig. 1. We firstly apply our discounting factor to our model of the opponent's concession function (Equation 1), to create a function which gives us an estimate of the discounted utility of our opponent's offers at any point in the negotiation session. Therefore, the discounted utility function is given by:

$$U_d(t) = (U_0 + e^a t^b)e^{-\delta t},\tag{3}$$

where δ is the discounting factor.

If the opponent's future concession is as predicted, then our agent cannot achieve a utility higher than the maximum on the discounted opponent concession curve (given by Equation 3) for the time period at which an agreement can be reached. By this we mean that parts of the curve that represent times in the past, or times which are beyond the negotiation deadline should be ignored. Therefore, the next step is to solve:

$$t^* = \underset{t_{\text{now}} \leq t \leq t_{\max}}{\arg\max} ((U_0 + e^a t^b)e^{-\delta t}),\tag{4}$$

where t_{now} is the current time.

By solving Equation 4, our agent has identified the time t^* at which the discounted utility to our agent of our opponent's offers is likely to be maximised. Given this, our agent chooses a target utility level, which matches the estimated utility of the opponent's offer (without any discounting) at that time. This is given by:

$$U_{o,\max} = U_o(t^*)\tag{5}$$

We now have a point in time t^* at which we expect to reach agreement, and a target utility level $U_{o,\max}$ that we should use at that time. We then choose the value of β such that the concession curve (Equation 2) passes through $[t^*, U_{o,\max}]$. By rearranging Equation 2, we find β as follows:

$$\beta = \frac{\log(t^*)}{\log\left(\dfrac{1 - U_{o,\max}}{1 - U_{\min}}\right)}\tag{6}$$

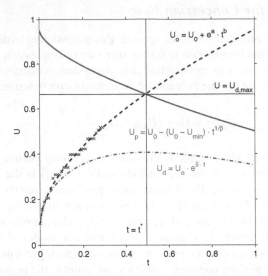

Fig. 1 Setting our Concession Rate. The crosses represent the best offers made by the opponent. The estimated future concession of the opponent is shown by the dashed (undiscounted) and dot-dashed (discounted) lines. The maxima on the dot-dashed line, indicated by the vertical line, represents the time (t^*) at which the expected discounted utility of the opponent's offers is maximised. The horizontal line shows the expected undiscounted utility of the opponent's offer $(U_{d,max})$ at that time. The solid curve is our planned concession.

A limitation of this strategy is that, at the beginning of a negotiation, the curve fitting is performed using a small number of points, and therefore the curve may not accurately reflect the future concession of the opponent. Thus, we set upper and lower bounds on our actual concession parameter (β) so that the agent does not use a concession strategy that is too extreme, either by conceding too quickly at the beginning of the session (thereby forming a low utility agreement), or by playing too tough (thereby conceding too little).

Having shown our concession strategy, we now discuss how the strategy produces a multi-issue offer at the utility level selected by the concession strategy.

3   Negotiating over Multiple Issues

In the previous section we showed how our agent can choose its desired utility level at any point during the negotiation session. However, in a multi-issue negotiation, there are likely to be a number of different offers at any given utility level. Our agent is indifferent between these offers, since they all result in the same utility. However, the opponent has a different utility function to ours. Therefore, our aim, and the basis of our negotiation strategy, is to select the offer which maximises the utility to the opponent whilst maintaining our desired utility level. The reason for this is that the opponent is more likely to accept offers with a higher utility, and, from a performance perspective, such outcomes are more Pareto-efficient. The challenge here is to try to do so without knowing the opponent's utility function.

In this section, we consider issues to belong to one of two classes: *ordered*, and *unordered*. Ordered issues have an ordering that is common to and known by both agents, though the agents may have different preferences over the issue values. Conversely, unordered issues do not have a common ordering.

In domains which consist solely of ordered issues, our agent does not attempt to learn the opponent's utility function. Instead, we use a Pareto-search approach to selecting an offer. In domains with unordered issues, we extend this approach by using Bayesian learning to learn the opponent's utility functions for the unordered issues.

To this end, in Section 3.1, we describe our approach in a domain without unordered issues, and in Section 3.2, we consider domains with such issues.

3.1 Domains without Unordered Issues

In this section, we present our approach to selecting the package with a given utility that we consider to be closest to the best offer that we have seen from our opponent. At this point, we only consider ordered issues.

Our strategy is based on the Pareto-search approach developed in [3, 5]. We consider our agent's utility function to be a mapping from a multi-dimensional space (in which there is a dimension representing each ordered issue) to a real value which represents the utility of the outcome. Our strategy treats integer based issues in the same way as continuous issues. For example, for a domain which consists of only one linear and one triangular issue, this multi-dimensional space is shown graphically in Fig. 2.

Now, by taking a cross-section of the utility space, we can construct an iso-utility space, which is a multi-dimensional space, with the number of dimensions equal to the number of ordered issues. This space represents all packages which result in a particular utility for our agent.

Moreover, the iso-utility space that is chosen at a particular time is the one which represents our current desired utility level (as decided by our concession strategy, which we detailed in Section 2). Based on the work of [5], we use projection to find the point on our iso-utility space which is closest to the best offer our agent has received from its opponent.

Specifically, in terms of closeness between two offers $[v_1, v_2, ..., v_n]$ and $[y_1, y_2, ..., y_n]$, we use the Euclidean distance, that is:

$$\sqrt{\sum_{i=1}^{n} \left(\left(\frac{v_i - y_i}{\text{range}_i} \right)^2 \right)}, \qquad (7)$$

where range_i is the range of values allowed for issue i. The reason that we divide by the range is to ensure that the scale of the issue's values does not affect the distance

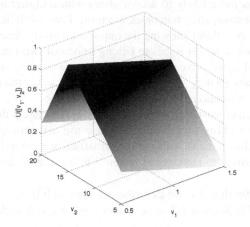

Fig. 2 Multi-dimensional space representing the utility of outcomes in a domain with one linear issue and one triangular issue. Values v_1 and v_2 are the values of the two issues. $U([v_1, v_2])$ is the utility of the offer represented by those values.

measurement. In doing so, all of the issues are considered to have equal weightings. Otherwise, the distance calculation could be excessively biased by an issue with a large range relative to the other issues.

3.2 Domains with Unordered Issues

In order for our strategy to negotiate in domains with unordered issues, we need to make some modifications to it. In particular, we cannot treat the unordered issues as further dimensions in our space, since they cannot be ordered based on their similarity. To address this problem, we continue to create an iso-utility space to represent the ordered issues. However, to handle the additional complexity of unordered issues, we create an iso-utility space for each combination of the unordered issues. As an example, consider a domain with one linear, one triangular and one discrete issue. The discrete issue can take the values *'red'*, *'green'* or *'blue'*.

For each combination of the unordered issues, we create a multi-dimensional space, and use the iso-utility projection method (described in Section 3) to find a solution for each of those combinations. We demonstrate this projection in Fig. 3, where the opponent's offer $[1.35, 7.0, v_3]$ (we write v_3 to represent the value of the unordered issue 3, since it does not affect the projection) is projected to give the solutions $[1.17, 15.0, \text{red}]$, $[1.25, 14.8, \text{green}]$ and $[1.26, 14.3, \text{blue}]$. For some combinations, the maximum overall utility available from the package with the best values for the ordered issues may be lower than our current utility level. In this case, there will not be a solution which contains this combination.

Once a solution has been found for each combination of the unordered issues, it is necessary for our agent to choose one of these solutions as its offer. Our agent

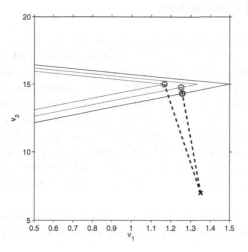

Fig. 3 Projection of a point representing an opponent's offer (at $[1.35, 7.0, v_3]$, marked with a cross) onto an iso-utility space for a domain with one linear, one triangular and a discrete unordered issue. The projections are shown for each of the three discrete combinations.

is indifferent between each of them, as they all belong to the same iso-utility space, resulting in them having an identical utility. However, their utility to the opponent may vary, and in order to negotiate efficiently, we should choose the one which maximises the utility to the opponent.

Now, since we assume that the opponent's utility function is unknown, we need a way to estimate the utility of the opponent. In our work, this is done using the approach taken by [4], using Bayesian updating to learn the preferences of the opponent in a scalable manner.

As the agent receives offers from the opponent, it uses Bayes' rule to update the probability of each hypothesis. It then evaluates the solution for each unordered combination using our model of the opponent's utility function, in order to obtain an estimate of the utility of the offer to the opponent. The overall solution that is chosen is then the one which maximises the opponent's utility according to our model of its utility function.

In order to ensure that our approach remains computationally tractable even in domains with large outcome spaces, we limit the number of combinations of unordered issues that we perform the iso-utility projection method for. Specifically, we choose a maximum of 1000 such combinations, by identifying those which maximise the sum of our utility and our opponent's utility (according to our model of the opponent's utility function).

4   Results and Conclusions

In this work, we have described our negotiating agent, IAMhaggler, covering both the process that our agent uses to set its concession rate, and the way in which it selects an offer with a given utility. In doing so, we have combined and extended several existing approaches, and designed a new adaptive strategy.

Our agent finished in 3rd place in the 2010 competition (according to the official results[1]). After the competition, we identified some minor implementation problems with our agent. By fixing these problems and then re-running the competition setup, our agent was able to finish in 2nd place. Since the competition domains did not include a discounting factor, we performed a further re-run in which we used the competition domains but with a discounting factor ($\delta = 1$). In this re-run, our agent finishes in 1st place.

We plan to extend our work by investigating how the agent can better adjust its parameters (such as the reservation value) based on its knowledge of the domain.

Reference

1. Baarslag, T., Hindriks, K., Jonker, C.M., Kraus, S., Lin, R.: The first automated negotiating agents competition (ANAC 2010). In: Ito, T., et al. (eds.) New Trends in Agent-Based Complex Automated Negotiations. SCI, vol. 383, pp. 113–135. Springer, Heidelberg (2010)
2. Faratin, P., Sierra, C., Jennings, N.R.: Negotiation decision functions for autonomous agents. Robotics and Autonomous Systems 24(3), 159–182 (1998)
3. Faratin, P., Sierra, C., Jennings, N.R.: Using similarity criteria to make issue trade-offs in automated negotiations. Artificial Intelligence 142(2), 205–237 (2002)
4. Hindriks, K., Tykhonov, D.: Opponent modelling in automated multi-issue negotiation using bayesian learning. In: Proc. of the 7th Int. Joint Conference on Autonomous Agents and Multiagent Systems, vol. 1, pp. 331–338 (2008)
5. Somefun, D.J.A., Gerding, E.H., La Poutré, J.A.: Efficient methods for automated multi-issue negotiation: Negotiating over a two-part tariff. Int. Journal of Intelligent Systems 21(1), 99–119 (2006)

---

[1] http://mmi.tudelft.nl/negotiation/images/0/03/Results.pdf

AgentFSEGA: Time Constrained Reasoning Model for Bilateral Multi-issue Negotiations

Liviu Dan Şerban, Gheorghe Cosmin Silaghi, and Cristian Marius Litan

Abstract. This paper presents AgentFSEGA experience at the Automated Nego-tiation Agents Competition 2010. AgentFSEGA is a time-constrained negotiation strategy that learns the opponent's profile out of its moves. Having at the baseline the Bayesian negotiation strategy [3], AgentFSEGA considers the negotiation time as a resource, being prepared to concede more as time passes. While AgentFSEGA performs well on relatively large domains, we prove that the performance of the negotiation strategy does not downgrade significantly even for small, engineered, domains not suited for a learning strategy.

1 Introduction

A bilateral negotiation is a game with two agents, having a common interest in cooperating in order to reach an agreement, but conflicting interests concerning the particular way of achieving it [1], [5].

The challenge in developing negotiating agents is to find strategies which react fast at opponent's actions, learn well opponent's preferences, better use the opponent's profile when proposing bids, and use time effectively during negotiation.

We designed AgentFSEGA - described in this paper - for the purpose of ANAC 2010, which uses strategies having above-mentioned properties. AgentFSEGA is built on the grounds offered by the Bayesian agent [3], already implemented in Genius before ANAC. We modified it to accommodate for the time restrictions of the ANAC game and to improve the efficiency of the learning procedure. AgentF-SEGA is an instantiation of the general framework for building negotiation agents described in [7].

Like other learning-based reasoning models, at a given negotiation step our agent performs two actions: analyzes the previous action of the opponent and selects and

Liviu Dan Şerban · Gheorghe Cosmin Silaghi · Cristian Marius Litan
Babeş-Bolyai University
Str. Theodor Mihali 58-60, 400591, Cluj-Napoca, Romania
e-mail: {Liviu.Serban,Gheorghe.Silaghi,Cristian.Litan}@
        econ.ubbcluj.ro

T. Ito et al. (Eds.): New Trends in Agent-Based Complex Automated Negotiations, SCI 383, pp. 159–165.
springerlink.com   © Springer-Verlag Berlin Heidelberg 2012

offers the next bid. AgentFSEGA reasoning model uses a concession-based strategy which reduces the own utility of the offered bid as the negotiation time passes, in order to avoid the situation that negotiation ends without an agreement.

This paper is organized according to the two main actions performed by AgentFSEGA. Section 2 shortly introduces the Bayesian learning strategy. Section 3 described the details of the AgentFSEGA implementation. Section 4 presents the experience of AgentFSEGA during the ANAC game and develops a discussion about the observed strengths and weaknesses of our proposed learning strategy. Finally, section 5 concludes the paper.

2 The Bayesian Learning Setup

In general, at each negotiation step, a learning-based agent performs two actions:

- analyze the incoming opponent's proposal and update the locally stored opponent's profile
- select and propose the next bid

Learning the opponent's profile consists mainly of approximating its utility function $U(v)$. As the search space for the opponent profile is infinite, different learning models propose various techniques to reduce it and to make it computer tractable via simplification. Bayesian learning - which stays at the foundation of AgentFSEGA - does not suppose to learn the precise weights w_k. The literature [2] argues that learning the ranks of the weights is typically sufficient to significantly increase the efficiency of the outcome. In this way, the search space for the weights reduces to $n!$ possible rankings of the preferences for the n issues. For each individual utility function $U_k(v_k)$, agent's preferences on the value of the k-th issue can be encoded starting from three basic function shapes:

- downhill shape: minimal issue values are preferred over the rest of values and the function decreases linearly with an increase of the value of the issue;
- uphill shape: maximal issue values are preferred over the rest of values and the function increases linearly with a decrease of the value of the issue;
- triangular shape: a specific issue value in the interior of the issue range is most valued, and the function linearly decreases at the left and at the right of this issue.

A function with a unique extreme point on an interval can be approximated with a linear combination of *basis functions*[1] from the three types presented above. Therefore, given that the domain of an issue k has m_k values, one can draw m_k functions with the maximal issue value in each of the m_k values and linearly combine these functions to obtain the individual utility function U_k. This approach is very suitable for discrete issue domains, while continuous domains might be discretized

---

[1] In mathematics, a basis function is an element of a particular basis for a function space. Every function in the function space can be represented as a linear combination of basis functions, just as every vector in a vector space can be represented as a linear combination of basis vectors (G. Strang, "Wavelets", American Scientist, Vol. 82, 1994, pp. 250-255.).

(i.e. split in m equal subintervals). Thus, for an issue k, the search for the individual utility function U_k is in fact reduced to learning some numerical values $0 \leq p_i \leq 1$, $i = \overline{1, m_k}$, which allows to compose U_k as a linear combination of triangular functions.

From the discussion above, we note that learning the opponent profile means selecting the proper ranking of weights and assigning some probabilistic values for each triangular valuation function, for each issue value, and for all issues. Therefore, in the Bayesian learning setup, a user profile is a matrix that associates probabilities for each hypothesis $h_j \in H$, where $H = H_w \times H_1 \times H_2 \times \ldots H_n$ is the hypothesis space. One row of this matrix represents the probabilities $(p_w, p_1, \ldots, p_n)$ associated for a hypothesis h, and the matrix has $M = n! \times m_1 \times m_2 \times \cdots \times m_n$ lines (the total number of possible hypotheses). Bayesian learning starts with some random initialization of the agent profile matrix. At each step, this matrix is updated using the Bayes rule. If b_t is the newest received bid, then

$$P(h_j|b_t) = \frac{P(h_j)P(b_t|h_j)}{\sum_{k=1}^{M} P(h_k)P(b_t|h_k)} \tag{1}$$

Hindriks & Tykhonov [3] shows how to compute eq. 1 in the setup above described. It is worth noting that if the opponent agent is a rational one, after a large number of proposed bids, the agent profile matrix converges.

The second action that a negotiating agent performs in each step is *to select and propose the next bid*. For each possible own bid $b = (v_1, v_2, \ldots, v_n)$ the agent can estimate the utility that this bid is giving to the opponent:

$$U(b) = \sum_{j=1}^{M} P(h_j) \sum_{k=1}^{n} w_k U_k(v_k) \tag{2}$$

A smart strategy is to select and propose the bid that maximizes the utility U_b of the opponent, while the own utility U_{own} stays as close as possible to the own target negotiation objective τ:

$$b = \underset{b' \in \{x||U_{own}(x)-\tau| \leq \delta\}}{\arg\max} U(b') \tag{3}$$

3 AgentFSEGA - the Time Constrained Reasoning Model

The core of our negotiation strategy consists in adapting the agent behavior to the duration of the negotiation. Usually, a learning scheme takes time to converge to some consistent opponent's profile, therefore the agent should allow several bilateral bids exchange rounds without much own utility concession and concede only when the total negotiation time T_{max} is about to elapse. By that time, the Bayesian learning is more likely to already converge to the opponent profile; therefore, our agent has better chances to propose acceptable deals for the opponent.

AgentFSEGA reasoning model splits total negotiation time in three intervals $[0,t_1]$, $[t_1,t_2]$ and $[t_2,T_{max}]$ and for each interval it applies different bidding strategies. The intervals' lengths had been selected based on experimentation. We considered: $t_1 = 0.85 * T$ and $t_2 = 0.95 * T$, where $T = T_{max}$.

Regardless of the interval, the bid selection strategy computes two values: U_{min} and U_{max} which are minimum and maximum acceptable own utilities for the next bid. Then, the agent selects from the interval $[U_{min}, U_{max}]$ the bid with the highest utility for the opponent, based on the opponent learned profile. U_{max} is its own highest utility of a bid, which was not been previously offered, and U_{min} is calculated differently for each interval.

In the first interval (I_1) from $[0, t_1]$: the agent performs only very small concessions, this time is used to learn the opponent's profile. Supposing that b_t is the last bid proposed by the opponent, then at step $t+1$ AgentFSEGA will propose a bid b_{t+1} that not only maximizes the opponent's utility, but also ensures that our agent does not attain an utility lower than the one implied by the opponent's offered bid b_t. On I_1, U_{min} is calculated as:

$$U_{min} = \max\{U_{max} - \delta, U_{own}(b_t)\} \tag{4}$$

On this time interval, the negotiation game may end only with the opponent accepting our offered bid; AgentFSEGA would never accept a bid offered by the opponent.

In the second interval (I_2) from $[t_1, t_2]$ the conditions for selecting our next bid are relaxed to the usual $U_{min} = U_{max} - \delta$ (see [3]). Our agent accepts the offer b_t only if $U_{own}(b_t) > U_{own}(b_{t+1})$.

We introduced the second interval I_2, because we noticed that, without this interval, the agent switches very abrupt from a powerful agent that does not concede to an agent that concedes very quickly. If a smart agent is our opponent, this agent can simply learn this behavior, thus our agent shall never win in this case. By performing the transition between the strong-holding agent behavior of I_1 to the standard behavior of I_3, via some "mixed" behavior in I_2, we harden the job of an opponent to learn our strategy.

On the last interval (I_3) (from $[t_2, T]$) the agent concedes, up to a reservation utility u_r. The lower limit will be given by:

$$U_{min} = \min\{\max(u_r, U_{max} - \delta), e^{c+d(T_{max}-t)}\} \tag{5}$$

where $c = \ln(u_r)$, $d = \frac{\ln(\frac{u_{0min}}{u_r})}{(1-0.95)\cdot T_{max}}$, and u_{0min} is the last lower bound before entering in the time interval I_3. Notice that for $t = T_{max}$, then

$$e^{c+d(T_{max}-t)} = u_r \tag{6}$$

which is the minimum of the given exponential function. For $t = 0.95 \cdot T_{max}$, then

$$e^{c+d(T_{max}-t)} = u_{0min} \tag{7}$$

Table 1 AgentFSEGA against ANAC2010 agents. Columns represents the opponent agents. Each row represents a role played by AgentFSEGA.

AgentFSEGA	Agent K	Nozomi	AnAgent
SON buyer	0.882 / 0.587	0.877 / 0.612	0.972 / 0.137
SON seller	0.847 / 0.622	0.847 / 0.622	0.965 / 0.175
Itex-Cypress Itex	0.653 / 0.703	0.548 / 0.732	0.576 / 0.776
Itex-Cypress Cypress	0.670 / 0.721	0.588 / 0.759	0.588 / 0.759
England-Zimbabwe England	0.700 / 0.871	0.611 / 0.924	0.586 / 0.946
England-Zimbabwe Zimbabwe	0.664 / 0.935	0.664 / 0.935	0.596 / 0.955

which is the maximum of the given exponential function. The rest of the conditions for selecting our next bid, or accepting an offered bid, remain the same as before. One may note that on I_3, at the end of the negotiation, the agent does not accept an agreement below its reservation utility u_r. This can be:

- the utility of the bid that maximizes the estimated opponent's utility on the whole domain;
- the utility given by the first offer of the opponent;
- another of its past offers.

Any of these choices should give the agent an acceptable positive utility, ensuring as well that the opponent will accept it, thus avoiding the negotiation to end without an agreement.

4 Experiments and Results - AgentFSEGA at ANAC2010

In this section we presents the way AgentFSEGA behaved during ANAC 2010. Fully experimentation of AgentFSEGA against non-time-constrained agents previously implemented in GENIUS can be found in [7]. Previous to ANAC 2010 competition (as described in [7]) we tested AgentFSEGA against ABMP [4] and against the standard Bayesian [3] agents, on the Service-oriented negotiation (SON) domain [2] and Zimbabwe - England domain [6].

In this paper, we present the results of AgentFSEGA against *Agent K*, *Nozomi* and *AnAgent*. All these three agents were in the ANAC 2010 competition, and we took their implementation from the Genius repository after ANAC took place. We considered the following domains: SON, Itex-Cypress and England-Zimbabwe. Results are presented in table 1. Each cell of the table represents the utility of AgentFSEGA (when playing the role specified on the row) against the utility of the opponent.

Note that AgentFSEGA performs best on the SON domain. SON is a huge domain with about 810000 possible agreements, uniformly distributed on the search space, depicting a problem that happens quite often in the real life. England-Zimbabwe domain has only 576 possible agreements, however with a stronger opposition between the payoffs received by the negotiation parties. In this respect, this

Table 2 AgentFSEGA relative results scored at ANAC2010, compared with the winning agent of the domain

Domain - role	AgentFSEGA average utility	Domain winner average utility	relative result
Itex-Cypress Itex	0.533	0.598	0.891
Itex-Cypress Cypress	0.511	0.61	0.838
England-Zimbabwe England	0.712	0.886	0.804
England-Zimbabwe Zimbabwe	0.667	0.863	0.773

domain is cumbersome, as there are very few mutually acceptable agreements. Itex-Cypress has 180 possible agreements, with the strongest opposition between the negotiating parties. We note that on bigger domains AgentFSEGA performs well: weaker the opposition, better the results. That is because at every round, AgentFSEGA computes and proposes a bid. If the number of possible bids is small, this leads to AgentFSEGA conceding quicker. But, even on such domains, the defeat of AgentFSEGA is not too strong.

In table 2 we present the relative results of AgentFSEGA, expressed in percentages of the best result obtained by the winning agent, on a specific domain, in the ANAC 2010 competition. One may notice that in both domains AgentFSEGA scores relatively good results compared with the winner of the domain.

Unfortunately, AgentFSEGA did not run on the third domain - Travel. This third domain has 188160 possible aggreements with a weak opposition between them. This means that the utility difference between two possible agreements in both profiles is very small. The number of possible agreements over the 7 issues of the domain leads to almost 1 billion hypothesis for AgentFSEGA to estimate in the Bayesian space. This is 10 times bigger than the case of SON, with almost 100 million hypotheses. The failure of AgentFSEGA on the Travel domain is due to the impossibility of the Bayesian learner to initialize the hypothesis search space, as the number of alternatives is very large. Further optimization is required such that the Bayesian learner to accommodate only a subset of the hypothesis space, at the beginning of each negotiation.

The conclusion of the ANAC competition is that AgentFSEGA performs better when the search space increases. Of-course with the limitation that, after a certain threshold, it cannot accommodate a larger number of hypotheses in the agent memory.

5 Conclusions and Future Work

This paper presents AgentFSEGA experience at the ANAC 2010 competition. AgentFSEGA is a classical learning agent that derives the profile of the opponent out of its past moves. Taking as baseline the Bayesian agent [3], AgentFSEGA further improves the Bayesian strategy in order to accommodate the negotiation time constraint. Observing the number of rounds of the negotiation sessions in ANAC 2010, it seems that the majority of the agents designed for the competition exploited the same resource: negotiation should finish in a given time-frame.

It is noteworthy that the learning scheme works well, under the condition that the hypothesis space is not small. However, even on such tricky domains, AgentFSEGA is not strongly defeated. As the learning scheme is memory consuming, then on very large exponential domains further optimization is required, in order to make the agent scalable. The experience of AgentFSEGA within ANAC 2010 further proves that the strategies of learning opponents' profiles are worth investigating, as they lead to consistent negotiation results on realistic assumptions.

Acknowledgements. This work is supported by the Romanian Authority for Scientific Research under project PNII IDEI_2452.

Reference

1. Baarslag, T., Hindriks, K., Jonker, C., Kraus, S., Lin, R.: The first automated negotiating agents competition (ANAC 2010). In: Ito, T., et al. (eds.) New Trends in Agent-Based Complex Automated Negotiations. SCI, vol. 383, pp. 113–135. Springer, Heidelberg (2010)
2. Faratin, P., Sierra, C.: Jennings, N. Using similarity criteria to make issue trade-offs in automated negotiations 142(2) (2002)
3. Hindriks, K., Tykhonov, D.: Opponent modelling in automated multi-issue negotiation using bayesian learning. In: AAMAS 2008: Proceedings of the 7th International Joint Conference on Autonomous Agents and Multiagent Systems, IFAAMAS, pp. 331–338 (2008)
4. Jonker, C.M., Robu, V., Treur, J.: An agent architecture for multi-attribute negotiation using incomplete preference information. Autonomous Agents and Multi-Agent Systems 15(2), 221–252 (2007)
5. Li, C., Giampapa, J., Sycara, K.: A review of research literature on bilateral negotiations. Tech. Rep. CMU-RI-TR-03-41, Robotics Institute, Carnegie Mellon University (2003)
6. Lin, R., Oshrat, Y., Kraus, S.: Investigating the benefits of automated negotiations in enhancing people's negotiation skills. In: AAMAS 2009: Proceedings of The 8th International Conference on Autonomous Agents and Multiagent Systems, IFAAMAS, pp. 345–352 (2009)
7. Cosmin Silaghi, G., Dan Şerban, L., Marius Litan, C.: A Framework for Building Intelligent SLA Negotiation Strategies Under Time Constraints. In: Altmann, J., Rana, O.F. (eds.) GECON 2010. LNCS, vol. 6296, pp. 48–61. Springer, Heidelberg (2010)

Agent Smith: Opponent Model Estimation in Bilateral Multi-issue Negotiation

Niels van Galen Last

Abstract. In many situations, a form of negotiation can be used to resolve a problem between multiple parties. However, one of the biggest problems is not knowing the intentions and true interests of the opponent. Such a user profile can be learned or estimated using biddings as evidence that reveal some of the underlying interests. In this paper we present a model for online learning of an opponent model in a closed bilateral negotiation session. We studied the obtained utility during several negotiation sessions. Results show a significant improvement in utility when the agent negotiates against a state-of-the-art Bayesian agent, but also that results are very domain-dependent.

1 Introduction

Whenever two or more parties are in a conflict, the most common way to resolve the problem is by negotiating towards a solution. Negotiation is not limited to scenarios involving a conflict between parties and we can generalize the concept of negotiation as the interaction between two or more parties who are trying to find a solution to a situation in which everybody is satisfied. The negotiation process is a widely-studied topic in various research areas such as mathematics, economics and artificial intelligence.

Coinciding with the increasing availability of internet to people over the world, online businesses are becoming more popular than ever before. With the increasing competition, the e-commerce businesses are trying to make things as simple as possible for their customers. For example, offering different possibilities for making

Niels van Galen Last
Man-Machine Interaction, Faculty of EEMCS,
Delft University of Technology,
Mekelweg 4, 2628CD Delft,
The Netherlands
e-mail: n.a.vangalenlast@student.tudelft.nl

T. Ito et al. (Eds.): New Trends in Agent-Based Complex Automated Negotiations, SCI 383, pp. 167–174.
springerlink.com

electronic payments, semantic search through the product catalogs and personalized recommendations. However, human involvement is still required in the buying phase. The process of buying a product can be viewed as negotiation between the buyer and the seller. This type of negotiation during which two parties are trying to come to an agreement is known as bilateral negotiation.

During a negotiation session there are two important aspects to consider. First, to anticipate the opponents interests and motivation, because of their great influence on the opponents actions. When the other party reveals all his interests, it is much more convenient to reach an agreement. Unfortunately this type of negotiation is very uncommon. In this paper we focus on closed, bilateral negotiation. The algorithm that we present in this paper tries to estimate an opponent's preference model during a negotiation session. The negotiation in which this negotiation session takes place is called the party domain. The idea is that two agents are organizing a party and both have a different take (i.e., different preference profiles) on what is important to them. However, there is a maximum amount of money they can spend, and if the issues combined exceed the maximum, utility of the offer will be zero for both parties. There is also a time constraint on the negotiation, meaning if the two parties fail to reach an agreement before this time is up, both agents will receive a utility of zero.

For a graduate course on Artificial Intelligence and negotiation we developed the agent presented in this paper. This agent was later submitted to AAMAS 2010 to participate in the Automated Negotiating Agents Competition (ANAC) [1].

2 Negotiation Model

In the design of the overall strategy of the negotiation agent several constraints had to be taken into account. First of all, an agreement must be reached within three minutes. If the agents fail to reach an agreement within this time span, both agents receive a utility of zero. Another constraint is set by the cost of an offer. For example, in the party domain the price of a bid must not exceed 1200 euros. It would be not be beneficial for an agent to propose or accept such an offer for it has a utility of zero for both agents.

Our goal was for the agent to propose offers on or as close as possible to the Pareto efficient frontier. This requires not only the agent's own preference profile, but also (a decent estimation) of the opponents profile. Since the environment was closed negotiation (the agent does not have direct access to the opponents profile, this reflects most real life negotiation situations), the agent must make an educated guess on the utility and weights of the opponent profile. This is only possible under the assumption of negotiating with a rational opponent. Therefore the agent assumes that the first bid of the opponent will be the maximum utility bid of the agent and that subsequent bids slowly concede towards the agent's own preferences.

Constructing the opponent preference profile is a difficult problem, because several assumptions have to be made about the opponents behavior. The most important one is that the first offer the opponent makes is the maximum utility offer. If the opponent is playing rationally, values with small weights shall be adjusted most often.

Issues that are important to the opponent shall not be adjusted as often. This information is used together with the opponent's consecutive offers to construct an opponent preference profile model.

For offering bids we created the following protocol. In the first round the agent always offers his maximum utility bid. The reason for this is twofold. First, this is the maximum utility the agent can ever achieve, if the opponent accepts this bid our agent has achieved the highest possible utility. Second, this enables the opponent to make a proper estimate of the maximum utility values, which will help the opponent in knowing where to concede towards. Subsequent bids are made based on the constructed opponent preference profile, slowly conceding towards the opponents preferences. When the time is almost up, the agent takes on a strategy implemented to come quickly to an agreement. This is done by proposing offers that the opposing party has previously made with the highest utility values for our agent.

The agent accepts an offer under the following circumstances. In general, any bids made by the opponent which have a utility value over the 0.9 threshold are accepted. If the opposing party makes an offer that exceeds the utility value of our agent's last offered bid the agent accepts as well. The reason for this is that the order in which the agent makes his bids follows a monotone decreasing shape. Hence if the opponents makes a better offer than the upcoming bid of our agent, the utility exceeds the utility of our next bid and it is wise to accept this offer. The last situation in which the agent accepts is when time is almost running out, in this case the agent accept bids at a much lower threshold: 0.7.

3   Experimental Results

In this section we discuss the results of several negotiation sessions. As our first opponent we selected the state-of-the-art Bayesian agent presented in [6]. This agent is also designed for the GENIUS [5] environment, but adopted a Bayesian approach to the opponent modeling. Second, we discuss the ANAC results.

3.1   Bayesian Agent

The initial implementation of the agent sampled the entire bidspace with a utility higher than zero and ordered these in order of highest utility first. Based on this list the first bid contains the highest utility. This was advantageous for our agent, but sometimes resulted in no agreement at all, which means the agent scored a utility of zero. The performance was improved by taking on a more temporal approach. The agent's threshold for acceptance also decreases over time, to a 0.7 in the last 10 seconds of the negotiation session and the agent will propose the highest bids that the opponent already has proposed. It is likely for a rational opponent to accept one of those bids, since he previously proposed them.

To improve the agent's selection of bids to offer to the opponent, we implemented a mechanism that constructs an opponent model. This represents the importance and preference of each of the items for each issue. This model is used to make bids that are not only beneficial for us, in the sense that they are close to the Pareto optimal frontier, but it also accounts for the opponents' preference. The to test the performance of the improved agent, a tournament was created against all readily available agents and against itself where each session consists of five rounds.

The first tournament consisted of negotiation sessions between our agent and the Bayesian agent, using party preference profiles 4, 8, 20 and 24 with five sessions. These party profiles come with the GENIUS environment and were selected at random. This experiment was performed twice: our agent makes the first bid and the Bayesian agent makes the first bid. The results are shown in table 1. Figure 1 shows one of the sessions from the tournament in which a pareto efficient agreement was obtained. In this figure the Bayesian agent corresponds to Agent A (green line) and Agent Smith corresponds to Agent B (blue line), it corresponds to the seventh session in Table 1. From this we can see that the average utility of Agent Smith, given that he starts with the first bid, is 0.872, whereas the average utility of the Bayesian agent is 7.96. The Bayesian agent obtains an average utility of 0.818 (or 0.750 if we take the zero utility into account) when making the first bid. Again, the average of Agent Smith is much higher with a utility of 0.873 (0.801 if the zero utility is taken into account). These results show that the average utility of our agent is about 8% higher than the utility of the Bayesian agent.

Table 1 Smith Agent (A) vs. Bayesian Agent (B)

Profile A	Profile B	Smith vs. Bayes			Bayes vs. Smith		
		UtilityA	UtilityB	Outcome	UtilityA	UtilityB	Outcome
profile 4	profile 8	0.840	0.857	Lose	0.840	0.926	Win
profile 4	profile 20	0.885	0.801	Win	0.785	0.896	Win
profile 4	profile 24	0.920	0.878	Win	0.875	0.900	Win
profile 8	profile 4	0.926	0.840	Win	0.857	0.840	Lose
profile 8	profile 20	0.914	0.849	Win	0.726	0.958	Win
profile 8	profile 24	0.966	0.765	Win	0.874	0.825	Lose
profile 20	profile 4	0.896	0.682	Win	0.822	0.825	Win
profile 20	profile 8	0.911	0.794	Win	0.822	0.903	Win
profile 20	profile 24	0.829	0.592	Win	0.759	0.648	Lose
profile 24	profile 4	0.900	0.875	Win	0.878	0.920	Win
profile 24	profile 8	0.825	0.874	Lose	0.765	0.966	Win
profile 24	profile 20	0.648	0.759	Lose	0.000	0.000	Lose

The second tournament was a negotiation session between two instances of our agent, again using profiles 4, 8, 20 and 24 with five sessions total. The results are shown in table 2.

In this case you can see that A wins most of the time, this is because the agents will slowly work to each other, in most of the cases it will take the maximum time possible. In this case player B has to make a BATNA bid, which player A will accept. Because agent A wins most of the time there can be concluded that the starting agent has an advantage. This can also be seen from the average utilities that are obtained by the players. The average of the agent that makes the first offer is 0.857, whereas the average of the other agent is 0.829.

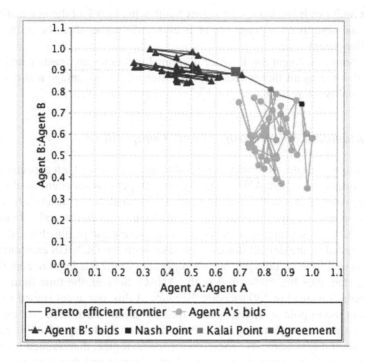

Fig. 1 Example of a negotiation session ending in a Pareto efficient agreement

Table 2 Smith Agent (A) vs. Smith Agent (B)

Profile A	Profile B	UtilityA	UtilityB	Outcome
profile 4	profile 8	0.840	0.926	B Win
profile 4	profile 20	0.885	0.801	A Win
profile 4	profile 24	0.920	0.878	A Win
profile 8	profile 4	0.857	0.840	A Win
profile 8	profile 20	0.889	0.867	A Win
profile 8	profile 24	0.853	0.850	Tie
profile 20	profile 4	0.847	0.803	A Win
profile 20	profile 8	0.879	0.863	A Win
profile 20	profile 24	0.804	0.633	A Win
profile 24	profile 4	0.878	0.920	B Win
profile 24	profile 8	0.850	0.852	Tie
profile 24	profile 20	0.785	0.696	A Win

Based on these results one can conclude that most of the negotiation sessions are won by our agent, where the winner is determined by the agent that obtained the highest utility given that the session ended in an agreement. Most the sessions the agreement results were on the Pareto optimal frontier with an advantage for our agent. This leads us to the following brief summary of the results:

- Agent Smith vs Bayesian agent: Agent Smith wins 62.5% of the negotiation sessions and obtains an average utility is about 8% higher than the utility of the Bayesian Agent.
- Agent Smith vs Agent Smith: Pareto optimal outcomes are reached and results show that the agent that makes the first offer has an advantage and reachers a higher utility on average.

3.2 Automated Negotiating Agents Competition

The preliminary results were very good and therefore we had high expectations for the performance during the ANAC. During this competition our agent would negotiate against other agents in different domains. Unfortunately, the results turned out to be overly optimistic, since the agent could not manage to consistently obtain good utilities. There are several arguments that can explain these results. First, the agent was only tested in the default domains bundled with the GENIUS environment. It performed well in these domains but was never put to the test in more complex domains like the ones that were used in the ANAC. Second, the time limit was set to 120 seconds instead of 180 seconds. Because of this, our agent started to accept offers from the opponents much sooner then he should have done. As explained before, we tested our agent only in the Party domain and during ANAC it was subjected to several new domains. Therefore, it seems logical that unintentionally the agent was in some way optimized for the Party domain and therefore was lacking the strategy or generality to succeed in other domains.

4 Related Work

The principles behind negotiation can be applied to many situations and, therefore, is pursued in various fields. Roots can be found in areas such as game theory [8, 13], but it is also applied social sciences, economics and (distributed) artificial intelligence. However, every field is working on different aspects of the negotiation process and have different constraints. For example, researchers in game theory are creating protocols that specify the boundaries of strategies that an agent can take during a negotiation session. Such a protocol can, for instance, prescribe that no agents are allowed to deceive others during an encounter [9]. The reasoning is that when an agent puts all his cards on the table the outcome of negotiation will be better for both agents because they can more easily work towards agreements much closer to the Pareto efficient frontier. This work contrasts the work discussed in Section 2, because these situations are almost never occurring in the real world. Our work focusses more at the negotiation process itself and not only on the outcome. Like [3], the main interests is on the dynamics and experimental results. we concentrate on estimating a model of the opponent [11, 10] and making slow concessions (cf. [4]) but without relying on a specific learning algorithm such as [7, 12, 2].

5 Conclusion

In this paper we presented an agent model for estimating the opponents interests profile during closed bilateral negotiation sessions. The agent was developed assuming a rational opponent and experiments have shown that the best results are obtained when our agent is using playing rational. In order to get the best results against agents who play rationally, our agent also has to negotiate with a rational strategy. Hence, to reach a Pareto optimal outcome, the agent needs to be open about our preferences to the opponent and assume that the opponent plays rational as well. During a negotiation session, our agent will try to estimate an opponent profile using the information that the opponent implicitly gives, regarding his preferences, through his bid proposals. Then our agent will slowly make concessions based on the likelihood of the opponents profile as he is trying to come to an agreement. We created a tournament environment for our agent to negotiate against an agent that tries to learning our preferences using a Bayesian approach. Results show that 62.5% of the negotiation sessions end with a higher utility for our agent than the Bayesian agent.

The agent was also submitted to the Automated Negotiating Agents Competition, however most of the domains our agents was subjected to, the results were not very good. The most important reason is that the time constraint for our agent was set to two minutes instead of three, resulting in accepting offers that are not very good. Also, our agent was only tested in one domain and it apparently there are some accidental optimizations made for this domain, resulting in bad performance in others. However, these results provide much useful feedback for further improvements.

Acknowledgements. We would like to thank Catholijn Jonker and Tim Baarslag for their useful comments.

Reference

1. Baarslag, T., Hindriks, K., Jonker, C., Kraus, S., Lin, R.: The First Automated Negotiating Agents Competition (ANAC 2010). In: Ito, T., et al. (eds.) New Trends in Agent-Based Complex Automated Negotiations. SCI, vol. 383, pp. 113–135. Springer, Heidelberg (2010)
2. Coehoorn, R., Jennings, N.: Learning on opponent's preferences to make effective multi-issue negotiation trade-offs. In: ... of the 6th International Conference on ... (January 2004)
3. Faratin, P., Sierra, C., Jennings, N.R.: Negotiation decision functions for autonomous agents. Robotics and Autonomous Systems 24(3-4), 159–182 (1998)
4. Fatima, S., Wooldridge, M., Jennings, N.: Optimal negotiation strategies for agents with incomplete information. In: Intelligent Agents VIII (January 2002)
5. Hindriks, K., Jonker, C., Kraus, S., Lin, R.: Genius: negotiation environment for heterogeneous agents. In: Proceedings of The 8th ... (January 2009)
6. Hindriks, K., Tykhonov, D.: Opponent modelling in automated multi-issue negotiation using bayesian learning. In: Proceedings of the 7th International Joint Conference on Autonomous Agents and Multiagent Systems, vol. 1, pp. 331–338 (2008)

7. Mudgal, C., Vassileva, J.: Bilateral negotiation with incomplete and uncertain information: A decision-theoretic approach using a model of the opponent. In: Cooperative Information Agents IV-The Future of Information Agents in Cyberspace, pp. 1–43 (2004)
8. Rosenschein, J., Genesereth, M., University, S.: Deals among rational agents. Citeseer (January 1984)
9. Rosenschein, J.S., Zlotkin, G.: Rules of encounter: designing conventions for automated negotiation among computers. MIT Press, Cambridge (1994)
10. Soo, V.W., Hung, C.A.: On-line incremental learning in bilateral multi-issue negotiation. In: Proceedings of the First International Joint Conference on Autonomous Agents and Multiagent Systems: part 1, p. 315 (2002)
11. Zeng, D., Sycara, K.: How can an agent learn to negotiate? Intelligent Agents III Agent Theories (January 1997)
12. Zeng, D., Sycara, K.: Bayesian learning in negotiation. International Journal of Human-Computers Studies (January 1998)
13. Zlotkin, G., Rosenschein, J.: Negotiation and task sharing among autonomous agents in cooperative domains. In: Proceedings of the Eleventh International . . . (January 1989)

Author Index

Vassilis G. Kaburlasos

Towards a Unified Modeling and Knowledge-Representation based
on Lattice Theory

Studies in Computational Intelligence, Volume 27

Editor-in-chief
Prof. Janusz Kacprzyk
Systems Research Institute
Polish Academy of Sciences
ul. Newelska 6
01-447 Warsaw
Poland
E-mail: kacprzyk@ibspan.waw.pl

Further volumes of this series
can be found on our homepage:
springer.com

Vol. 10. Andrzej P. Wierzbicki, Yoshiteru
Nakamori
Creative Space, 2005
ISBN 3-540-28458-3

Vol. 11. Antoni Ligęza
*Logical Foundations for Rule-Based
Systems,* 2006
ISBN 3-540-29117-2

Vol. 12. Jonathan Lawry
*Modelling and Reasoning with Vague Con-
cepts,* 2006
ISBN 0-387-29056-7

Vol. 13. Nadia Nedjah, Ajith Abraham,
Luiza de Macedo Mourelle (Eds.)
Genetic Systems Programming, 2006
ISBN 3-540-29849-5

Vol. 14. Spiros Sirmakessis (Ed.)
Adaptive and Personalized Semantic Web, 2006
ISBN 3-540-30605-6

Vol. 15. Lei Zhi Chen, Sing Kiong Nguang,
Xiao Dong Chen
*Modelling and Optimization of
Biotechnological Processes,* 2006
ISBN 3-540-30634-X

Vol. 16. Yaochu Jin (Ed.)
Multi-Objective Machine Learning, 2006
ISBN 3-540-30676-5

Vol. 17. Te-Ming Huang, Vojislav Kecman,
Ivica Kopriva
*Kernel Based Algorithms for Mining Huge
Data Sets,* 2006
ISBN 3-540-31681-7

Vol. 18. Chang Wook Ahn
Advances in Evolutionary Algorithms, 2006
ISBN 3-540-31758-9

Vol. 19. Ajita Ichalkaranje, Nikhil
Ichalkaranje, Lakhmi C. Jain (Eds.)
*Intelligent Paradigms for Assistive and
Preventive Healthcare,* 2006
ISBN 3-540-31762-7

Vol. 20. Wojciech Penczek, Agata Półrola
*Advances in Verification of Time Petri Nets
and Timed Automata,* 2006
ISBN 3-540-32869-6

Vol. 21. Cândida Ferreira
*Gene Expression on Programming: Mathematical
Modeling by an Artificial Intelligence,* 2006
ISBN 3-540-32796-7

Vol. 22. N. Nedjah, E. Alba, L. de Macedo
Mourelle (Eds.)
Parallel Evolutionary Computations, 2006
ISBN 3-540-32837-8

Vol. 23. M. Last, Z. Volkovich, A. Kandel (Eds.)
Algorithmic Techniques for Data Mining, 2006
ISBN 3-540-33880-2

Vol. 24. Alakananda Bhattacharya, Amit Konar,
Ajit K. Mandal
Parallel and Distributed Logic Programming,
2006
ISBN 3-540-33458-0

Vol. 25. Zoltán Ésik, Carlos Martín-Vide,
Victor Mitrana (Eds.)
*Recent Advances in Formal Languages
and Applications,* 2006
ISBN 3-540-33460-2

Vol. 26. Nadia Nedjah, Luiza de Macedo Mourelle
(Eds.)
Swarm Intelligent Systems, 2006
ISBN 3-540-33868-3

Vol. 27. Vassilis G. Kaburlasos
*Towards a Unified Modeling and Knowledge-
Representation based on Lattice Theory,* 2006
ISBN 3-540-34169-2

Vassilis G. Kaburlasos

Towards a Unified Modeling and Knowledge-Representation based on Lattice Theory

Computational Intelligence and Soft Computing Applications

With 47 Figures and 31 Tables

 Springer

Prof. Vassilis G. Kaburlasos, Ph.D.
Department of Industrial Informatics
Division of Computing Systems
Technological Educational Institution of Kavala
65404 Kavala
Greece
E-mail: vgkabs@teikav.edu.gr

ISSN print edition: 1860-949X
ISSN electronic edition: 1860-9503
ISBN 978-3-642-07058-7   e-ISBN 978-3-540-34170-3

Springer is a part of Springer Science+Business Media
springer.com
© Springer-Verlag Berlin Heidelberg 2006
Softcover reprint of the hardcover 1st edition 2006

The use of general descriptive names, registered names, trademarks, etc. in this publication
does not imply, even in the absence of a specific statement, that such names are exempt from
the relevant protective laws and regulations and therefore free for general use.

Cover design: deblik, Berlin

Foreword

What can be said at all can be said clearly, and what we cannot talk about
we must pass over in silence.

(Wittgenstein 1975)

Various modeling paradigms are proposed under computational intelligence and soft computing. In the aforementioned context lattice theory
emerges as a sound foundation for unifying diversity rigorously. Novel
perspectives and tools are presented here towards improving both analysis
and design in practical applications. Hopefully the material in this volume
is not only illuminating and useful but also inspiring.

To my family

Preface

By 'model' we mean a mathematical description of a world aspect. Models are important because they enable employment of the vast resources in mathematics in order to improve analysis and design in real world applications. A model is useful to the extent it generalizes accurately.

The development of a model, namely modeling, is close to an *art* since a model should be both 'detailed enough', to accurately describe phenomena of interest, and 'simple enough', to be amenable to rigorous analysis.

Originally, modeling has been formalized rigorously in the Euclidean space R^N – Note that the *totally ordered* lattice R of real numbers has emerged historically from the conventional measurement process of successive comparisons. With the proliferation of computers, a variety of modeling paradigms emerged under *computational intelligence* (CI) and *soft computing* (SC). However, an advancing technology is fragmented in different domains, where domain specific tools are developed. An effective unification would be in the interest of all parts involved.

A major reason for the development of different tools in different domains is the need to cope with disparate types of data including vectors, Boolean data, symbols, images, text, graphs, (fuzzy) sets, etc. It appears that an effective unification should also be pursued in the data domain.

This research monograph proposes a unified approach for knowledge representation and modeling beyond R^N as well, based on *lattice theory*. The emphasis is on clustering, classification, and regression applications.

Currently lattice theory is not in the mainstream in CI-SC. Nevertheless several research groups around the world, working separately, employ lattice theory *explicitly* in specific domains including neural computing, mathematical morphology, information retrieval, inference systems, etc. In addition, lattice theory is employed *implicitly* in a number of popular computational intelligence paradigms as shown below.

To the best of our knowledge the material in this monograph is the first effort to present in a single volume various applications of lattice theory in computational intelligence and soft computing. The material here is multi-disciplinary based mainly on our on-going research published regularly in major scientific journals and conferences. Relevant work by other authors is also presented extensively and comparatively. The presentation here is concise. Pointers to more detailed publications are shown. Novel results and perspectives are also presented.

The essential prerequisite for understanding the material below is a familiarity with general computational intelligence practices. Lattice theory is useful but not necessary since all the required mathematics is covered in this volume. The material here is structured in four parts.

Part I describes the context. More specifically, chapter 1 outlines the origins of this work; chapter 2 reviews extensively relevant literature.

Part II presents the mathematical background as well as algorithms. In particular, chapter 3 summarizes basic lattice theory; chapter 4 shows the relevance of the aforementioned tools; chapter 5 shows the capacity of lattice theory for knowledge representation; chapter 6 formulates the modeling problem as a function approximation problem; chapter 7 presents several algorithms for clustering, classification, and regression.

Part III shows applications and interesting connections. More specifically, chapter 8 demonstrates applications of the proposed models on numeric data, whereas chapter 9 demonstrates applications on non-numeric data; chapter 10 describes connections with established paradigms.

Part IV includes discussions and the conclusion. In particular, chapter 11 discusses practical implementation issues; chapter 12 summarizes advantages and limitations of our work.

The Appendices include supportive material. More specifically, Appendix A summarizes useful mathematical definitions and notions; Appendix B presents mathematical proofs; Appendix C shows convenient geometric interpretations on the plane.

Acknowledgements

Carl G. Looney has directed me to lattice theory during my Ph.D. work on neural computing applications at the University of Nevada, Reno. My long collaboration with Vassilis Petridis at Aristotle University of Thessaloniki has given my work a direction towards practical system modeling. I had useful discussions with Theodore Alevizos at Technological Educational Institution of Kavala regarding computing issues. Substantial has been the mathematical contribution of Athanasios Kehagias. Several colleagues and graduate students have contributed by carrying out creative experiments; especially significant at different stages has been the contribution of Stelios Papadakis, Ioannis Athanasiadis, Al Cripps, and Spyros Kazarlis.

Contents

Lists of Figures and Tables

Acronyms and Symbols

Acronym	Meaning
AI	artificial intelligence
AN	ammonium nitrate
ARMA	autoregressive moving average
ART(MAP)	adaptive resonance theory (MAP)
ART1	ART 1
ART2	ART 2
BCP	Bayesian combined precictor
CALFIN (algorithm)	algorithm for CALculating a FIN
CART	classification and regression tree
CBR	case-based reasoning
CDF	cumulative distribution function
CI	computational intelligence
CInd	circularity index
CNF	conjunctive normal form
CTP	computational theory of perceptions
CW	computing with words
FBF	fuzzy basis functions
FCA	formal concept analysis
FIN	fuzzy interval number
FINkNN	FIN k nearest neighbor
FIS	fuzzy inference system
FLBN	fuzzy logic-based neural network
FLN(MAP)	fuzzy lattice neurocomputing (MAP)
FLNff	FLN first fit
FLNmtf	FLN max tightest fit
FLNotf	FLN ordered tightest fit
FLNsf	FLN selective fit
FLNtf	FLN tightest fits
FLR	fuzzy lattice reasoning
FLSTN	fuzzy lattice state transition net
FNN	fuzzy neural network
GA	genetic algorithm

g.l.b.	greatest lower bound
grARMA	granular ARMA
grSOM	granular self organizing map
GTM	generative topographic map
HC	hard computing
HCI	human-computer interaction
hw	hyperwords
IC	integrated circuit
IM (algorithm)	algorithm for calculating an Inclusion Measure
KL	Kullback-Leibler
kNN	k nearest neighbor
KSOM	Kohonen's self organizing map
L-fuzzy set	lattice fuzzy set
LR	Left Right (fuzzy number representation)
LSE	least square error
l.u.b.	least upper bound
ML	machine learning
MLP	multiplayer perceptron
mod	modulo
MOS	metal-oxide-semiconductor
NBC	naïve Bayes classifier
O_3RTAA	a multi-agent decision-support software system
PAC	probably approximately correct
PAREST (algorithm)	algorithm for PARameter ESTimation
pdf	probability density function
POL	percentage of sugar in fresh root weight
poset	partially ordered set
RAAM	recursive auto-associative memory
RBF	radial basis function
RPM	rotations per minute
RW	roots weight
SC	soft computing
SOM	self organizing map
SOMM	self organizing mixture model
SVM	support vector machines
TSK	Takagi-Sugeno-Kang
VC	Vapnik-Chervonenkis
VLSI	very large scale integration
Voting σ-FLNMAP	ensemble of σ-FLNMAP modules
WWW	world wide web

Symbol	Meaning
$\leq$	order, relation
$<$	strict order, relation
$<<$	order(s) of magnitude less, relation
$\leq^{\partial}$ (or $\leq^{-1}$, or $\geq$)	dual of $\leq$
$\|$	incomparability, relation
$\wedge$	meet/infimum/g.l.b. operator in a lattice
$\vee$	join/supremum/l.u.b. operator in a lattice
$'$	complement (unary) operator
$\overline{\wedge}$	meet operator in a hidden lattice
$\underline{\vee}$	join operator in a hidden lattice
$\subseteq$	set-inclusion, relation
$\cap$	set-intersection, operator
$\cup$	set-union, operator
$\varnothing$	empty set
$\cong$	isomorphic, relation
$\sim$	equivalence, relation
$\times$	direct (Cartesian) product
$:=$	equal, by definition
$:\Leftrightarrow$	equivalent, by definition
$\otimes$	symmetric difference, operator
$\lfloor . \rfloor$	floor function
$\| \cdot \|$	real number 'norm' of its vector argument
$\| \cdot \|$	vector 'absolute value' of its vector argument
$\exists x$	there exists a x
$\forall x$	for all x
$x \in A$	x is in set A
B	set of balls in (L,d)
C	set of continuous functions
F	set of FINs
$\mathfrak{F}$	set of functions $f: R \rightarrow R$
$\mathcal{F}$	set of non-negative functions in $\mathfrak{F}$
G	set of (directed) graphs in $\mathcal{M}$
$\mathcal{G}$	set of integrable functions in $\mathcal{F}$
$\mathcal{G}^{*}$	set of equivalence classes of zero distance in $\mathcal{G}$

I	complete lattice unit interval [0,1]
I	greatest element in a poset
M	set of generalized intervals
M	set of Mamdani type FISs
$\mathcal{M}$	master-graph
N	set of natural numbers
O	least element in a poset
Q	totally ordered lattice of rational numbers
R	totally ordered lattice of real numbers
R	binary relation between two sets
S	set of TSK type FISs
U	unit (N-dimensional) hypercube
x	vector
Δ	interval in τ(L)
Φ	set of generalized spheres
TRN	training data set
TST	testing data set
d_p	Minkowski metric, also known as L_p metric
D_{crit}, D_a	diagonal size in σ-FLN(MAP)
F_a	a-cut of a fuzzy set
F_c	collection of all families that represent a class c
F_n	set of fuzzy numbers
$F_+/F_0/F_-$	set of positive/trivial/negative FINs
$\mathbf{I}_{[a,b]}$	set of non-decreasing functions on interval $[a,b]$
K_{ij}	label
L_1	Hamming (city-block) distance
L_2	Euclidean distance
n_V	number of voters in Voting σ-FLNMAP
N_s	number of senses
N_w	number of words
P_p	set of parametric models with p parameters
T_d	distance threshold
W_{ij}	FIN weight in grSOM
ρ_a	vigilance parameter in σ-FLNMAP
$\aleph_0$	cardinality of Q
$\aleph_1$	cardinality of R
$\aleph_2$	cardinality of Pow(R)
A^{-1}	inverse of A
e^h	basis vector in linear space $\mathbf{M}^h$
$\mathbf{M}^h$	set of generalized intervals of height h
S^∂	the dual of statement S

x^+	vector 'positive part' of vector x
x^-	vector 'negative part' of vector x
X^c	complement of X
$M_+^h / M_0^h / M_-^h$	set of positive/trivial/negative generalized intervals
R_0^+	set of non-negative real numbers
$(a]$	principal ideal set $\{x: x \le a\}$
$[b)$	principal filter set $\{x: x \ge b\}$
$\hat{f}$	optimal estimate for f
$Pow(S)$	power set of set S
$[u]^a$	a-cut of fuzzy set u
$a(L)$	sets of atoms in lattice $(L, \le)$
$\tau(L)$	set of intervals in lattice $(L, \le)$
$\mathcal{C}_{\tau(L)}$	set of classes in lattice $(L, \le)$
$\mathcal{F}_{\tau(L)}$	set of families of intervals in lattice $(L, \le)$
$O(n^2)$	complexity order n^2
$Q(f)$	quotient of family/class f
$(P, \le)$	poset
$(P, \le)^\partial$	dual of $(P, \le)$
$(L, \le)$	(complete) lattice
$(L^\partial \times L, \le)$	lattice of generalized intervals
$(\tau(L), \le)$	lattice of intervals
$(M^h, \le)$	lattice of generalized intervals of height h
(a,b)	element of lattice $(L \times L, \le)$
$[a,b]$	element of lattice $(L^\partial \times L, \le)$
$[O,I]$	largest element in complete lattice $(\tau(L), \le)$
$[I,O]$	least element in complete lattice $(\tau(L), \le)$
$[a,b]^h$	generalized interval of height h
(L,d)	(linear) metric space L with metric d
$[S, F]$	algebra, F is a set of operations on set S
(X, μ)	fuzzy set $\mu: X \to [0,1]$
$(L, \le, \mu)$	fuzzy lattice
(Ω, S, P)	probability space
arg max/min	argument which corresponds to a *max/min* value
ball(c,r)	set $\{x \in L: d(x,c) \le r\}$, where (L,d) is a metric space
card(S)	cardinality of set S
dim(x)	dimension of vector x
$f = \{w_i\}_{i \in I}$	family f of (lattice) intervals w_i

$\{F_a\}_{a \in [0,1]}$	fuzzy number, F_a is a a-cut
$\mathcal{G} = (\mathcal{V}, \mathcal{E})$	graph $\mathcal{G}$ with vertices in $\mathcal{V}$ and edges in $\mathcal{E}$
$median(x)$	median of the entries in vector x
$N(\mu, \sigma)$	normal pdf with mean μ and standard deviation σ
$P(M/D)$	Bayesian conditional of model M given data D
$sphere(c,r)$	sphere with center at c and radius $r \geq 0$
$h(t)$	neighborhood function in KSOM
$M(t)$	number of intervals learned by a σ-FLN
$m_A(f)$	fuzzy membership of f in A
$m_i(t)$	reference (weight) vector in KSOM
$\mu^h_{x_1,x_2}(x)$	generalized interval (box function) of height h
$(\Delta_i, g(\Delta_i))$	pair of a datum $\Delta_i \in \tau(L)$ and its category $g(\Delta_i)$
d: $L \times L \rightarrow R_0^+$	(pseudo)metric
d_h: $M^h \times M^h \rightarrow R_0^+$	metric in M^h
$d_H([u]^a, [v]^a)$	Hausdorf metric between intervals/a-cuts $[u]^a$ and $[v]^a$
d_K: $F \times F \rightarrow R_0^+$	(pseudo)metric
d_p: $S \times S \rightarrow R_0^+$	Minkowski metric for an integer parameter p
$diag$: $\tau(L) \rightarrow R_0^+$	diagonal of an interval
f: $L \rightarrow K$	function/model, where $(L, \leq)$ and $(K, \leq)$ are lattices
F: $(0,1] \rightarrow M$	FIN, in particular $F = \bigcup_{h \in (0,1]} \{[a(h), b(h)]^h\}$
g: $a(L) \rightarrow K$	category function
m_h: $R \rightarrow R_0^+$	(integrable) mass function
me: $\mathcal{F}_{\tau(L)} \rightarrow \mathcal{F}_{\tau(L)}$	maximal expansion
v: $L \rightarrow R$	positive valuation function
k: $L \times L \rightarrow [0,1]$	a specific inclusion measure
s: $L \times L \rightarrow [0,1]$	a specific inclusion measure
θ: $L^{\partial} \rightarrow L$	isomorphic function
σ: $L \times L \rightarrow [0,1]$	a general inclusion measure function
σ_K: $F_+ \times F_+ \rightarrow [0,1]$	a specific inclusion measure
ϕ: $\tau(L) \rightarrow L \times L$	composite injective monotone map
φ: $B \rightarrow \Phi$	injective monotone map
$d_S(x, [a,b])$	distance between a point x and an interval $[a,b]$
$M/\sim$	quotient (set) of set M
R: $A \rightarrow C$	rule R 'if A then C'

PART I: THE CONTEXT

Part I describes the origins of this work.
It also includes a relevant literature review.

1 Origins in Context

This chapter summarizes the chronicle of this work in context

The origins of this work are traced to an application of Carpenter's fuzzy Adaptive Resonance Theory neural network, or fuzzy-ARTMAP for short, in health care databases towards medical diagnosis (Kaburlasos 1992). Note that other researchers, later, also pursued medical pattern classification using fuzzy-ARTMAP (Downs et al. 1996). However, the work in Kaburlasos (1992) broke new ground by introducing lattice theory in neural computing as explained next.

Fuzzy-ARTMAP operates by conditionally enlarging hyperboxes in the unit hypercube. It was realized that the set of hyperboxes is lattice-ordered; hence improvements were sought using lattice theory. A naive theory of perception was proposed in the unit hypercube by introducing novel tools such as an 'inclusion measure' function for computing a fuzzy degree of inclusion of a hyperbox into another one; moreover, the notion 'fuzzy lattice' was introduced in R^N (Kaburlasos 1992).

Subsequent work extended the applicability domain of fuzzy-ARTMAP to a Cartesian product $L = L_1 \times \ldots \times L_N$ involving disparate 'complete' lattices. A series of Fuzzy Lattice Neurocomputing (FLN) models was launched and effective applications were demonstrated in pattern recognition (Kaburlasos and Petridis 1997, 2000; Petridis and Kaburlasos 1998, 1999), intelligent control in surgery (Kaburlasos et al. 1999), and text classification (Petridis and Kaburlasos 2001; Kehagias et al. 2003).

Retaining the basic tools of a FLN model, interest gradually shifted to machine learning (Kaburlasos et al. 1999, 2002; Kaburlasos 2004a; Petridis and Kaburlasos 2003). A breakthrough analysis of fuzzy numbers using lattice-ordered 'generalized intervals' further turned interest to fuzzy inference systems (Kaburlasos 2002; Kaburlasos and Kehagias 2006a, 2006b). Optimization was pursued using genetic algorithms (Kaburlasos et al. 2003). Recent work has proposed a granular extension of Kohonen's Self-Organizing Map to a linguistic data domain (Kaburlasos et al. 2005; Kaburlasos and Papadakis 2006). Apart from an introduction of effective

Vassilis G. Kaburlasos: *Origins in Context*, Studies in Computational Intelligence (SCI) **27**, 3–4 (2006)
www.springerlink.com

techniques, a most significant contribution of our work has been the elevation of lattice theory in computational intelligence – soft computing.

Lately there is a growing interest in lattice theory applications in various domains including machine learning (Mitchell 1997), knowledge representation (Ganter and Wille 1999), image processing (Braga-Netto 2001; Goutsias and Heijmans 2000), neural computing (Ritter and Urcid 2003), mathematical morphology (Sussner 2003; Maragos 2005), probabilistic inference (Knuth 2005), etc. Unique characteristics of the lattice theoretic techniques here include (1) complete and/or non-complete lattices, (2) lattices of either finite- or infinite- cardinality, and (3) positive valuation functions. Moreover, our techniques retain a unifying character.

Various authors currently employ lattice theory explicitly as explained above. In addition, lattice theory is employed implicitly. For instance, it is known that the family of all partitions of a set (also called 'equivalence relations') is a lattice when partitions are ordered by refinement. It turns out that popular information-processing frameworks, including probability theory, rough set theory, etc. employ equivalence relations. Furthermore, lattice theory also emerges implicitly in popular computational intelligence paradigms as shown below.

Currently, it is 'mathematics' which remains the application domain of lattice theory par excellence. Note that, historically, mathematical lattices emerged as a spin off of work on formalizing propositional logic. More specifically, in the first half of the nineteenth century, Boole's attempt to formalize propositional logic led to the concept of Boolean algebras. While investigating the axioms of Boolean algebras at the end of the nineteenth century Peirce and Schröder introduced the lattice concept. Independently Dedekind's research on ideals of algebraic numbers led to the same concept (Grätzer 1971). Nevertheless, it was Garrett Birkhoff's work in the mid-1930s that started the general development of lattice theory. Birkhoff showed that lattice theory provides a unifying framework for hitherto unrelated developments in many mathematical disciplines including linear algebra, logic, probability, etc. The development of lattice theory by eminent mathematicians including Dedekind, Jónsson, Kurosh, Malcev, Ore, von Neumann, Tarski and Birkhoff contributed a new vision of mathematics. Note that the introduction of lattice theory has met resentment and hostility. Nevertheless, lattice theory is expected to play a leading role in the mathematics of the twenty-first century (Rota 1997).

This research monograph focuses on the utility of lattice theory for unified modeling and knowledge-representation in the context of computational intelligence – soft computing.

2 Relevant Literature Review

> This chapter stakes out the domain of interest

The interest here is in models implementable on a computing device for driving a practical application. The models of interest are to be induced empirically from training data by novel *learning* algorithms. According to some authors the problem of learning is at the very core of both biological and artificial intelligence (Heisele et al. 2002). In the latter sense the aforementioned algorithms are *intelligent*.

2.1 Classic Modeling and Extensions

Classic modeling typically regards physical phenomena. In particular, principles/laws of physics, biology etc. are invoked to derive parametric algebraic expressions that quantify a functional relation between variables of interest. In particular, *dynamic* models employ differential/difference equations whose solutions are analog signals (functions of time).

Analog signals are measured by sampling them at discrete times t_k; hence *discrete* signals emerge. Continuous and discrete time signals have been studied jointly in the time domain or in the Fourier-/Laplace-/Z-transform domain. The celebrated *Sampling Theorem* prescribes the sampling rate for reconstructing a (band limited) continuous-time signal $f(t)$ from its samples $f(t_k)$ (Oppenheim et al. 1997). A discrete-time signal becomes a *digital* one when stored on a digital device using finite word length. Digital signals are hard to study rigorously due to round-off errors; however, a longer word length has practically enabled the treatment of digital signals as discrete ones.

Classic modeling can be traced back to work by Newton and Gauss. As soon as a classic model is chosen the next task is to fit it to measured data in order to optimally estimate the model parameters in a LSE sense.

Classic modeling ultimately rests upon the conventional measurement process, which is carried out by comparing successively an unknown quan-

Vassilis G. Kaburlasos: *Relevant Literature Review*, Studies in Computational Intelligence (SCI) **27**, 5–18 (2006)
www.springerlink.com

tity to a known prototype. For instance, an unknown *length* is measured by comparing it successively to a known prototype (e.g. a 'meter') as well as to subdivisions of it. The quotient and remainder of a measurement jointly define a real number; that is how the set R of real numbers emerges.

The development of a classic model requires a deep understanding of what is being modeled; nevertheless classic modeling might be time-consuming. Moreover, classic modeling is restricted to similar problems. For instance a heat transfer model (based on Fourier transform analysis) may be suitable for modeling mechanical vibrations but neither for industrial assembly nor for human cognition. With the accelerating proliferation of ever more-sophisticated, computer-supported systems in manufacturing, services, etc. a need arises for comparable model development. A useful modeling alternative is an empirical one described next.

Based, first, on a set of input-output training data samples ($x_i \in R^N$, $y_i \in R^M$), i=1,...,n and, second, on a *learning* algorithm including a *cost* function, an *empirical model* induces a function f: $R^N \rightarrow R^M$ from the training data so as to minimize the corresponding *cost* function while maintaining a capacity for generalization. Actually, a function is induced by selecting it 'optimally' within a parametric family of functions. The difference with classic modeling is that no specific 'first principles (laws)' need to be adhered to. Instead, modeling is pursued by mathematical curve fitting – Note that the problem of fitting a mathematical model to measured data is ubiquitous in science and engineering (Rust 2001).

It turns out that empirical modeling is especially convenient when *little*, if anything at all, is known about the data generating law. An additional advantage is that useful knowledge may be induced regarding the data generating law. However, a demand still remains to accommodate non-numeric data including text, symbols, structured /linguistic data, etc.

A popular practice to meet, in part, the aforementioned demand is to consider multidimensional feature extraction for multivariate modeling in R^N (Mardia et al. 1979). For instance, in digital image processing, multi-dimensional models have been considered for capturing subjective impressions (Martens 2002). However, an increase of N in R^N increases complexity exponentially. Moreover, flat (real number) vectors cannot express (*human*) *intention*. An enhanced modeling paradigm is needed.

With the proliferation of digital computers new modeling paradigms emerged under *hybrid models*. *Hybrid system theory* is defined as the modeling, analysis, and control of systems that involve the interaction of both discrete state systems, represented by finite automata (Kohavi 1978), and continuous state dynamics, represented by differential equations. Hybrid models have been used for meeting high-performance specifications of sophisticated, human-made systems in manufacturing, communication networks, auto pilot design, chemical processes, etc. (Antsaklis 2000). Hybrid systems may also arise from the hierarchical organization of complex systems, where higher levels in the hierarchy require less detailed models than the lower levels (Saridis 1983).

Formal methods for hybrid systems have been proposed including mathematical logics based on general topology (Davoren and Nerode 2000). Stability analysis for hybrid systems may involve energy (Lyapunov) function techniques. Verification of hybrid systems has also been pursued by computing *reachable sets*, which satisfy safety specifications (Tomlin et al. 2003).

Hybrid system theories provide a foundation for *embedded* computing. The latter is computing for implementing and controlling complex interactions among physical system components (Sastry et al. 2003). Formal modeling methods for certain embedded systems may involve power sets of general input and output values in a function domain and range, respectively (Sifakis et al. 2003). Nevertheless, in the aforementioned context an employment of common-sense *semantics* is not straightforward. A comprehensive modeling paradigm is required.

2.2 Other Modeling Paradigms

This section presents additional modeling paradigms starting from simpler ones, which pertain to the physical world, then proceeding to more sophisticated ones towards human-oriented modeling.

Uncertainty and other non-classic models

A popular model to accommodate uncertainty is the Kalman filter regarding sequential state estimation of dynamical systems (Kalman 1960). The derivation of classic Kalman filter is based on the following simplifying assumptions: (1) linearity of the model, and (2) Gaussianity of both the dynamic noise in the process equation and the measurement noise in the measurement equation. Under the aforementioned assumptions derivation of the Kalman filter leads to an elegant algorithm that propagates the mean vector and covariance matrix of the state estimation error in an iterative manner; moreover, the Kalman filter is optimal in the Bayesian setting.

The *extended* Kalman filter has been introduced for state estimation in non-linear dynamical systems (Haykin and de Freitas 2004). However, many of the state estimation problems encountered in practice, in addition

to non-linear, are also non-Gaussian thereby limiting the practical usefulness of Kalman filtering in R^N.

Empirical models of uncertainty have been proposed. For instance, multivariate statistical models have demonstrated a capacity for industrial
process monitoring, abnormal situation detection, and fault diagnosis
(Kourti 2002). Moreover, statistical regression models have been popular
in economic forecasting (Pindyck and Rubinfeld 1991). Furthermore, probabilistic models of uncertainty are popular in various contexts. For instance, in robotics a probabilistic model may assign a distribution over all
possible outcomes for improved decision-making under uncertainty (Thrun
2002). Probabilistic techniques have also enabled switching between different predictive (neural) modules in time series classification and segmentation problems (Petridis and Kehagias 1998).

A popular non-classic modeling paradigm regards graphical models
(Edwards 2000; Jordan and Sejnowski 2001) as well as Bayesian networks. Advantages of the latter include representation of more than pairwise relationships between variables, resistance to overfitting, robustness
in the face of noisy data. Furthermore, Bayesian networks are interpretable
and can incorporate a priori knowledge as a prior distribution. The capacity of Bayesian networks was demonstrated in biology (Hartemink et al.
2002), information retrieval in the Web (Rowe 2002), etc. However, the
usefulness of Bayesian approaches to practical problem solving has been
questioned (Hearst 1997).

Bayesian networks are sometimes combined with alternative technologies. For instance, Bayesian information fusion from different sensors has
been presented towards industrial quality prediction in the context of neural network model combination (Edwards et al. 2002). Moreover, Bayesian
network models have been proposed for supervised learning based on multimodal sensor fusion (Garg et al. 2003). Furthermore, Bayesian and
Dempster-Shafer decision-making has been employed comparatively
(Buede and Girardi 1997).

Additional paradigms for coping with various types of uncertainty include interval computation (Moore 1979; Tanaka and Lee 1998), qualitative information (El-Alfy et al. 2003), fuzzy set theory (Klir and Folger
1988; Zimmermann 1991), etc.

Knowledge discovery

Apart from inducing a 'data generating law' there might also be an interest
in inducing knowledge from the data. *Knowledge discovery* is the automated induction of fundamental, simple, yet useful properties and principles from vast pools of data using computer techniques (Quinlan 1986,

1990; Motoda et al. 1991; Fujihara et al. 1997). Data clustering algorithms (Jain et al. 1999) are popular for knowledge discovery. Moreover, knowledge-modeling procedures (Chandrasekaran et al. 1992) as well as knowledge management models have been proposed (Hou et al. 2005).

Knowledge discovery from databases, also known as *data mining* (Cios et al. 1998), bridges several technical areas including databases, human-computer interaction, statistical analysis, and machine learning (ML). The latter is central to the data mining process. Nevertheless, first-generation machine learning algorithms, including decision trees, rules, neural networks, Bayesian networks, and logistic regression, have significant limitations; for instance, they cannot cope with disparate types of data, they fail to allow (interactive) user guidance, etc. (Mitchell 1999).

Of particular interest is 'computational learning theory' (Kearns and Vazirani 1994). The latter is not human motivated. However, it can deal with uncertainty by *probably approximately correct* (PAC) learning a set of functions (Valiant 1984; Blumer et al. 1989; Haussler 1989; Angluin et al. 1992; Kearns et al. 1994).

Statistical learning and classification were considered by a number of authors (McLachlan 1992; Hummel and Manevitz 1996; Vidyasagar 1997; Vapnik 1998, 1999; Jain et al. 2000; Herbrich 2002). Cluster analysis, also known as 'unsupervised learning' in the literature of pattern recognition (Duda et al. 2001), pursues natural groupings in the data, e.g. in a database. Using a well-defined distance, or a similarity measure, between the data an unsupervised scheme may organize objects in a population as clusters of subpopulations (Jain et al. 1999). This organization is unbiased in the sense that no a priori assumptions, common to most statistical methods, are required. However, clustering methods neither find meaningful sub-groupings within a class nor explain their answers the way an expert can.

Knowledge acquisition has been demonstrated beyond space R^N in structured data domains for instance in biology (Conklin et al. 1993), etc. Lately, data mining in the World Wide Web (WWW) poses additional challenges since the WWW is a complex and highly dynamic information source involving, as well, (semi-) structured data. Hence, a fundamentally different knowledge discovery paradigm is needed.

Towards human-oriented modeling

Most of the aforementioned models regard the physical world. However, human-oriented models increasingly appear including knowledge discovery models above. Additional human-oriented models are described next.

Neural computing models have been proposed originally in cognition and psychology (Rumelhart et al. 1986). Additional models were considered lately regarding consciousness, emotion, etc. (Sun 1997; Taylor and Fragopanagos 2005). Symbolic behavior was also considered (Omori et al. 1999). Computational models have also been presented inspired from psychology (Chaib-draa 2002) regarding *cognitive maps* based on the *personal construct theory* (Kelly 1992).

Human-computer interaction (HCI) is a domain where non-classic models appear. For instance, HCI user-models have been proposed involving multimodal human cognitive, perceptual, motor, and affective factors (Duric et al. 2002; Pantic and Rothkrantz 2003). Furthermore, human-computer communication was modeled via speech using both speech-waveforms and a hidden Markov model (HMM) (O'Shaughnessy 2003). A modeling framework has been proposed in software for multimodal HCI based on 'abstract modalities' (Obrenovic and Starcevic 2004). Additional models have been proposed for user interaction with a software platform (Liu et al. 2003). Note that a 'long expected' HCI theory is currently missing. However, the expected benefits of the aforementioned theory can be substantial (Long 1991; Fukuda et al. 2001).

Modeling is typically carried out in the Euclidean space R^N involving N-dimensional vectors of numbers. An extension of conventional modeling in R^N considers multiple sensor modalities (Nakamura 2002). Multiple modalities have also been proposed for a single sense, i.e. vision (Hiruma 2005). A more sophisticated extension towards human-oriented modeling involves non-numeric data, for instance in database applications (Palopoli et al. 2003), in face recognition (Brunelli and Poggio 1993), etc. Furthermore, with the ever-increasing scale and complexity of human-made systems non-numeric data emerge (Cassandras et al. 2001; Giles et al. 1999). Various system models have been proposed for processing real numbers at a lower level and symbolic data at a higher level (Song et al. 2000). Likewise, a practical vision system was proposed partitioned into two computing stages, first, for 'front-end' processing using sampled image data and, second, for 'high-level' interpretation using symbolic data (Burt 2002). Moreover, a translation of (time) signals into symbols, the latter are spatial patterns, has been studied (Principe et al. 2001). A need for modeling based as well on non-numeric data also appears in other areas as described in the following.

Classification as well as clustering models have been proposed involving non-numeric data including symbols and structures (Chandrasekaran and Goel 1988). Furthermore, multimedia modeling is a fairly new interdisciplinary area whose fast growth is not currently matched by comparable data processing technologies. In the aforementioned context semantic multimedia retrieval has been described as a promising (multimodal) pattern recognition problem (Naphade and Huang 2002). There is also a need for modeling in hypermedia applications, the latter are collections of hypermedia documents organized into a hypertext net. However, a difficulty arises from the need to model effectively general navigational structures, interactive behaviors, and security policies (Díaz et al. 2001).

An ontology has been defined as a (non-numeric) formal specification of a shared conceptualization (Chandrasekaran et al. 1999; Swartout and Tate 1999). Ontologies have been employed for conceptual modeling (Richards and Simoff 2001) and they are expected to be a critical part of the Semantic Web for enabling computers to online process information (Berners-Lee et al. 2001; Berners-Lee 2003). Note also that structural data (graphs) have been used for representing text as well as semantics and logic (Hsinchun Chen et al. 1996; Wolff et al. 2000; Lloyd 2003).

A number of specialized tools have been developed for processing non-numeric data. For instance, regarding ontologies, computational tools have been proposed for matching concepts for similarity (Rodríguez and Egenhofer 2003). In another context, an effective distance measure has been introduced to express similarity in heterogeneous databases (Dey et al. 2002). Graph-based metrics were used for measuring the semantic similarity between pairs of objects in heterogeneous databases (Palopoli et al. 2003). Non-numeric data tools have also been employed for both calculating similarity (Berretti et al. 2000) and for semantic retrieval in multimedia databases (Yoshitaka and Ichikawa 1999). They can also be useful for classification and various information retrieval tasks (Berztiss 1973; Salton 1988; Chakrabarti et al. 1999; Green 1999; Cook and Holder 2000; Berretti et al. 2001; Cook et al. 2001).

In various information processing applications a need emerges for integrating information (Hearst et al. 1998) especially in multimedia applications (Yoshitaka and Ichikawa 1999; Chang and Znati 2001; Meghini et al. 2001). In the aforementioned applications, useful models should have a capacity to jointly process disparate data types.

Non-numeric data are typically dealt with in applications by transforming them to numeric data; see, for instance, both the 'animal identification' data and the 'DNA promoter' in Tan (1997). Sometimes numeric data are converted to nominal data; see, for instance, the 'cylinder bands' data example in Evans and Fisher (1994). A general methodology for dealing, in principle, with disparate types of data would be valuable.

The general practice for non-numeric data is to 'expert' transform them to R^N using, for example, kernel functions in support vector machines (SVM) or in radial basis function (RBF) networks, transition probabilities in (hidden) Markov models, etc. However, a data transformation to R^N typically abandons original data semantics. Then, a solution is pursued based on 'algebraic semantics'; the latter quoted term means sound mathematical techniques (in R^N). Finally, a solution computed in R^N is transformed back to the original data domain where it is interpreted. A capacity to carry out computations beyond R^N could be valuable for developing useful models while retaining original problem semantics all along. A number of authors have pursued computation beyond R^N as outlined in the following.

Various algebraic approaches have been proposed involving non-numeric data in various information processing contexts (Güting et al. 1989; Fuhr and Rölleke 1997; Goshko 1997; Rus and Subramanian 1997; Halmos and Givant 1998; Marcus 2001; Kobayashi et al. 2002). Moreover, a novel notion of computationalism was proposed, which embodies interaction, semantics and other real-world constraints (Scheutz 2002).

Many non-numeric data processing techniques have lately been subsumed under *computational intelligence* discussed next.

2.3 Computational Intelligence - Soft Computing

There is a tendency in science and technology to commit to a specific methodology. However, in real-world problems, a pluralistic coalition of methodologies is often required (Zadeh 2001). *Soft computing* (SC) is an evolving collection of methodologies with tolerance for imprecision, uncertainty, and partial truth in order to achieve robustness and low cost in practice (Bonissone et al. 1999). Note that the term *hard computing* (HC) is used to denote (traditional) analytical methods (Dote and Ovaska 2001). Here, SC is used as a synonym with *computational intelligence* (CI). The potential and broad scope of CI was shown early (Zurada et al. 1994); furthermore, various CI-SC technologies have been published extensively (Pedrycz 1998; Fogel et al. 1999; Dote and Ovaska 2001; Kecman 2001).

Using the computer as 'enabling technology', CI pursues technological improvements inspired from nature. The core methodologies of CI include *neural networks* inspired from brain, *fuzzy (inference) systems* (FIS) inspired from natural language, and *evolutionary computing* inspired from Darwinian evolution theory. Additional methodologies include probabilistic computing, machine learning, cognitive artificial intelligence, etc. SC methodologies are frequently used for designing devices with high machine intelligence quotient (MHIQ). An interesting connection with conventional artificial intelligence (AI) is shown next.

Conventional AI has flourished as a result of the proliferation of digital computers. The history of AI is characterized by towering expectations and stagnant progress. It is interesting that if John McCarthy, the father of AI, were to coin a new phrase for Artificial Intelligence today he would probably use *Computational Intelligence* (Andresen 2002). Therefore, it appears that modern CI may subsume conventional AI.

Among core CI methodologies the ability to learn was, originally, a unique advantage of neural networks. The latter are also known for their capacity for fault tolerant, massively parallel, distributed data processing and generalization. The resurgence of the field of neural networks during the 1980s was marked by the promotion of both binary Hopfield- (Hopfield 1982) and backpropagation networks (Rumelhart et al. 1986). An encyclopaedia of neural networks was published lately (Arbib 2003).

Neural networks are typically employed for learning various non-linear input-output maps in clustering, classification and regression applications (Fritzke 1994; Jianchang Mao and Jain 1996; Bishop 1997; Looney 1997; Card et al. 1998; Setiono et al. 2002). Specialized neural clustering techniques have been used for function approximation (González et al. 2002). However, conventional neural networks can neither justify their answers nor deal with uncertainty (Browne and Sun 2001).

Another core CI methodology is fuzzy systems, which can deal with ambiguity. Moreover, fuzzy systems can explain their answers using rules, and they have demonstrated practical advantages in knowledge representation (Pedrycz and Reformat 1997). It was shown that a non-linear map implemented by a neural network can be approximated to any degree of accuracy by a fuzzy system (Wang and Mendel 1992a). However, original fuzzy systems were short of a capability for learning (Werbos 1993).

Fuzzy neural networks (FNN) were proposed for combining the aforementioned advantages of neural networks and fuzzy systems (Lin and Lee 1996; Fuessel and Isermann 2000; Jia-Lin Chen and Jyh-Yeong Chang 2000; Zhang et al. 2000; Castro et al. 2002; Rutkowski and Cpalka 2003). Further improvements were pursued including the following ones.

Recurrent fuzzy neural networks were proposed for reducing the size of the input space and for cognitive (fuzzy decision tree) decision-making (Lee and Teng 2000). Furthermore, multimodal fuzzy data fusion algorithms were proposed (Tahani and Keller 1990; Dayou Wang et al. 1998) as well as various distance- and similarity- functions between fuzzy sets (Setnes et al 1998b; Chatzis and Pitas 1999, 2000). The effectiveness of alternative distances than the Euclidean one in soft computing was also demonstrated (Karayiannis and Randolph-Gips 2003).

A number of improved fuzzy systems were proposed for non-linear system modeling and rule induction by clustering towards classification and regression (Klir and Yuan 1995; Pedrycz 1996; Chen and Xi 1998; Setnes 2000; Papadakis et al. 2005). The inputs to a fuzzy system may be vectors as well as fuzzy numbers (Chen 2002; Boukezzoula et al. 2003).

SC has been especially successful in industrial modeling applications, in data /Web mining, etc. (Lu et al. 1997; Mitra et al. 2002; Pal et al. 2002). Evolutionary algorithms are often used for tuning performance. For instance, evolutionary algorithms can significantly improve the performance of artificial neural network by optimizing connection weights, architectures, learning rules, and input features (Yao 1999).

Machine learning (ML) is defined as the computation of a mathematical rule that generalizes a relation from a finite sample of data (Mitchell 1997). ML techniques are often employed for classification. Note that neural networks have been used as ML techniques, and interesting studies were reported regarding the size of either a network or a training data set that give a valid generalization (Baum and Haussler 1989; Ehrenfeucht et al. 1989). The Vapnik-Chervonenkis (VC) dimension of various learning machines including neural networks was also studied (Koiran and Sontag 1998). Note that the VC-dimension is defined as the size of the largest data set which can be *shattered* by a hypothesis space of a family of models (Mitchell 1997), where 'shattered' means that all the corresponding subsets, or equivalently *dichotomies* (Cover 1965), can be produced.

Another ML algorithm is linear regression. However, linear regression is limited because of its inherent linearities. One approach in ML towards non-linear regression is support vector machines (SVM) based on *kernel functions* (Cristianini and Taylor 2000). The latter functions are used to map, non-linearly, an input space to a high-dimensional feature space. Hence, non-linear relations in the input space may become linear ones in the feature space where powerful (linear) algorithms are employed. The difficulty in the SVM approach is to find correct kernel functions that transform a non-linear problem into an easily solvable linear problem in more dimensions. Based on Mercer kernels (from support vector machines) the classic perceptron has been fuzzified (Chen and Chen 2002).

Non-linear regression in ML can be pursued using non-linear regressors, e.g. neural networks (Fine 1999). The difficulty of this method is that, usually, many computer cycles are required for computing locally optimal solutions. The Vapnik-Chervonenkis theory relates the complexity of the class of functions used for approximation, in either linear- or nonlinear-regression, with the expected quality of generalization.

Machine learning as well as other CI methodologies are often used for black-box prediction. However, in many applications, mere black-box prediction is not satisfactory, and understanding the data is important. Major approaches to an optimal extraction of logical rules based on neural networks, decision trees, machine learning, and statistical methods have been reported (Duch et al. 2004). Note also that, for a number of authors, the main criterion for selecting a CI methodology is practical effectiveness (Joshi et al. 1997; Duch et al. 2004).

It has been argued that despite their brilliant successes modern science and technology cannot perform simple human tasks such as driving in city traffic, etc. because humans employ perceptions rather than measurements (Zadeh 2001). It is this human capacity that existing scientific theories do not possess. Therefore, computing with words (CW) was proposed as a fuzzy-logic-based methodology that provides a basis for a Computational Theory of Perceptions (CTP) towards introducing human capacities to measurement-based theories (Zadeh 1996, 1997, 1999).

Disparate types of data including non-numeric/structured data representations, symbols, etc. naturally emerge in applications (Dawant and Jansen 1991; Holder and Cook 1993; Sugeno and Yasukawa 1993; Yao et al. 1995; Wiesman and Hasman 1997; Corridoni and Del Bimbo 1998; Boinski and Cropp 1999; Li and Biswas 2002; Huang et al. 2000). Moreover, in an artificial intelligence (AI) context a symbolic measurement process has been proposed (Goldfarb and Deshpande 1997). Until lately the capacity of AI to deal with symbols was unmatched by conventional CI methodologies. For instance, even in knowledge discovery applications (Bengio et al. 2000), conventional neural-fuzzy systems typically deal only with numeric and/or linguistic (fuzzy) data, and they cannot easily accommodate additional types of data such as structured data (graphs), etc. Nevertheless, other data representations than 'flat vectors' may arise in practice. Hence, lately, a number of neural networks have been presented for dealing with structured data (Sperduti and Starita 1997; Frasconi et al. 1998; Weber and Obermayer 2000; Carrasco and Forcada 2001; Petridis and Kaburlasos 2001; Hagenbuchner et al. 2003; Hammer et al. 2004).

Other non-numeric data in applications may include linguistic (fuzzy) data as was explained above, matrices (Seo and Obermayer 2004), complex numbers (Hirose 2003), etc. Non-numeric data processing in CI is often carried out in space R^N after a suitable data transformation as explained in section 2.2. Moreover, (fuzzy) logic or probabilistic techniques may be used for dealing with non-numeric data (Chen and Zhang 2003).

It turns out that most CI methodologies are applicable to one of (1) Boolean lattices of zeros and ones, (2) the N-dimensional Euclidean space R^N, where one dimension corresponds to the totally-ordered lattice of real numbers, and (3) lattices of fuzzy numbers. A unifying modeling framework is expected to be valuable in CI.

This work proposes mathematical *lattice theory* as a sound, unifying foundation for analysis and design of useful systems in computational intelligence (CI). Practical applications of lattice theory in various domains by a number of authors are shown next.

2.4 Lattice Theory in Practice

The word 'lattice' is used occasionally to denote a two-dimensional grid of objects; for instance, lattices of filters are popular in signal processing applications. In the context of this work by lattice we mean a *mathematical lattice* (Rutherford 1965; Birkhoff 1967; Grätzer 1971; Gierz et al. 1980; Davey and Priestley 1990). Currently, lattice theory is not in the mainstream in computational intelligence (CI). Nevertheless several research groups around the world, working separately, employ lattice theory in specific application domains as explained next.

Originally, lattice theory has been popular in applied mathematics, where it was used to unify the study of different algebraic systems including groups, rings, linear (sub)spaces, etc. (Birkhoff 1967). It was pointed out that lattice theory fits partitions (of a set) like a shoe (Rota 1997). Hence, lattice theory may be employed where 'set partitions' are employed as shown below. Furthermore, a number of authors have employed explicitly (1) the partial ordering, (2) the algebra, and/or (3) the unifying capacity, of (mathematical) lattices as explained in the following.

Moray (1987) has employed lattices for describing acquisition of simple mental models. Various conceptual models have been proposed based on lattices (Kent 2000; Fang and Rousseau 2001). Sahami (1995) employed lattices for rule extraction. The significance of partially (lattice) ordered data in the context of *Behavioral Finance* was acknowledged (Simon 1955). Moreover, lattice theory was employed by pioneers of quantum mechanics (Birkhoff and von Neumann 1936) and electrical communication theory (Shannon 1953).

Lattice theory has been a tool for studying logic (Birkhoff 1967; Margaris 1990; Xu et al. 2003). In the introduction of fuzzy set theory, Zadeh (1965) pointed out that 'fuzzy sets (over a universe of discourse) constitute a distributive lattice with 0 and 1'. Moreover, Goguen (1967) explained how a L(lattice)-fuzzy set generalizes the notion of a fuzzy set. Gaines (1978) and Edmonds (1980) employed lattice theory for 'decision-making under ambiguity' in the context of fuzzy logic. Furthermore, lattice theory based quantum logics are of interest in theoretical physics (Riecanova 1998; Pykacz 2000).

In computer science, finite state machines were studied using lattices (Liu 1969). Moreover, various types of data structures including lists, trees, frames can be represented as lattices (Scott 1976). Ontology metadata, the latter are data about data, can be modeled in a lattice (Sowa 2000) also for Semantic Web applications. Elementary lattice operations were used for studying multisource data flow problems (Masticola et al. 1995).

Moreover, lattices can be used for studying properties of algorithms (Kohavi 1978; Ait-Kaci et al. 1989). For instance, if a partial order relation is assumed for an implication relation then the meaning of a declarative program can be computed more efficiently (Wuwongse and Nantajeewarawat 2002). Lattices have also been used for measuring rigorously functional (semantic) distances between software specifications (Jilani et al. 2001).

Databases are a favorite application domain for lattice theory. For instance, a lattice of rules was employed for the analysis of techniques regarding rule execution termination in a knowledge base (Baralis et al. 1998). An authorization model for digital library protection has employed a (lattice) order relation for representing a credential-type hierarchy (Adam et al. 2002). Complete lattices, in particular, were employed in deductive databases (Lu et al. 1996).

Formal concept analysis (FCA), that is a lattice theory-based field of applied mathematics (Ganter and Wille 1999), is based on complete lattice analysis. In this context several techniques have been proposed for knowledge acquisition, classification, and information retrieval in databases (Carpineto and Romano 1996; Priss 2000; Bloehdorn et al. 2005).

Lattices are popular in mathematical morphology including image processing applications (Bloch and Maitre 1995; Dougherty and Sinha 1995; Sussner and Ritter 1997; Goutsias and Heijmans 2000; Heijmans and Goutsias 2000; Braga-Neto 2001; Nachtegael and Kerre 2001; Sussner and Graña 2003). Also, lattice theory has been employed for unifying morphological and fuzzy algebraic systems in image processing (Maragos 2005).

Neural computing is typically carried out in space R^N. However, there is no evidence that biological neurons operate in R^N. Rather, there is evidence that biological neurons carry out lattice- *meet* (min) and *join* (max) operations. Hence, lattice algebra was employed for modeling both biological neurons (Ritter and Urcid 2003) and associative memories (Ritter et al. 1998, 1999).

In machine learning, lattice theory was originally used in *version spaces* (Mitchell 1997). Moreover, a unification was shown including a Boolean lattice of logical assertions and a lattice of questions (Knuth 2005). A lattice representation of marginals/conditionals in probabilistic decision-making was also considered (Lin 2005). Furthermore, pattern recognition and classification have been pursued using Steinbuch's Lernmatrix based on the lattice operations of *join* and *meet* (Sánchez-Garfias et al. 2005).

In system modeling, mathematical lattices were used to study hybrid (control) systems; the latter are systems in which the state set possesses both continuous and discrete components. In particular, the lattice of hybrid in-block controllable partition machines was both defined and investigated (Caines and Wei 1998). Moreover, mathematical lattices were employed for control in discrete event systems (Cofer and Garg 1996).

Lattices appear implicitly in a number of neural network models including fuzzy-ART (Carpenter et al. 1991b), the min-max neural networks (Simpson 1992, 1993) as well as in the extension neural networks (Wang 2005). All aforementioned neural networks operate in the lattice of hyperboxes in R^N. Note that lattices of hyperboxes in R^N were explicitly employed in neural computing first in Kaburlasos (1992); nevertheless the latter work was primarily oriented towards medical diagnosis rather than towards a theoretical substantiation.

The rationale for an employment of lattices in the context of this work is straightforward: With the advent of computers, lattice-ordered data including N-dimensional vectors, hyperboxes, hyperspheres, (fuzzy) sets, events in a probability space, propositional statements, graphs, etc. have proliferated. A popular practice is to transform disparate types of data to a single data domain; the latter is usually R^N. Our proposal is to develop algorithms applicable in a general lattice data domain. Hence, disparate types of data can be treated, in principle, by the same algorithm. Moreover, since the Cartesian product of lattices is a lattice it follows that disparate types of data can be treated jointly in a product lattice. Note also that a single lattice element may be a whole procedure (of a kind).

Previous work has pursued neurocomputing in the framework of *fuzzy lattices* (Kaburlasos and Petridis 1997, 2000, 2002; Petridis and Kaburlasos 1998, 1999, 2001; Kaburlasos et al. 1999), where a fuzzy lattice stems from a conventional one by fuzzifying the crisp partial order relation. This work presents, more generally, a rigorous unification in modeling and knowledge-representation based on novel lattice theoretic tools.

A single most important novelty between the employment of lattice theory here and elsewhere is that positive valuation functions are used here. Moreover, lattice theory is typically employed elsewhere in data domains of finite cardinality. This book shows lattices of cardinality $\aleph_1$, where $\aleph_1$ is the cardinality of the totally ordered lattice R of real numbers.

Finally, lattice theory is proposed here for the design of improved intelligent systems (Turing 1950; Searle 1980; Bains 2003). The prototype for intelligence in our study is the human mind. Even though differential equations may model brain activity at a physiological (neuronal) level in R^N, a different approach is required for modeling higher-level brain faculties. In this context lattice theory is proposed with the far-reaching objective to develop models of the mind (Grossberg 1976b; Penrose 1991; von Neumann 2000).

PART II: THEORY AND ALGORITHMS

Part II includes novel mathematical theory and tools.
It also presents algorithms for clustering, classification, and regression.

3 Novel Mathematical Background

This chapter presents mathematical tools

The objects of mathematical interest here are *(partially) ordered sets*. Some well-known facts are summarized next.

3.1 Partially Ordered Sets

A *binary relation* R between two sets P and Q is defined as a subset of the Cartesian product $P \times Q$. Instead of $(p,q) \in R$ we often write pRq. If $P=Q$, we speak of a *binary relation on the set* P. The symbol R^{-1} denotes the *inverse relation* to R, i.e. $qR^{-1}p :\Leftrightarrow pRq$. A specific binary relation is defined next.

Definition 3-1 A binary relation R on a set P is called *(partial) order*, if it satisfies the following conditions, for $x,y,z \in P$

B1. xRx	*(Reflexive)*
B2. xRy and $yRx \Rightarrow x = y$	*(Antisymmetry)*
B3. xRy and $yRz \Rightarrow xRz$	*(Transitivity)*

Condition B2 can be replaced by the following equivalent condition
B2′. xRy and $x \neq y \Rightarrow$ not yRx *(Antisymmetry)*

Instead of xRy we can write $x \leq y$ and say x 'is contained in' y, or x 'is a part of' y, or x 'is less than or equal to' y. If $x \leq y$ and $x \neq y$, one writes $x < y$, and says that x 'is less than' or 'properly/strictly contained in' y. Likewise, for R^{-1} we can use the symbols $\geq$ and $>$.

A *partially ordered set* (or, *poset*, for short) is a pair $(P, \leq)$, where P is a set and $\leq$ is an order relation on P. More than one orders, e.g. $\leq_1$ and $\leq_2$, may be defined on a set P. Examples of *posets* include the following.

Example 3-2 Let $Pow(S)$ be the power set, i.e. the set of subsets, of some set S; in this case, let $x \leq y$ mean that x is a subset of y.

Vassilis G. Kaburlasos: *Novel Mathematical Background*, Studies in Computational Intelligence (SCI) **27**, 21–34 (2006)
www.springerlink.com

Example 3-3   Consider the set of all real functions defined on the set R of real numbers; in this case, let $f \leq g$ mean that $f(x) \leq g(x)$ for every x in R.

Example 3-4   The set R of real numbers with the usual order relation $\leq$ is a poset. In particular, the poset $(R, \leq)$ is called a *chain* or, equivalently, *totally ordered* set (Halmos 1960; Birkhoff 1967).

A chain is formally defined in the following.

Definition 3-5   A *chain* or, equivalently, *totally ordered* set, is a poset $(P, \leq)$ such that given $x, y \in P$ it is either $x \leq y$ or $y \leq x$.

Two different elements in a chain are *comparable* in the sense that it is either $x \leq y$ or $y \leq x$. The posets in Examples 3-2 and 3-3 above are not chains: They contain pairs of elements x, y which are *incomparable*, i.e. neither $x \leq y$ nor $y \leq x$, symbolically $x \| y$.

The notion of 'immediate superior' in a poset is defined next.

Definition 3-6   By 'a covers b' in a poset $(P, \leq)$, it is meant that $a > b$, but $a > x > b$ for no $x \in P$.

A finite poset $(P, \leq)$ can be represented by a *line diagram*, also called *Hasse diagram*, where the elements of P are depicted by small circles on the plane such that two elements $a \in P$ and $b \in P$ (respectively, 'above' and 'below' on the plane) are connected by a line segment if and only if a covers b. Fig.3-1 shows a Hasse diagram.

Definition 3-7   The *least* element of a subset X of P in a poset $(P, \leq)$ is an element $a \in X$ such that $a \leq x$ for all $x \in X$. Likewise, the *greatest* element of a subset X of P is an element $b \in X$ such that $b \geq x$ for all $x \in X$.

The least element, if it exists, in a poset will be denoted by O; likewise, the greatest element will be denoted by I.

Let $(P, \leq)$ be a poset with least element O. Every $x \in P$ which covers O, if such x exists, is called *atom*; for instance, the sets $\{a\}$, $\{b\}$, $\{c\}$ in Fig.3-1 are atoms. Given a poset, various sets can be defined as follows.

Definition 3-8   If $(P, \leq)$ is a poset and $a, b \in P$ with $a \leq b$, we define interval $[a,b]$ as the set $[a,b] := \{x \in P: \ a \leq x \leq b\}$.

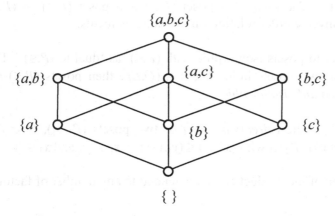

Fig.3-1   Hasse diagram for the power set of a set $S= \{a,b,c\}$.

Definition 3-9   If $(P,\le)$ is a poset and $a,b \in P$, the set $(a] := \{x \in P: x \le a \}$ is called *principal ideal* (generated by a); moreover, the set $[b):= \{x \in P: x \ge b \}$ is called *principal filter* (generated by b).

Functions between posets are presented next.

Definition 3-10   A function $\varphi: P \to Q$ between posets $(P,\le)$ and $(Q,\le)$ is *order-preserving*, if $x \le y \Rightarrow \varphi(x) \le \varphi(y)$ for $x,y \in P$. If, in addition, φ fulfills the converse implication $x \le y \Leftarrow \varphi(x) \le \varphi(y)$, then φ is *order-embedding*. A bijective (one-to-one) order-embedding is called *(order-) isomorphism*.

When there is an (order-) isomorphism between two posets $(P,\le)$ and $(Q,\le)$, then $(P,\le)$ and $(Q,\le)$ are called *(order-)isomorphic*, symbolically $(P,\le) \cong (Q,\le)$. Extensions of an order relation are shown next.

The Duality Principle for posets: The inverse $\ge$ of any order relation $\le$ is itself an order relation. The order $\ge$ is called the *dual* order of $\le$, symbolically $\le^{\partial}$, or $\le^{-1}$, or $\ge$.

We obtain the dual statement S^{∂} of an order-theoretic statement S, if we replace the symbol $\le$ in S by the symbol $\ge$. It follows that S holds for an ordered set if and only if S^{∂} holds for the dual ordered set. The Duality Principle above is used to simplify definitions and proofs.

Definition 3-11   The *dual* of a poset $(P,\leq)$ is a poset $(P,\leq)^{\partial} = (P,\leq^{\partial})$ defined by the inverse order relation on the same elements.

We shall refer to posets isomorphic with $(P,\leq)^{\partial}$ as 'dual to $(P,\leq)$ '. Furthermore, if for two posets holds $(P,\leq) \cong (Q,\leq)^{\partial}$, then posets $(P,\leq)$ and $(Q,\leq)$ are called *dually isomorphic*.

Definition 3-12   The (*direct*) *product* of two posets $(P_1,\leq_1)$, $(P_2,\leq_2)$ is defined as poset $(P_1 \times P_2,\leq)$ with $(x_1,x_2) \leq (y_1,y_2) :\Leftrightarrow x_1 \leq_1 y_1$ and $x_2 \leq_2 y_2$.

The definition of the product can be extended to any number of factors.

3.2 Elements from Lattice Theory

This section summarizes relevant elements from standard lattice theory (Rutherford 1965; Birkhoff 1967; Davey and Priestley 1990).

Let $(P,\leq)$ be a poset and $X \subseteq P$. An *upper bound* of X is an element $a \in P$ with $x \leq a$ for every $x \in X$. The *least upper bound* (l.u.b.), if it exists, is the unique upper bound contained in every other upper bound; the l.u.b. is also called *supremum* or *lattice join* of X and denoted by $\sup X$ or $\vee X$. The notions *lower bound* of X and *greatest lower bound* (g.l.b.) of X are defined dually; the g.l.b. is also called *infimum* or *lattice meet* of X and denoted by $\inf X$ or $\wedge X$. If $X=\{x,y\}$, we will write $x \wedge y$ for $\inf X$ and $x \vee y$ for $\sup X$.

Definition 3-13   A *lattice* is a poset $(L,\leq)$ any two of whose elements have both a greatest lower bound (g.l.b.), denoted by $x \wedge y$, and a least upper bound (l.u.b.), denoted by $x \vee y$. A lattice $(L,\leq)$, or equivalently *crisp lattice*, is called *complete* when each of its subsets X has a l.u.b. and a g.l.b. in L.

Setting $X=L$ in definition 3-13 it follows that a nonvoid complete lattice contains both a *least element* and a *greatest element* denoted, respectively, by O and I. By definition, an *atomic* lattice $(L,\leq)$ is a complete lattice in which every element is a join of atoms.

We now define a *hidden lattice*. Let, $(P,\leq)$ be a partially ordered set, $(L,\leq)$ be a lattice, and $\psi: P \rightarrow L$ be an order-preserving map. For $x,y \in P$ consider an upper bound, say z, of x and y. If there exists a unique upper bound z_0 of x and y in P such that $\psi(z_0) \leq \psi(z)$ for all upper bounds $z \in P$ then

z_0 is called the *h-join* ($\underline{\vee}$) of x and y in P. The notion *h-meet* ($\overline{\wedge}$) of $x,y \in P$ is defined dually. The definition of the notion *hidden lattice* follows.

Definition 3-14   A partially ordered set $(P,\leq)$ is called *hidden lattice* if there exists a lattice $(L,\leq)$, namely *underlying lattice*, and an order-preserving map $\psi: P \rightarrow L$, namely *corresponding map*, such that any two elements $x,y \in P$ have both a *h-join* $x\underline{\vee}y$ and a *h-meet* $x\overline{\wedge}y$.

The above definition for a lattice is called here *set theoretic*. There is an equivalent, *algebraic* definition for a lattice based on the binary operations $\wedge$ and $\vee$ as shown in the following (Birkhoff 1967) – For definitions of *algebra* and *operation* see in Appendix A.

Theorem 3-15   Any system with two binary operations which satisfy L1-L4 below is a lattice, and conversely.

L1. $x \wedge x = x$, $x \vee x = x$	(*Idempotent*)
L2. $x \wedge y = y \wedge x$, $x \vee y = y \vee x$	(*Commutative*)
L3. $x \wedge (y \wedge z) = (x \wedge y) \wedge z$, $x \vee (y \vee z) = (x \vee y) \vee z$	(*Associative*)
L4. $x \wedge (x \vee y) = x \vee (x \wedge y) = x$	(*Absorption*)

Moreover, $x \leq y$ is equivalent to each of the conditions
$$x \wedge y = x \text{ and } x \vee y = y \qquad (Consistency)$$

The binary operations of meet ($\wedge$) and join ($\vee$) in lattices have various algebraic properties (Birkhoff 1967); for instance, both operations $\wedge$ and $\vee$ are *monotone*: $y \leq z \Rightarrow x \wedge y \leq x \wedge z$ and $x \vee y \leq x \vee z$, for all x. Moreover, in a complete lattice, it holds: $O \wedge x = O$, $O \vee x = x$, $x \wedge I = x$, and $x \vee I = I$, for all x.

The dual of a lattice is a lattice. Furthermore, the dual of a complete lattice is a complete lattice, with meets and joins interchanged. The *direct product* of two lattices is a lattice, more specifically $(L_1,\leq_1) \times (L_2,\leq_2) = (L_1 \times L_2, \leq)$, where $(a_1,a_2) \leq (b_1,b_2) \Leftrightarrow a_1 \leq_1 b_1$ and $a_2 \leq_2 b_2$.

Any chain is a lattice, in which $x \wedge y$ is simply the smaller and $x \vee y$ is the larger of x and y. Not every lattice is complete: For instance, the chain R of real numbers (in their natural order) is not a complete lattice. However, the power set $Pow(S)$ of a set S is a complete lattice, where the empty set { } is the least element O, moreover S itself is the greatest element I (Fig.3-1).

Definition 3-16   A *sublattice* $(S,\leq)$ of a lattice $(L,\leq)$ is another lattice such that both $S \subseteq L$ and $a,b \in S \Rightarrow a \wedge b \in S$ and $a \vee b \in S$.

Given $a \leq b$ in a lattice $(L, \leq)$, the closed interval $[a,b]$ is a sublattice. A subset $S \subseteq L$ is called *convex sublattice* when $a, b \in S$ imply $[a \wedge b, a \vee b] \subseteq S$.

Definition 3-17   A lattice is *distributive* if and only if any of the following equivalent identities hold, for all x, y, z

L5. $x \wedge (y \vee z) = (x \wedge y) \vee (x \wedge z)$ L5'. $x \vee (y \wedge z) = (x \vee y) \wedge (x \vee z)$

The *complement* of an element x in a lattice $(L, \leq)$ with least (greatest) element O (I) is another lattice element y such that $x \wedge y = O$ and $x \vee y = I$; a lattice $(L, \leq)$ is called *complemented* if all its elements have complements.

Definition 3-18   A complemented distributive lattice is called *Boolean*.

The posets in Examples 3-2, 3-3, and 3-4 are also lattices; more specifically, power set *Pow(S)* is a Boolean lattice. Moreover, the closed interval $[0,1]$ of real numbers, in its natural order, is a complete lattice $([0,1], \leq)$; nevertheless, the lattice $(R, \leq)$ of real numbers is not complete because it lacks both a least and a greatest element.

Definition 3-19   A *Boolean algebra* is an algebra with two binary operations $\wedge$, $\vee$ and one unary operation $'$ satisfying L1-L8.

L6. $x \wedge x' = O,   x \vee x' = I$

L7. $(x')' = x$

L8. $(x \wedge y)' = x' \vee y',   (x \vee y)' = x' \wedge y'$

The Duality Principle for lattices: We obtain the dual statement of a lattice theoretic statement, if we replace symbols $\leq$, $\vee$, $\wedge$, O, I by symbols $\geq$, $\wedge$, $\vee$, I, O, respectively.

Definition 3-20   Let $(L, \leq)$ and $(M, \leq)$ be lattices, furthermore let θ: $L \rightarrow M$ be a function. Then function θ is called

(i) *join-morphism*: if $\theta(x \vee y) = \theta(x) \vee \theta(y)$, $x, y \in L$.

(ii) *meet-morphism*: if $\theta(x \wedge y) = \theta(x) \wedge \theta(y)$, $x, y \in L$.

A function θ is called (*lattice-*) *morphism* if θ is both *join-morphism* and *meet-morphism*. A bijective (one-to-one) morphism is called a (*lattice-*) *isomorphism*.

It turns out that in a Boolean lattice, each element x has one and only one complement denoted by x'. Under function θ: $x \rightarrow x'$ it follows that a Boolean lattice is dually isomorphic with itself (Birkhoff 1967).

Taski's celebrated fixedpoint theorem is shown next.

Theorem 3-21   Let $(L,\leq)$ be a complete lattice and θ: $L{\rightarrow}L$ a *monotone* function on L into L, i.e. $x{\leq}y \Rightarrow \theta(x){\leq}\theta(y)$. Then $\exists a$ in L such that $\theta(a) = a$.

3.3 Positive Valuation Functions

Consider the following function in a lattice.

Definition 3-22   A *valuation* v in a lattice $(L,\leq)$ is a real function v: $L{\rightarrow}R$ which satisfies $v(x)+v(y) = v(x{\wedge}y)+v(x{\vee}y)$, $x,y{\in}L$. A valuation is called *monotone* iff $x{\leq}y$ implies $v(x){\leq}v(y)$, and *positive* iff $x{<}y$ implies $v(x){<}v(y)$.

Valuations have been used in mathematical lattice theory for studying limiting processes similar to those of real analysis. The most important valuations on lattices are *measure* and *probability* functions on fields of sets (Birkhoff 1967). The utility of positive valuation functions in the context of this work is that they enable manipulating sensibly general lattice elements by operating on real numbers.

Even though partial ordering relations and/or lattices have been considered in the literature, nevertheless positive valuation functions are typically not considered (Goguen 1967; Scott 1976; Mitchell 1997; Ganter and Wille 1999; Braga-Neto 2001; Ritter and Urcid 2003; Sussner 2003; Maragos 2005). The employment of positive valuations is an instrumental novelty of this work. A simple positive valuation function in shown next.

Example 3-23   Consider finite-dimensional vector space R^N, lattice-ordered by letting $(x_1,...,x_N) \leq (y_1,...,y_N)$ mean $x_k \leq y_k$, for $k=1,...,N$. Then, any linear function $c(\mathbf{x})= c_1x_1+...+c_Nx_N$ is a valuation. This valuation is positive if and only if all c_k are positive numbers.

3.3.1 Metric and inclusion measure functions

A monotone valuation function v: $L{\rightarrow}R$ in a general lattice $(L,\leq)$ implies a *pseudometric* d: $L{\times}L{\rightarrow}R_0^+$ given by $d(x,y)=v(x{\vee}y)-v(x{\wedge}y)$, $x,y{\in}L$. If the monotone valuation v is, in addition, positive then the distance $d(x,y)=v(x{\vee}y)-v(x{\wedge}y)$ is a *metric* (Rutherford 1965; Birkhoff 1967) – For definition of a (*pseudo*)*metric* see in Appendix A.

A positive valuation can also be used for defining a degree of inclusion of a lattice element into another one based on the following definition.

Definition 3-24 Let $(L,\leq)$ be a complete lattice with least and greatest elements O and I, respectively. An *inclusion measure* in $(L,\leq)$ is a mapping $\sigma: L\times L\rightarrow[0,1]$, which satisfies the following conditions for $u,w,x\in L$.

C0. $\sigma(x,O)= 0, x\neq O$

C1. $\sigma(x,x)=1, \forall x\in L$

C2. $u \leq w \Rightarrow \sigma(x,u) \leq \sigma(x,w)$ (*Consistency Property*)

C3. $x\wedge y < x \Rightarrow \sigma(x,y) < 1$

For non-complete lattices condition C0 above is dropped.

Conditions C0-C3 are interpreted as follows. Condition C0 shows that the degree of inclusion of a complete lattice element (other than the least one) to the least element is zero. Condition C1 requires that a lattice element is fully included into itself. Condition C2, required as well by other authors in a different context (Sinha and Dougherty 1993, 1995), stipulates the common sense *Consistency Property* for an inclusion measure function. Finally, condition C3 requires that (1) when x and y are *incomparable*, then x is included in y to a degree less than 1, and (2) when y is strictly included in x, then x is included in y to a degree less than 1.

Based on equivalence relation $x\wedge y < x \Leftrightarrow y < x\vee y$ (Birkhoff 1967) it follows that condition C3 can, equivalently, be replaced by

C3′. $y < x\vee y \Rightarrow \sigma(x,y) < 1$

Conditions C1 and C2 jointly imply $u \leq w \Rightarrow \sigma(u,u) \leq \sigma(u,w) \Rightarrow \sigma(u,w)=1, u,w\in L$; hence $\sigma(x,I)=1, \forall x$ in a complete lattice L.

An inclusion measure (σ) can be used to quantify partial set inclusion. In this sense $\sigma(x,y)$ is similar to alternative definitions proposed in the literature for quantifying a degree of inclusion of a (fuzzy) set into another one (Klir and Folger 1988; Sinha and Dougherty 1993, 1995; Young 1996; Popov 1997; Setnes et al. 1998b; Fan et al. 1999; Burillo et al. 2000; Chatzis and Pitas 2000; Paul and Kumar 2002; Cornelis et al. 2003; Kehagias and Konstantinidou 2003). However, the aforementioned 'alternative' definitions typically involve only overlapping (fuzzy) sets otherwise the corresponding inclusion index equals zero. Whereas, definition 3-24 is more general, since it applies to any lattice, not only to a lattice of (fuzzy) sets. Indeed, $\sigma(x,y)$ can be interpreted as the fuzzy degree to which x is less than y; therefore the notation $\sigma(x\leq y)$ is sometimes used in place of $\sigma(x,y)$. The following theorem shows two inclusion measures in a lattice based on a positive valuation function.

Theorem 3-25   The existence of a positive valuation function $v: \mathsf{L} \rightarrow \mathsf{R}$ in a lattice $(\mathsf{L}, \leq)$ is a sufficient condition for inclusion measure functions

(a) $k(x,u) = \dfrac{v(u)}{v(x \vee u)}$, and   (b) $s(x,u) = \dfrac{v(x \wedge u)}{v(x)}$.

The proof of Theorem 3-25 is given in Appendix B.

Practical decision-making typically boils down to computing either a distance or a similarity function. Lattice theory provides both aforementioned functions based on a positive valuation.

Consider a hidden lattice P with underlying lattice $(\mathsf{L}, \leq)$ and corresponding map ψ. Let v be a positive valuation in L. Then both a metric and an inclusion measure can be defined in hidden lattice P as follows:

(1) a metric is defined by $d(x,y) = d(\psi(x \underline{\vee} y), \psi(x \underline{\wedge} y)) = v(\psi(x \underline{\vee} y)) - v(\psi(x \underline{\wedge} y))$,

and (2) an inclusion measure is defined by $\sigma(x \leq y) = v(\psi(y))/v(\psi(x \underline{\vee} y))$.

All the analysis above was carried out with regards to lattice $(\mathsf{L}, \leq)$. However, $(\mathsf{L}, \leq)$ could be the Cartesian product of N lattices, i.e. $\mathsf{L} = \mathsf{L}_1 \times \ldots \times \mathsf{L}_N$; each one of lattices $(\mathsf{L}_i, \leq)$, $i \in \{1, \ldots, N\}$ is called here *constituent lattice*. The product-lattice meet $(\wedge)$ and join $(\vee)$ are, respectively, given by

$x \wedge y = (x_1, \ldots, x_N) \wedge (y_1, \ldots, y_N) = (x_1 \wedge y_1, \ldots, x_N \wedge y_N)$, and
$x \vee y = (x_1, \ldots, x_N) \vee (y_1, \ldots, y_N) = (x_1 \vee y_1, \ldots, x_N \vee y_N)$.

The next proposition shows how a valuation function can be defined in a product-lattice from valuation functions in the constituent lattices.

Proposition 3-26   If $v_1, \ldots, v_N$ are valuations in lattices $(\mathsf{L}_1, \leq), \ldots, (\mathsf{L}_N, \leq)$, respectively, then function $v: \mathsf{L}_1 \times \ldots \times \mathsf{L}_N \rightarrow \mathsf{R}$ given by $v = v_1 + \ldots + v_N$ is a valuation in the product lattice $(\mathsf{L}, \leq)$, where $\mathsf{L} = \mathsf{L}_1 \times \ldots \times \mathsf{L}_N$.

The proof of Proposition 3-26 is given in Appendix B.

We remark that should all valuations $v_1, \ldots, v_N$ be *monotone* so would be valuation $v = v_1 + \ldots + v_N$. If at least one of the monotone valuations $v_1, \ldots, v_N$ is, in addition, a *positive valuation* then v is a positive valuation.

Positive valuation functions $v_1, \ldots, v_N$ in lattices $(\mathsf{L}_1, \leq), \ldots, (\mathsf{L}_N, \leq)$, respectively, imply metric distances $d_i: \mathsf{L}_i \times \mathsf{L}_i \rightarrow \mathsf{R}_0^+$ given by $d_i(x_i, y_i) = v(x_i \vee y_i) - v(x_i \wedge y_i)$, $i = 1, \ldots, N$. In conclusion, the following Minkowski metrics are implied $d_p(\mathbf{x}, \mathbf{y}) = [d_1(x_1, y_1)^p + \ldots + d_n(x_N, y_N)^p]^{1/p}$, $p \geq 1$, $x_i, y_i \in \mathsf{L}_i$, $i = 1, \ldots, N$ in product lattice $(L, \leq) = (L_1, \leq) \times \ldots \times (L_N, \leq)$, where $\mathbf{x}=(x_1, \ldots, x_N)$ and $y=(y_1, \ldots, y_N)$. For $L_1=\ldots=L_N=\mathbf{R}$ and $v_1(x)=\ldots=v_N(x)=x$ it follows $d_p(\mathbf{x}, \mathbf{y})= [|x_1-y_1|^p+\ldots+|x_N-y_N|^p]^{1/p}$. Metric d_p is known as L_p metric. In particular, the L_1 metric in lattice $\mathbf{R}^N$ equals $d_1(\mathbf{x}, \mathbf{y})= |x_1 - y_1| + \ldots + |x_N - y_N|$; the latter is known as Hamming distance or city-block distance. Moreover, L_2 metric equals the Euclidean distance $d_2(\mathbf{x}, \mathbf{y})= \sqrt{(x_1 - y_1)^2 + \ldots (x_N - y_N)^2}$. Furthermore, L_∞ equals $d_\infty(\mathbf{x}, \mathbf{y})= \max\{d_1(x_1, y_1), \ldots, d_N(x_N, y_N)\}$.

3.3.2 Fuzzy lattice extensions

The notion *fuzzy lattice* is introduced next to extend a crisp lattice's $(L, \leq)$ ordering relation to all pairs $(x, y) \in L \times L$. Such an extended relation may be regarded as a fuzzy set on the universe of discourse $L \times L$. In this work a fuzzy set is denoted by a pair (X, μ), where X is the universe of discourse and μ is a membership real function $\mu: X \to [0,1]$.

Definition 3-27 A *fuzzy lattice* is a triple $(L, \leq, \mu)$, where $(L, \leq)$ is a crisp lattice and $(L \times L, \mu)$ is a fuzzy set such that $\mu(x, y)=1$ if and only if $x \leq y$.

The collection of fuzzy lattices is called *framework of fuzzy lattices* (Kaburlasos and Petridis 1997, 2000; Petridis and Kaburlasos 1998, 1999). A fuzzy lattice is different than a *L-fuzzy set* (Goguen 1967); the latter is a mapping from a universe of discourse onto a lattice and as such it is a generalization of the notion 'fuzzy set'. Furthermore, a fuzzy lattice is different than a *type-2 fuzzy set*; the latter maps a universe of discourse to the collection of either conventional fuzzy sets or of intervals to deal with ambiguity in practical applications (Karnik et al. 1999; Liang and Mendel 2000; Mendel and John 2002).

Several authors have employed a notion 'fuzzy lattice' in mathematics emphasizing algebraic properties of lattice ideals, etc. (Yuan Bo and Wu Wangming 1990; Ajmal and Thomas 1994); recent extensions have been presented (Tepavcevic and Trajkovski 2001). Definition 3-27 is different from a synthesis of fuzzy multivalued connectives (Kehagias 2002, 2003).

The motivation for definition 3-27 is to enable a comparison of incomparable lattice elements. More specifically, to every $(x, y) \in L \times L$ a real number $\mu_P(x, y) \in [0,1]$ is attached to indicate the degree of inclusion of x in y. When $x \leq y$ then the fuzzy relation μ holds to the maximum degree (i.e. 1) between x and y; but μ may also hold to a lesser degree between x and y even when $x\|y$ (i.e. x and y are incomparable). Hence μ can be understood as a *weak (fuzzy) partial order* relation. In particular, μ possesses a weak form of transitivity: when both $\mu(x,y)=1$ and $\mu(y,z)=1$, then we also have $\mu(x,z)=1$; but if either $\mu(x,y)\neq1$ or $\mu(y,z)\neq1$, then $\mu(x,z)$ can take any value in $[0,1]$. The motivation for definition 3-27 is similar to the motivation for introducing a 'fuzzy lattice' in mathematics (Nanda 1989; Chakrabarty 2001), or a fuzzified *zeta* function in Bayesian probability theory (Knuth 2005). However, the objective here is, ultimately, a unifying applicability. A fuzzy lattice can be obtained from a crisp lattice as follows.

Proposition 3-28   Let $(L,\leq)$ be a crisp lattice and σ: $L\times L\rightarrow[0,1]$ be an inclusion measure on $(L,\leq)$. Then $(L,\leq,\sigma)$ is a fuzzy lattice.

The proof of Proposition 3-28 is given in Appendix B.

3.4 Useful Isomorphisms in a Complete Lattice

This section presents an inclusion measure in the set of intervals in a complete product-lattice. To simplify notation a convention is proposed regarding the elements of lattices $(L\times L,\leq)$ and $(L^{\partial}\times L,\leq)$, where $(L,\leq)$ is a lattice. More specifically, an element of lattice $(L\times L,\leq)$ is denoted by (a,b), whereas an element of lattice $(L^{\partial}\times L,\leq)$ is denoted by $[a,b]$. Likewise, if $(L\times L,\leq)$ is the product of *constituent* lattices $(L_1,\leq),...,$ $(L_N,\leq)$, an element of lattice $(L\times L,\leq)$ will be denoted by $(a_1,b_1,...,a_N,b_N)$ with $(a_i,b_i)\in L_i\times L_i$, $i=1,...,N$, whereas an element of lattice $L^{\partial}\times L$ will be denoted by $[a_1,b_1,...,a_N,b_N]$ with $[a_i,b_i]\in L_i^{\partial}\times L_i$, $i=1,...,N$.

Consider a function τ, which maps a complete lattice $(L,\leq)$ to the collection of non-empty intervals including the empty set. The following proposition introduces another complete lattice.

Theorem 3-29   Let $(L,\leq)$ be a complete lattice. Then the set $\tau(L)$ of intervals in $(L,\leq)$ is a complete lattice with largest and least intervals denoted by $[O,I]$ and $[I,O]$, respectively. The lattice ordering relation $[a,b]\leq[c,d]$ is equivalent to '$c\leq a$ and $b\leq d$'. For two intervals $[a,b]$, $[c,d]$ their *lattice join* is given by $[a,b]\vee[c,d]=[a\wedge c,b\vee d]$. Moreover their *lattice meet* is given by either $[a,b]\wedge[c,d]=[a\vee c,b\wedge d]$ if $a\vee c\leq b\wedge d$, or $[a,b]\wedge[c,d]=[I,O]$ otherwise.

The proof of Theorem 3-29 is given in Appendix B.

Of particular interest is the subset $a(L)$ of $\tau(L)$ that includes trivial intervals (singletons). Since relation $[I,O]<[a,b]<[x,x]$ holds for no $[a,b]\in\tau(L)$, a singleton $[x,x]$ *covers* $[I,O]$; i.e. the elements of $a(L)$ are *atoms*.

It turns out that (1) a positive valuation v: $L\rightarrow R$, and (2) an isomorphism θ: $L^\partial\rightarrow L$ are sufficient conditions for the existence of an inclusion measure in the lattice $(\tau(L),\leq)$ of intervals. In particular, an inclusion measure σ can be defined in $(\tau(L),\leq)$ according to the following algorithm.

Algorithm IM: An inclusion measure in a complete lattice of intervals
1. A positive valuation v: $L\rightarrow R$ in a complete lattice $(L,\leq)$ implies a positive valuation V in the complete lattice $(L\times L,\leq)$ given by $V(a,b) = v(a)+v(b)$. Hence, it implies an inclusion measure in $(L\times L,\leq)$ given by $\sigma((a,b),(c,d))=k((a,b)\leq(c,d))=V(c,d)/V((a,b)\vee(c,d))=V(c,d)/V(a\vee c,b\vee d)$.
2. Consider lattice $(L^\partial\times L,\leq)$, namely *lattice of generalized intervals*. Furthermore, let θ: $L^\partial\rightarrow L$ be an isomorphism.
3. Map $\theta\times Id_A$: $L^\partial\times L\rightarrow L\times L$, where Id_A is the identity map in L, implies isomorphism $(L\times L,\leq) \cong (L^\partial\times L,\leq)$.
4. Consider the subset L_t of $L^\partial\times L$ defined by $L_t=\{[a,b]: a,b\in L$ and $a\leq b\}\cup\{[I,O]\}$. It holds $(\tau(L),\leq) \cong (L_t,\leq)$.
5. Consider the injective monotone map i: $\tau(L)\rightarrow L^\partial\times L$, given by $i(\Delta)= [\wedge\Delta,\vee\Delta]$, where $\Delta\in\tau(L)$, $\wedge\Delta$ is the greatest lower bound of Δ, and $\vee\Delta$ is the least upper bound of Δ in the complete lattice $(L,\leq)$.
6. Consider composite injective monotone map $\phi=(\theta\times Id_A)\circ i$: $\tau(L)\rightarrow L\times L$.
7. An inclusion measure $\sigma(x\leq y)$ in $(\tau(L),\leq)$ is given by $\sigma([a,b]\leq[c,d]) =$

$$k(\phi([a,b]),\phi([c,d])) = k((\theta(a),b) \leq (\theta(c),d)) = \frac{v(\theta(c)) + v(d)}{v(\theta(a \wedge c)) + v(b \vee d)}.$$

Relations between various sets of interest above are summarized next

$$\tau(L) \cong L_t \subseteq L^\partial\times L \cong L\times L \qquad\qquad (ISO)$$

The composite injective monotone map ϕ above implies that if $[a,b]\leq[c,d]$ in $\tau(L)$ then $\phi([a,b])\leq\phi([c,d])$ in $L\times L$.

The *diagonal* of a lattice interval computes the size of a lattice interval.

Definition 3-30   Let v be a positive valuation in a lattice $(L,\leq)$, and let $(\tau(L),\leq)$ be the lattice of intervals. The *diagonal of an interval* $[a,b]$ is a real function *diag*: $\tau(L)\rightarrow R_0^+$ given by $diag([a,b])=v(b)-v(a)$.

Proposition 3-31 It holds $diag([a,b]) = d(a,b) = \max\limits_{x,y\in[a,b]} d(x,y)$.

The proof of Proposition 3-31 is given in Appendix B.

Let $(L,\leq)$ be the product of lattices $(L_1,\leq),\ldots,(L_N,\leq)$ with positive valuations $v_1,\ldots,v_N$, respectively. Consider an interval $[a,b]=[a_1,b_1]\times\ldots\times[a_N,b_N]$. Then $diag([a,b]) = \sum_{i=1}^{N}d(a_i,b_i)$, i.e. the *diagonal* of a lattice interval, is a Hamming distance. In particular for an atom $t_r=[a,a]$ it follows that $diag(t_r) = diag([a,a])=v(a)-v(a)=0$; that is, the size of an atom equals zero. Furthermore, assuming $v(O)=0$, it follows that the diagonal of the least interval $[I,O]$ in the complete lattice $\tau(L)$ equals $diag([I,O])=v(O)-v(I)=-v(I)$.

3.5 Families of Lattice Intervals

In this section we consider finite collections of lattice intervals.

Definition 3-32   Let $(L,\leq)$ be a lattice. A *family f of lattice intervals* $f=\{w_j\}_{j\in J}$ is a finite collection of lattice intervals.

The set of families of intervals in a lattice $(L,\leq)$ is denoted by $\mathcal{F}_{\tau(L)}$.

Our interest is in describing 'closed and bounded' subsets of a lattice using families of intervals. The aforementioned description is feasible for lattices of practical interest with countable cardinality. It is also feasible for the uncountably infinite set R of real numbers due to the following theorem from mathematical topology (Chae 1995).

Heine-Borel Theorem 3-33   A subset of R is *compact* if and only if it is *closed* and *bounded*.

We remark that a set K is called *compact* if every *open cover* of K admits a finite subcover, where definitions for *open (sub)cover* /*closed* /*bounded* sets are given in Appendix A.

A family $f=\{w_j\}_{j\in J}\in\mathcal{F}_{\tau(L)}$ implies a fuzzy set on the lattice-of-intervals universe of discourse $\tau(L)$ as shown in the following.

Definition 3-34   The *membership function of a family* $f=\{w_j\}_{j\in J}$ *(with respect to an inclusion measure σ in $\tau(L)$)* is a real function m_f: $\tau(L)\to[0,1]$ given by $m_f(x):=\max\limits_{j\in J}\sigma(x\leq w_j)$.

The terms *class* and *category* are used interchangeably in this work to denote the set-union of the intervals in a family.

Definition 3-35   Let $(L,\leq)$ be a lattice. A *class* or, alternatively, *category*, c in L is defined as $c=\bigcup\limits_{j\in J} w_j$, where $\{w_j\}_{j\in J}$ is a family in $\mathcal{F}_{\tau(L)}$.

The set of classes in a lattice $(L,\leq)$ is denoted by $\mathcal{C}_{\tau(L)}$. Apparently a class c is a subset of L. Note that two different families of intervals $f_1=\{u_i\}_{i\in I}$ and $f_2=\{w_j\}_{j\in J}$ may represent the same class, i.e. $\bigcup\limits_{i\in I}u_i=\bigcup\limits_{j\in J}w_j$.

Let F_c be the collection of all families that represent a class c. We define that a family $\{p_m\}\in F_c$ is 'smaller than or equal to' another family $\{q_n\}\in F_c$ symbolically $\{p_m\}\leq\{q_n\}$, if and only if $\forall p\in\{p_m\}\ \exists q\in\{q_n\}$ such that $p\leq q$. Moreover, we define $\{p_m\}<\{q_n\}$ if and only if it is $\{p_m\}\leq\{q_n\}$ and either (1) there is a p in $\{p_m\}$ and a q in $\{q_n\}$ such that $p<q$ or (2) the cardinality of family $\{p_m\}$ is strictly smaller than the cardinality of family $\{q_n\}$. It turns out that F_c is a partially ordered set with a greatest element.

Proposition 3-36   The collection F_c of families which represent a class $c=\bigcup\limits_{j} w_j$ has a greatest element, namely *quotient* of class c denoted by $Q(F_c)=Q(\{w_j\})$.

The proof of Proposition 3-36 is given in Appendix B.

We remark that quotient $Q(F_c)=Q(\{w_j\})$ maximizes the degree of inclusion of $x\in\tau(L)$ in class $c=\bigcup\limits_{j} w_j$.

A class $c\in\mathcal{C}_{\tau(L)}$ implies a fuzzy set as shown in the following.

Definition 3-37   *The membership function of a class c (with respect to an inclusion measure σ in $\tau(L)$)* is a real function m_c: $\tau(L)\to[0,1]$ given by $m_c(x):=m_{Q(c)}(x)$, where $Q(c)$ is the quotient family of class c.

4 Real-World Grounding

This chapter grounds lattice theory

Mathematical theories abound; however, many of them are not currently useful. On the other hand, popular practical techniques are often short of a sound mathematical basis. Hence, the latter techniques are short of instruments to optimize their application. It turns out that lattice theory is firmly grounded in the real world. This chapter illustrates how.

4.1 The Euclidean Space R^N

The Euclidean space R^N is the Cartesian product $R^N = R \times \ldots \times R$, where the set R of real numbers has emerged from the conventional measurement process of successive comparisons (Goldfarb and Deshpande 1997). People have used (and studied) real numbers for millennia, and essential properties of the set R are known. For instance, R is a (*mathematical*) *field*; moreover, there is a *metric* in R; furthermore, the cardinality $\aleph_1$ of R is larger than the cardinality $\aleph_0$ of the subset Q of rational numbers – For definitions of a *mathematical field, metric*, etc. see in Appendix A.

Practical decision-making is often based on the *total ordering* property of the non-complete lattice R. The *meet* ($\wedge$) and *join* ($\vee$) in the *chain* R are given, respectively, by $x \wedge y = \min\{x,y\}$ and $x \vee y = \max\{x,y\}$.

Mathematical studies were extended from R to R^N. The *linear space* $R^N = R \times \ldots \times R$ emerged 'par excellence' as the modeling domain in various applications. However, the product lattice $R^N = R \times \ldots \times R$ is not *totally ordered*; rather, R^N is *partially ordered*, as illustrated in Fig.4-1 on the plane for N=2, with partial ordering $(x_1,\ldots,x_N) \leq (y_1,\ldots,y_N) \Leftrightarrow x_1 \leq y_1,\ldots,x_N \leq y_N$. An interval in lattice R^N is a N-dimensional hyperbox, or *hyperbox* for short. The set of hyperboxes in R^N is a non-complete lattice.

A strictly increasing real function on R is a *positive valuation* function as shown in section 4.5, where a positive valuation is also defined in a lattice of intervals. Popular in applications is the 'unit hypercube' as explained in the following.

Vassilis G. Kaburlasos: *Real-World Grounding*, Studies in Computational Intelligence (SCI) **27**, 35–62 (2006)
www.springerlink.com

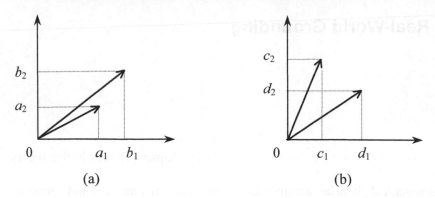

Fig.4-1   Geometric interpretations in the 2-dimensional lattice R^2.
(a) Two *comparable* elements with $(a_1,a_2) \leq (b_1,b_2)$.
(b) Two *incomparable* elements, symbolically $(c_1,c_2)\|(d_1,d_2)$.

4.2 Hyperboxes in R^N

A complete lattice, denoted by U, is defined in the N-dimensional unit hypercube with partial ordering $(x_1,...,x_N) \leq (y_1,...,y_N)$ $\Leftrightarrow$ $x_1 \leq y_1,...,x_N \leq y_N$. In the interest of simplicity the dimension (N) was suppressed in symbol U. Note that consideration of the lattice U, instead of R^N, is not restrictive by default, since all physical quantities have upper/lower bounds and suitable transformations to the unit hypercube can be found. Our interest is in learning sets of points using a finite collection of hyperboxes.

Lattice U is the product of N identical constituent lattices, these are the *complete* lattices (*chains*) I=[0,1] each with least element O=0 and greatest element I=1. It follows that U is a *complete* lattice with least element $(0,...,0)$ and greatest element $(1,...,1)$. Moreover, $\tau(U)$ is an *atomic* lattice because every hyperbox is the lattice-join of atoms 'min' and 'MAX'.

A number of computational intelligence schemes that learn by computing hyperboxes have been proposed including min-max neural networks (Simpson 1992, 1993; Gabrys and Bargiela 2000) as well as adaptive resonance theory (ART) inspired neural networks (Georgiopoulos et al. 1994; Healy and Caudell 1997; Cano Izquierdo et al. 2001; Anagnostopoulos and Georgiopoulos 2002; Parrado-Hernández et al. 2003; Castro et al. 2005). Moreover, in a machine-learning context, the class of axis-parallel rectangles was shown to be efficiently *probably approximately correct* (PAC) learnable (Blumer et al. 1989; Kearns and Vazirani 1994; Long and Tan 1998); other applications have also considered hyperboxes for learning (Salzberg 1991; Wettschereck and Dietterich 1995; Dietterich et al. 1997).

Note also that learning general lattice intervals beyond lattice R^N is carried out implicitly in Valiant (1984) as well as in Kandel et al. (1995) where conjunctive normal forms (CNF) are computed in a Boolean lattice.

None of the above algorithms employs lattice theory. This work engages explicitly lattice theory for learning hyperboxes in the unit hypercube U. Recall from section 3.4 that sufficient conditions for the existence of an inclusion measure σ in a *complete lattice* include (1) a *positive valuation* function v, and (2) an *isomorphic* function θ. It is shown next how the unit hypercube U satisfies the aforementioned sufficient conditions.

A strictly increasing function v_I: I→R is a valid *positive valuation* in I; for instance, $v_I(x)=x$, which implies positive valuation $v(a_1,...,a_N)= a_1+...+a_N$ in the product lattice U. Likewise, a strictly decreasing function θ_I: I→I is a valid *isomorphic function* in I; for instance, $\theta_I(x)=1-x$, which implies isomorphism $\theta(a_1,...,a_N)= (\theta_I(a_1),...,\theta_I(a_N))= (1-a_1,...,1-a_N)$ in the product lattice U. The choice of a 'good' isomorphic function θ and/or a 'good' positive valuation function v is problem dependent.

'Missing' data values in a *constituent* complete lattice I are replaced by the least element O=[1,0], whereas 'don't care' data values are replaced by the greatest element I=[0,1] in I. Note that a 'missing' value is interpreted as an *unknown* (*unavailable*) value for an attribute, whereas a 'don't care' value is interpreted as *all possible* attribute values.

4.3 Propositional (Boolean) Logic

The vertices of the unit hypercube is a popular data domain in computer science. More specifically, 'Boolean logic' involves true/false propositions, which can be represented as vertices in a unit hypercube. It is well known that propositions form a *Boolean algebra* (Birkhoff 1967).

In practice Boolean algebra is the basic tool for designing switches in the integrated circuits (ICs) manufacturing industry. In Computational Intelligence several schemes have been proposed for processing strings of 0s and 1s including ART1 neural networks (Carpenter and Grossberg 1987a), Hopfield networks (Hopfield 1982), morphological neural networks (Sussner 2003), etc. Moreover, a fuzzification of binary data was proposed for computing in a fuzzy (hyper)cube (Kosko 1997, 1998).

This work proposes the explicit employment of a Boolean lattice, where a positive valuation function (v) maps (1) 'bit 1' to a positive real number, and (2) 'bit 0' to 0; moreover, the corresponding isomorphic function (θ) maps 'bit 0' to 'bit 1' and vice versa.

4.4 A Probability Space

By definition a *probability space* is a triplet (Ω, S, P), where Ω is an abstract space, S is a σ-algebra of subsets of Ω, and P is a probability measure on S (Feller 1968; Doob 1994). A probability measure P is a *set function* defined for $E \in S$ and satisfying the following conditions:

P1. $P(E) \geq 0$

P2. $P(\bigcup_n E_n) = \sum_n P(E_n)$ for pairwise disjoint sets $E_n \in S$, $n=1,2,\ldots$

P3. $P(\Omega) = 1$

An element of S is called *(measurable) event*.

Probability theory has been popular for modeling uncertainty in applications. An additional reason for the popularity of probability theory is its capacity to handle non-numeric data. Theoretical connections between probability theory and lattice theory are well known (Birkhoff 1967). Recent perspectives have also been published (Knuth 2005).

The σ-algebra S in (Ω, S, P) is a lattice whose ordering relation is the conventional set-inclusion relation ($\subseteq$). The lattice *meet* ($\wedge$) and *join* ($\vee$) are the conventional set-intersection ($\cap$) and set-union ($\cup$), respectively. Moreover, S is a complete lattice with least and greatest elements O= $\varnothing$ and I= Ω, respectively. Furthermore, S is a *Boolean lattice*. A probability measure P is a *monotone valuation* in the lattice S of events. It follows that function $d(A,B)=P(A \cup B)-P(A \cap B)$ is a *pseudometric*. The latter has been employed to define a measure of equivalence between two propositions by $p(x \equiv y)=1-d(x,y)$, and ultimately to postulate equivalence, implication, and negation of propositions (Gaines 1978).

Relation $d(A,B)=0$ defines an *equivalence relation* in S. Let S be the set of equivalence classes in S. It turns out that S is a lattice; moreover, the probability measure P is a positive valuation in lattice S. There exists an isomorphic function θ in S, that is $\theta(X)=X^c$, where X^c denotes the complement of X. It follows that $(S, \subseteq, P(B)/P(A \cup B))$ is a fuzzy lattice.

A potentially useful algebra in a probability space (Ω, S, P) is described in the following. Let $\otimes$ be the *symmetric difference* operator $\otimes$: S×S→S defined by $A \otimes B = A \cup B - A \cap B$, where $A-B=A \cap B^c$. Then $(S, \otimes)$ is an *Abelian group* with identity element $e=\varnothing$ and inverse A^{-1} of $A \in S$ given by $A^{-1}=A$. Hence, equation $A \otimes X = C$ has a unique solution given by $X = C \otimes A^{-1}$. In this context, algorithms for classification and regression can be proposed (Kaburlasos and Petridis 1998).

4.5 FINs: Fuzzy Interval Numbers

This long section studies an extension of the set of fuzzy numbers, namely *fuzzy interval numbers*, or *FINs* for short, based on lattice theory.

Even though fuzzy sets can be defined on any universe of discourse, in practice fuzzy sets are typically defined on the real number universe of discourse R. More specifically *fuzzy numbers*, i.e. *normal* and *convex* fuzzy sets with an interval support, are considered. Recall that a fuzzy set is called *convex* if and only if the crisp set $\Gamma_a=\{x|\mu(x) \geq a\}$, namely *a-level set* or equivalently *a-cut*, is convex for a in $(0,1]$ (Zadeh 1965).

A fuzzy set F: R$\rightarrow$[0,1] is called a *fuzzy number* if it satisfies the following properties (Klir and Yuan 1995).

A1. It is normal (i.e. $\exists x_0$: $F(x_0)=1$)
A2. The a-cut $F_a= \{x$: $F(x) \geq a\}$ is a closed interval for all $a \in (0,1]$
A3. The support of F (i.e. the set $\{x$: $F(x) > 0\} = \bigcup_{a\in(0,1]} F_a$) is bounded

It is known that a fuzzy set can be advantageously represented in applications by its a-cuts (Uehara and Fujise 1993). It turns out that fuzzy numbers can be defined in terms of their a-cuts as follows.

Definition 4-1 A fuzzy number is a family of sets $\{F_a\}_{a\in[0,1]}$ which satisfy the following conditions.

F1. F_0=R
F2. For every $a,b \in [0,1]$ we have: $a \leq b \Rightarrow F_b \subseteq F_a$
F3. For every set $A \subseteq [0,1]$, letting $b=$ sup$A$, we have: $\bigcap_{a\in A} F_a = F_b$
F4. For every $a \in (0,1]$, F_a is a closed interval
F5. $\bigcup_{a\in(0,1]} F_a$ is a bounded interval
F6. $F_1 \neq \varnothing$

Instrumental in the study of FINs are *generalized intervals* presented in the following.

4.5.1 Generalized intervals

The notion 'interval' is basic in analysis and design (Moore 1979; Alefeld and Herzberger 1983; Kaufmann and Gupta 1985; Chae 1995; Tanaka et al. 2003). *Generalized intervals* (Kaburlasos 2002, 2004a; Petridis and Kaburlasos 2003; Kaburlasos and Papadakis 2006; Kaburlasos and Kehagias 2006a) are formally defined in the following.

Definition 4-2 A *positive generalized interval of height* $h \in (0,1]$ is a map $\mu^h_{x_1,x_2}(x)$: $R \rightarrow \{0,h\}$, given by $\mu^h_{x_1,x_2}(x) = \begin{cases} h, & x_1 \leq x \leq x_2 \\ 0, & otherwise \end{cases}$, where $x_1 \leq x_2$,; a *negative generalized interval of height* h is a map $\mu^h_{x_1,x_2}(x)$: $R \rightarrow \{0,-h\}$, given by $\mu^h_{x_1,x_2}(x) = \begin{cases} -h, & x_1 \geq x \geq x_2 \\ 0, & otherwise \end{cases}$, where $x_1 > x_2$, $h \in (0,1]$.

We remark that a generalized interval is a simple box function, either positive or negative. Generalized intervals will be useful for introducing a *metric* between *fuzzy interval numbers* (FINs) below. In the interest of simplicity a generalized interval will be denoted as $[x_1,x_2]^h$, where $x_1 \leq x_2$ ($x_1 > x_2$) for positive (negative) generalized intervals. In particular, a positive generalized interval $[x_1,x_1]^h$ is called *trivial*.

The set of positive (negative) generalized intervals of height h is denoted by M^h_+ (M^h_-). In particular, the set of *trivial* generalized intervals is denoted by M^h_0, where $M^h_0 \subset M^h_+$. The set of generalized intervals of height h is denoted by M^h, i.e. $M^h = M^h_- \cup M^h_+$. The set-union of all M^hs is the set M of *generalized intervals*, i.e. $M = \bigcup_{h \in (0,1]} M^h$. Our interest is in generalized intervals $[x_1,x_2]^h$ with $h \in (0,1]$ because the latter emerge from a-cuts of fuzzy numbers. An *ordering* relation is defined in $M^h \times M^h$ as follows.

R1. If $[a,b]^h$, $[c,d]^h \in M^h_+$ then: $[a,b]^h \leq [c,d]^h \Leftrightarrow [a,b] \subseteq [c,d]$

R2. If $[a,b]^h$, $[c,d]^h \in M^h_-$ then: $[a,b]^h \leq [c,d]^h \Leftrightarrow [d,c] \subseteq [b,a]$

R3. If $[a,b]^h \in M^h_-$, $[c,d]^h \in M^h_+$ then: $[a,b]^h \leq [c,d]^h \Leftrightarrow [b,a] \cap [c,d] \neq \varnothing$

In all other cases generalized intervals $[a,b]^h$ and $[c,d]^h$ are *incomparable*, symbolically $[a,b]^h \| [c,d]^h$.

Rule R1 indicates that a positive generalized interval $[a,b]^h$ is smaller than another one $[c,d]^h$ if and only if the interval support of $[a,b]^h$ is included in the interval support of $[c,d]^h$. Rule R2 indicates, dually, the converse for negative generalized intervals. Finally, rule R3 indicates that a negative generalized interval $[a,b]^h$ is smaller than a positive one $[c,d]^h$ if and only if their interval supports overlap.

The ordering relation above is illustrated geometrically on the plane in Fig.4-2. It turns out that the aforementioned ordering relation is a *partial ordering*; moreover, $(M^h, \leq)$ is a *mathematical lattice* (Kaburlasos 2004a; Kaburlasos and Kehagias 2006a). We point out that $(M^h, \leq)$ is not a complete lattice because the lattice of real numbers is not a complete lattice.

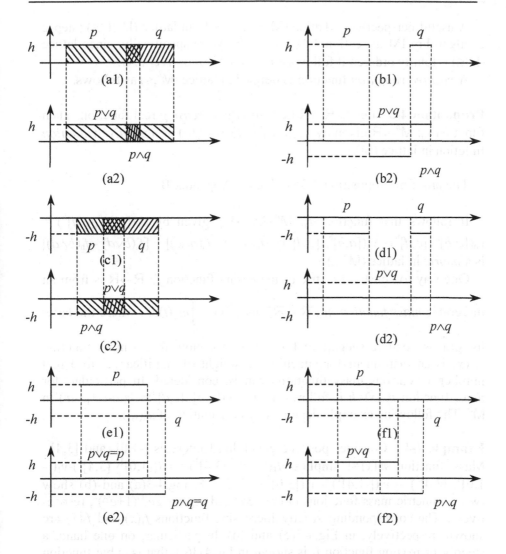

Fig.4-2   *Join* ($p \vee q$) and *meet* ($p \wedge q$) for all pairs (p,q) of generalized intervals of height *h*. Different fill-in patterns are used for partially overlapped generalized intervals.
(a) 'Intersecting' positive generalized intervals.
(b) 'Non-intersecting' positive generalized intervals.
(c) 'Intersecting' negative generalized intervals.
(d) 'Non-intersecting' negative generalized intervals.
(e) 'Intersecting' positive and negative generalized intervals.
(f) 'Non-intersecting' positive and negative generalized intervals.
In all other cases generalized intervals $[a,b]^h$ and $[c,d]^h$ are *incomparable*, symbolically $[a,b]^h \| [c,d]^h$.

A useful perspective in lattice $(\mathbf{M}^h,\leq)$ stems from lattice $(\mathbf{L}^\partial \times \mathbf{L},\leq)$ (step-2 of algorithm IM in section 3.4) for L = R. More specifically, since lattice $(\mathbf{R},\leq)$ is totally ordered it follows either $a \vee c \leq b \wedge d$ or $a \vee c > b \wedge d$.

A *positive valuation* function is defined in lattice $(\mathbf{M}^h,\leq)$ as follows.

Proposition 4-3 Let f_h: R→R be a *strictly increasing* real function. Then function v_h: $\mathbf{M}^h$→R given by $v_h([a,b]^h) = f_h(b) - f_h(a)$ is a *positive valuation* function in lattice $(\mathbf{M}^h,\leq)$.

The proof of Proposition 4-3 is given in Appendix B.

It follows that function d_h: $\mathbf{M}^h \times \mathbf{M}^h$→$\mathbf{R}_0^+$ given by $d_h([a,b]^h,[c,d]^h) = v_h([a,b]^h \vee [c,d]^h) - v_h([a,b]^h \wedge [c,d]^h) = [f_h(a \vee c) - f_h(a \wedge c)] + [f_h(b \vee d) - f_h(b \wedge d)]$ is a *metric* in lattice $(\mathbf{M}^h,\leq)$.

One way to construct a strictly increasing function f_h: R→R is from an integrable *mass function* m_h: R→$\mathbf{R}_0^+$ as: $f_h(x) = \int_0^x m_h(t)\mathrm{d}t$, where the latter integral is positive (negative) for $x>0$ ($x<0$). Note that a mass function $m_h(x)$ is an instrument for attaching 'a weight of significance' to a real number x. Various mass functions can be considered. In particular, for mass function $m_h(x)=h$ it follows metric $d_h([a,b]^h,[c,d]^h) = h(|a-c|+|b-d|)$ in $\mathbf{M}^h$. The following example demonstrates calculation of $d_h(.,.)$.

Example 4-4 Consider positive generalized intervals $[-1,0]^1$ and $[3,4]^1$. Mass function $m(x)=1$ implies $d_1([-1,0]^1,[3,4]^1) = v_1([-1,0]^1 \vee [3,4]^1) - v_1([-1,0]^1 \wedge [3,4]^1) = v_1([-1,4]^1) - v_1([3,0]^1) = 5+3 = 8$. Fig.4-3(a) and (b) show two symmetric mass functions $m_1(x) = 3x^2$ and $m_2(x) = 2e^{-x}/(1+e^{-x})^2$, respectively. The corresponding strictly increasing functions $f_1(x)$ and $f_2(x)$ are shown, respectively, in Fig.4-3(c) and (d). In particular, on one hand, a steeply increasing function f_1 is shown in Fig.4-3(c), that is cubic function $f_1(x)=x^3$; the computation of metric $d_1(.,.)$ using $m=m_1$ results in $d_1([-1,0]^1,[3,4]^1) = f_1([-1,0]^1 \vee [3,4]^1) - f_1([-1,0]^1 \wedge [3,4]^1) = 65+27 = 92$. On the other hand, a slowly increasing (saturated) function f_2 is shown in Fig.4-3(d), that is logistic function $f_2(x) = [2/(1+e^{-x})]-1$; the computation of metric $d_1(.,.)$ using $m=m_2$ results in $d_1([-1,0]^1,[3,4]^1) = f_2([-1,0]^1 \vee [3,4]^1) - f_2([-1,0]^1 \wedge [3,4]^1) \approx 1.4262+0.9052 \approx 2.3314$.

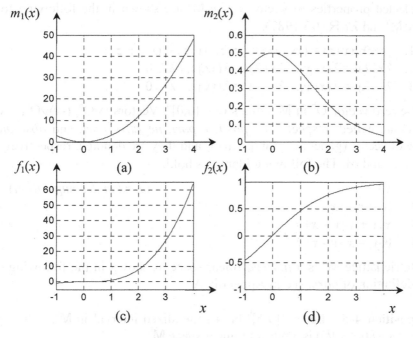

Fig.4-3   (a) Symmetric mass function $m_1(x) = 3x^2$.
(b) Symmetric mass function $m_2(x) = 2e^{-x}/(1+e^{-x})^2$.
(c) The *cubic* strictly increasing function $f_1(x) = x^3$ is an integral of mass function $m_1(x)$.
(d) The *logistic* strictly increasing function $f_2(x) = [2/(1+e^{-x})]-1$ is an integral of mass function $m_2(x)$.

The example above was meant to demonstrate that a different *mass function* could change substantially the distance between two intervals.

It further turns out that space M^h is a *real linear space* with

- *addition* defined as $[a,b]^h + [c,d]^h = [a+c,b+d]^h$.
- *multiplication* (by a real number k) defined as $k[a,b]^h = [ka,kb]^h$.

A generalized interval in linear space M^h is called *vector*, where the term *vector* is used here in the standard linear algebra sense (Itô 1987). A lattice-ordered linear space, such as space M^h in this work, is called *vector lattice* or *Riesz space*. Note that a theory of vector lattices has been introduced by Riesz (1928) and further developed by various authors (Vulikh 1967; Luxemburg and Zaanen 1971).

Selected properties in vector lattice M^h are shown in the following for $x,y,z \in M^h$ and $\lambda \in R$ (Itô 1987).

S1. $(x+z) \vee (y+z) = (x \vee y)+z$, $(x+z) \wedge (y+z) = (x \wedge y)+z$
S2. $\lambda x \vee \lambda y = \lambda(x \vee y)$, $\lambda x \wedge \lambda y = \lambda(x \wedge y)$, $\lambda \geq 0$
S3. $\lambda x \vee \lambda y = \lambda(x \wedge y)$, $\lambda x \wedge \lambda y = \lambda(x \vee y)$, $\lambda < 0$

The *zero vector* O^h in M^h equals $O^h = [0,0]^h$. Vectors $x \vee O^h$, $(-x) \vee O^h$, and $x \vee (-x)$ are called, respectively, *positive part*, *negative part*, and *absolute value* of vector (generalized interval) x, and they are denoted, respectively, by x^+, x^-, and $|x|$. The following identities hold:

I1. $x = x^+ - x^-$ (*Jordan decomposition*)
I2. $|x| = x^+ + x^-$
I3. $x \vee y + x \wedge y = x+y$
I4. $|x-y| = x \vee y - x \wedge y$

Metric lattice M^h is a *normed* linear space as shown in the following – For definition of *norm* see in Appendix A.

Proposition 4-5   Let $x = [a,b]^h$ be a generalized interval in M^h, $h \in (0,1]$. Then $\|x\| = h(|a| + |b|)$ is a norm in linear space M^h.

The proof of Proposition 4-5 is given in Appendix B.

Given strictly increasing function $f_h(x) = x$, from proposition 4-3 it follows $\|x\| = v_h(|x|)$. Moreover, it is well-known that a normed linear space S with norm $\|.\|$ implies metric $d(x,y) = \|x-y\|$, $x,y \in S$. Hence, based on Proposition 4-5, the norm-induced metric for two vectors $x = [a,b]^h$ and $y = [c,d]^h$ in M^h is given by $d(x,y) = \|x-y\| = \|[a-b,b-d]^h\| = h(|a-c| + |b-d|)$.

The dimension of linear space M^h equals two. A *basis* is selected in M^h: $[a,b]^h = [a,a+(b-a)]^h = [a,a]^h + [0,b-a]^h = a[1,1]^h + (b-a)[0,1]^h = ae_1^h + be_2^h$, $a,b \in R$. Hence, basis $(e_1^h, e_2^h) = ([1,1]^h, [0,1]^h)$ can be selected in M^h.

A subset C of a linear space is called *cone* if for all $x \in C$ and $\lambda > 0$ it follows $\lambda x \in C$ (Bertsekas et al. 2003). Two interesting cones in M^h include (1) the cone M_+^h of positive generalized intervals, and (2) the cone M_-^h of negative generalized intervals. Indeed, using the aforementioned basis $(e_1^h, e_2^h) = ([1,1]^h, [0,1]^h)$ it can be easily shown that (1) if $x \in M_+^h$ and $\lambda > 0$ then $\lambda x \in M_+^h$, and (2) if $x \in M_-^h$ and $\lambda > 0$ then $\lambda x \in M_-^h$.

4.5.2 Fuzzy interval numbers

Generalized intervals presented above are an instrument for analysis of *Fuzzy Interval Numbers* (*FINs*) defined in the following.

Definition 4-6 A *Fuzzy Interval Number* (*FIN*) is a function $F: (0,1] \rightarrow M$ such that (1) $F(h) \in M^h$, (2) either $F(h) \in M^h_+$ (*positive FIN*) or $F(h) \in M^h_-$ (*negative FIN*) for $h \in (0,1]$, and (3) $h_1 \leq h_2 \Rightarrow \{x: F(h_1) \neq 0\} \supseteq \{x: F(h_2) \neq 0\}$, where $h_1, h_2 \in [0,1]$.

We remark that a FIN was defined in Definition 4-6 via its range in a similar manner as the Lebesgue integral is defined (Chae 1995).

The set of FINs is denoted by **F**. More specifically, the set of *positive* (*negative*) FINs is denoted by $\mathbf{F}_+$ ($\mathbf{F}_-$); in particular, the set of *trivial* FINs is denoted by $\mathbf{F}_0$, where $\mathbf{F}_0 \subset \mathbf{F}_+$. Fig.4-4 shows examples of FINs including one negative FIN (F_n), one trivial FIN (F_t), and one positive FIN (F_p).

We point out that a FIN is an abstract mathematical notion. Various interpretations can be proposed for a general FIN. For instance, on the one hand, a positive FIN may be interpreted as a conventional fuzzy number; moreover a statistical interpretation is presented below for a positive FIN.

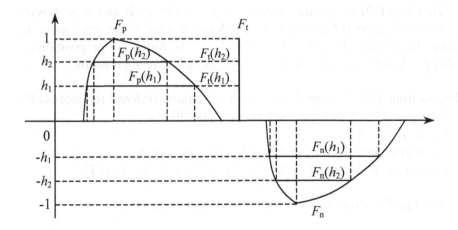

Fig.4-4   A positive FIN F_p, a trivial FIN F_t and a negative FIN F_n.

On the other hand, a negative FIN might be interpreted as an *intuitionistic (fuzzy) set* in the sense of Atanasov (Atanasov 1999). More specifically, the membership function of a negative FIN may denote the degree of certainty that a real number does *not* belong to a fuzzy set. FINs with negative membership functions might also be useful for interpreting significant improvements in fuzzy linear regression problems involving triangular fuzzy sets with negative spreads (Chang and Lee 1994). However, note that fuzzy sets with negative spreads are not regarded as fuzzy sets by some authors (Diamond and Körner 1997). In a different context, in particular in artificial intelligence (AI), note that expert system MYCIN, with strong influence in the design of commercial expert systems, has used *negative certainty factors* for medical diagnosis of certain blood infections (Buchanan and Shortliffe 1984). Negative FINs might find similar applications.

Under any interpretation, a FIN may be regarded as an *information granule* for dealing with ambiguity. Note that the term *granule* was introduced in fuzzy set theory to denote a clump of values drawn together by indistinguishability, similarity, proximity or functionality (Hirota and Pedrycz 1999; Zadeh 1999; Bortolan and Pedrycz 2002; Pedrycz 2002).

An ordering relation was introduced in the set F of FINs as follows.

Definition 4-7   Let $F_1, F_2 \in F$, then $F_1 \le F_2 \Leftrightarrow F_1(h) \le F_2(h)$, for h in (0,1].

That is, a FIN F_1 is smaller-than/equal-to a FIN F_2 if and only if every generalized interval $F_1(h)$ of F_1 is smaller-than/equal-to the corresponding generalized interval $F_2(h)$ of F_2 for $h \in (0,1]$. The following proposition establishes that the ordering relation $\le$ is a *partial ordering relation*.

Proposition 4-8   The set F of FINs is a *partially ordered set* because the ordering relation $\le$ satisfies the following conditions
(a) *reflexive*: $F_1 \le F_1$,
(b) *antisymmetric*: $F_1 \le F_2$ and $F_2 \le F_1 \Rightarrow F_1 = F_2$, and
(c) *transitive*: $F_1 \le F_2$ and $F_2 \le F_3 \Rightarrow F_1 \le F_3$, for F_1, F_2, F_3 in F.

The proof of Proposition 4-8 is given in Appendix B.

The following proposition introduces a pseudometric in F.

Proposition 4-9 Let F_1 and F_2 be FINs. A *pseudometric* d_K: $F \times F \rightarrow R_0^+$ is given by $d_K(F_1,F_2) = \int_0^1 d_h(F_1(h), F_2(h)) dh$, where $d_h(F_1(h),F_2(h))$ is a metric between generalized intervals $F_1(h)$ and $F_2(h)$.

The proof of Proposition 4-9 is given in Appendix B.

The reason that d_K is not a metric is that two FINs $F_1 \neq F_2$ could differ only on a set of measure zero; hence $d_K(F_1,F_2)=0$ does not necessarily imply that $F_1=F_2$. From a practical point of view the latter is unlikely to occur. In any case, from a pseudometric it is possible to obtain a true metric by a standard construction. Namely, we define an equivalence relation R on F as follows: F_1 and F_2 are equivalent if and only if they differ on a set of measure zero. Then d_K is a true metric on the quotient space F/R, i.e. d_K is a metric between the equivalence classes of R (Birkhoff 1967).

We remark that *negative* generalized intervals are used implicitly in the computation of d_K above. More specifically the computation of d_K is based on the metric $d_h(F_1(h),F_2(h))$ whose computation involves explicitly both positive and negative generalized intervals.

In the following, interest focuses mainly in the set F_+ of positive FINs with continuous membership functions including fuzzy numbers, intervals and real numbers. A conventional interval $[a,b]$, $a \leq b$ will be represented as $\bigcup_{h \in (0,1]} \{[a,b]^h\}$. Therefore, the set F_+ accommodates jointly *heterogeneous* data including (fuzzy) numbers and intervals (Hathaway et al. 1996; Shyi-Ming Chen 1997; Pedrycz et al. 1998; Shyi-Ming Chen et al. 2001; Bortolan and Pedrycz 2002; Paul and Kumar 2002).

Note that a metric, such as d_K, is typically useful in fuzzy- as well as in interval- regression analysis (Tanaka et al. 1982; Diamond 1988; Bandemer and Gottwald 1996; Diamond and Körner 1997; Yang and Ko 1997; Tanaka and Lee 1998). A similar metric, such as metric $d_K(.,.)$, has been presented between fuzzy sets elsewhere (Diamond and Kloeden 1994; Chatzis and Pitas 1995, 1999; Heilpern 1997) without reference to lattice theory as detailed in Kaburlasos (2004a). Other distance measures between fuzzy sets include Hamming, Euclidean, and Minkowski metrics (Klir and Folger 1988). However, all aforementioned distance functions are restricted to special cases, such as between fuzzy sets with triangular membership functions, between whole classes of fuzzy sets, etc. On the other hand, the metric d_K introduced above can compute a unique metric for any pair of fuzzy sets with arbitrary-shaped membership functions.

Various measure-theoretic considerations of fuzzy sets have been studied (Bollmann-Sdorra et al. 1993). Next we compare 'practically' metric d_K with another metric between convex fuzzy sets computed as follows

$$d_p(u,v) = (\int_0^1 d_H([u]^a,[v]^a)da)^{1/p}$$

whose calculation is based on the Hausdorf metric d_H between the a-cuts $[u]^a$ and $[v]^a$ of two fuzzy sets u and v, respectively (Ralescu and Ralescu 1984; Zwick et al. 1987; Diamond and Kloeden 1994; Heilpern 1997). Note that the Hausdorf metric $d_H(.,.)$ is a generalization of the distance between two points to two compact nonempty subsets (Zwick et al. 1987). For practical and theoretical purposes the membership functions of the fuzzy sets involved in the computation of $d_p(u,v)$ have been restricted to be *upper semicontinuous* (Diamond and Kloeden 1994). Advantages and disadvantages of the metric distances d_p and d_K are demonstrated comparatively in the following.

Metrics d_H and d_h produce quite different results in the space $\mathbb{R}^1$. More specifically, $d_H([a,b],[c,d])= \max\{|a-c|, |b-d|\}$, whereas using mass function $m(t)=1$ it follows that $d_h([a,b],[c,d])= |a-c|+|b-d|$. It turns out that the employment of d_h produces 'intuitive results', whereas the Hausdorf metric d_H may produce 'counter-intuitive' results as demonstrated next. For instance $d_H([1,2],[3,9])= \max\{2,7\}= 7$, moreover $d_H([1,2],[8,9])= \max\{7,7\}= 7$; whereas, $d_h([1,2],[3,9])= 2+7= 9$, moreover $d_h([1,2],[8,9])= 7+7= 14$. In words, metric d_h concludes that intervals [1,2] and [3,9] are closer to each other than intervals [1,2] and [8,9], the latter result is considered 'intuitive'; whereas, the Hausdorf metric d_H concludes that intervals [1,2] and [3,9] are as far from each other as intervals [1,2] and [8,9], the latter result is considered 'counter-intuitive'. Hence, a practical advantage of metric d_K is that it can capture sensibly the proximity of two FINs as demonstrated in the following example.

Example 4-10   Fig.4-5 illustrates the computation of metric $d_K(A,B)$ between FINs A and B (Fig.4-5(a)), where generalized intervals $A(h)$ and $B(h)$ are also shown for $h= 0.8$. To a height $h\in(0,1]$ there corresponds a value of metric $d_h(A(h),B(h))$ between generalized intervals $A(h)$ and $B(h)$ as shown in Fig.4-5(b). The area underneath the curve in Fig.4-5(b) equals the (metric) distance between FINs A and B. In particular, it was calculated $d_K(A,B)= 541.3$. The calculation of $d_K(.,.)$ based on generalized intervals also implies an enhanced capacity for *tuning* as shown below.

A metric between two N-tuples of FINs $A = (A_1,...,A_N)$ and $B = (B_1,...,B_N)$ can be computed using the following *Minkowski* metrics

$$d_p(A,B) = [d_K(A_1,B_1)^p + ... + d_K(A_N,B_N)^p]^{1/p}, \ p = 1,2,3,...$$

Note that various Minkowski metrics have been proposed regarding fuzzy sets (Bobrowski and Bezdek 1991; Hathaway et al. 2000).

Proposition 4-11 Let F_1 and F_2 be positive FINs. An *inclusion measure* function $\sigma_K: \mathbf{F}_+ \times \mathbf{F}_+ \to [0,1]$ is defined by $\sigma_K(F_1,F_2) = \int_0^1 k(F_1(h),F_2(h))dh$, where $k(F_1(h),F_2(h))$ is an inclusion measure function between positive generalized intervals $F_1(h)$ and $F_2(h)$.

The proof of Proposition 4-11 is shown in Appendix B.

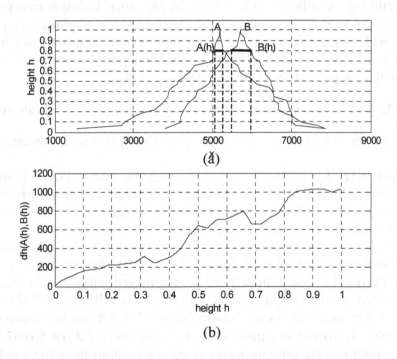

(å)

(b)

Fig.4-5   Computation of the metric distance $d_K(A,B)$ between FINs A and B.
(a) FINs A and B. Generalized intervals $A(h)$ and $B(h)$ are also shown for $h = 0.8$.
(b) The metric $d_h(A(h),B(h))$ between generalized intervals $A(h)$ and $B(h)$ is shown as a function of $h \in (0,1]$. The metric $d_K(A,B) = 541.3$ equals the area underneath the curve $d_h(A(h),B(h))$.

We remark that $k(F_1(h),F_2(h))$ in Proposition 4-11 can be calculated as shown in Theorem 3-25 based on a positive valuation function between positive generalized intervals shown in Proposition 4-3.

Since $\sigma_K(E,F)$ quantifies a degree of inclusion of FIN E in FIN F symbol $\sigma_K(E\leq F)$ might be used instead. An alternative tool for quantifying a degree of membership of FIN F in FIN E is using the fuzzy membership function $m_E(F)= 1/[1+d_K(E,F)]$ inspired from Krishnapuram and Keller (1993), where it is attributed to Zimmermann and Zysno. Note that using either function $\sigma_K(F,E)$ or function $m_E(F)$ it becomes feasible to compute a fuzzy degree of inclusion of a FIN F into another FIN E even when the interval supports of E and F do not overlap.

Equipped with inclusion measure $\sigma_K(.,.)$ the lattice F_+ of positive FINs becomes a *fuzzy lattice* $(\mathsf{F}_+,\leq,\sigma_K)$. Recall that metric d_K is also available in the lattice F_+. Both the inclusion measure σ_K and the metric d_K can be used for quantifying an 'affinity' of two FINs. In particular, inclusion measure $\sigma_K(F_1\leq F_2)$ can be used for quantifying a *fuzzy lattice degree of inclusion* of FIN F_1 in FIN F_2, whereas metric $d_K(F_1,F_2)$ can be used for quantifying the *proximity* of FINs F_1 and F_2. However, the employment of σ_K could be misleading as demonstrated in the following example.

Example 4-12   Fig.4-6 shows two pairs of FINs, namely 'E, F_1' (above) and 'E, F_2' (below). Using $\sigma_K(F_1\leq F_2)= \int_0^1 k(F_1(h), F_2(h))dh$ it was calculated both $\sigma_K(E\leq F_1)=0.1256$ and $\sigma_K(E\leq F_2)=0.3768$, that is FIN E is included more in FIN F_2 than it is in FIN F_1, as it might be expected intuitively by inspecting Fig.4-6. Nevertheless it is $\sigma_K(F_1\leq E) = 0.3140 = \sigma_K(F_2\leq E)$, that is FIN F_1 is included in FIN E as much as FIN F_2 is included in FIN E. The latter 'counter-intuitive' equality is due to the fact that the computation of $k(F_i(h),E(h))$, i=1,2 is based on the computation of $v(F_i(h)\vee E(h))$ i=1,2. Since the left-end of generalized interval $F_i(h)$ i=1,2 has no effect in the computation of $F_i(h)\vee E(h)$, it follows $\sigma_K(F_1\leq E) = \sigma_K(F_2\leq E)$. The aforementioned 'counter-intuitive' result can be amended using metric d_K instead. In particular it was computed $d_K(E,F_2) \approx 5.1667 \leq 5.8334 \approx d_K(E,F_1)$; the latter inequality is expected intuitively in Fig.4-6. It is interesting that fuzzy membership function $m_E(F_i)= 1/[1+d_K(E,F_i)]$, i=1,2 produces, as expected, the following 'intuitive inequality' $m_E(F_1) = 0.1463 < 0.1622 = m_E(F_2)$.

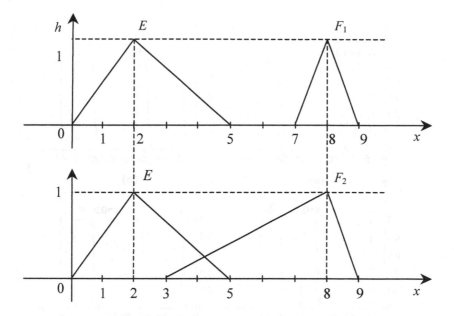

Fig.4-6   In the above figure it is $\sigma_K(F_1 \leq E) = 0.3140 = \sigma_K(F_2 \leq E)$; nevertheless it is $d_K(E, F_2) \approx 5.1667 \leq 5.8334 \approx d_K(E, F_1)$. Hence, metric d_K is more dependable than fuzzy inclusion measure σ_K for quantifying the 'relative position' between FINs.

In the following we extend addition and multiplication from M^h to **F**.

Definition 4-13   For $k \in R$, $F_1, F_2 \in F$, and $h \in (0,1]$
- *addition* $F_1 + F_2$ is defined as F_s: $F_s(h) = (F_1 + F_2)(h) = F_1(h) + F_2(h)$
- *multiplication* kF_1 is defined as *FIN* F_p: $F_p(h) = kF_1(h)$

It turns out that both F_+ and F_- are cones. Hence, if both FINs F_1 and F_2 are in cone F_+ (F_-) then $F_1 + F_2$ is in F_+ (F_-). Nevertheless, if $F_1 \in F_+$ and $F_2 \in F_-$ then $F_1 + F_2$ might not be a FIN.

Of particular interest are *convex* combinations $kF_1 + (1-k)F_2$, $k \in [0,1]$, where both F_1 and F_2 are positive FINs (in F_+); in the latter case it follows that $[kF_1 + (1-k)F_2] \in F_+$. Fig.4-7 shows two positive FINs A and B, respectively; moreover, the linear convex combination $kA + (1-k)B$ is shown for selected values of k, i.e. $k=0.8$, $k=0.6$, $k=0.4$ and $k=0.2$. Fig.4-7 demonstrates that the positive FIN '$kA + (1-k)B$' is progressively a combination of both the location and the shape of positive FINs A and B.

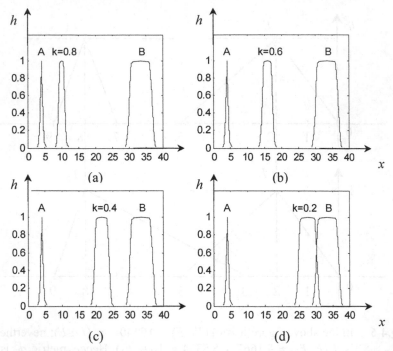

Fig.4-7   The convex combination $kA+(1-k)B$ for various values of k including (a) $k= 0.8$, (b) $k= 0.6$, (c) $k= 0.4$, (d) $k= 0.2$. Notice that the sum FIN '$kA+(1-k)B$' is progressively a combination of both the location and the shape of FINs A and B.

4.5.3 CALFIN: An algorithm for calculating a FIN

This section presents algorithm CALFIN in Fig.4-8 for computing a FIN from a population of samples (Kaburlasos 2002; Petridis and Kaburlasos 2003; Kaburlasos 2004a; Kaburlasos and Papadakis 2006).

In summary, algorithm CALFIN constructs a positive FIN according to the following procedure: Calculate the *median* number *median(x)* of N numbers in vector $x=[x_1,x_2,...,x_N]$. Insert number *median(x)* in a vector *pts*. Split vector x in two 'half vectors'; the latter include vector *x_left*, which contains the entries in vector x less-than *median(x)*, and vector *x_right*, which contains the entries in vector x larger-than *median(x)*. The afore-mentioned procedure is repeated recursively log_2N times until 'half vectors' are computed including a single number x_i, $i=1,...,N$; the latter are *median* numbers by definition.

Algorithm CALFIN

1. Let x be a vector of samples (real numbers), and let $dim(x)$ denote the dimension of x.

2. Order incrementally the numbers in vector x.

3. Initially vector pts is empty.

4. function $calfin(x)$

5. { while $(dim(x) \neq 1)$

6. $med:= median(x)$ [1]

7. insert med in vector pts

8. $x\_left:=$ elements in vector x less-than number $median(x)$

9. $x\_right:=$ elements in vector x larger-than number $median(x)$

10. $calfin(x\_left)$

11. $calfin(x\_right)$

12. endwhile

13. } //function $calfin(x)$

14. Sort vector pts incrementally.

15. Store in vector val, $dim(pts)/2$ numbers from 0 up to 1 in steps of $2/dim(pts)$ followed by another $dim(pts)/2$ numbers from 1 down to 0 in steps of $2/dim(pts)$.

Fig.4-8   Algorithm CALFIN computes a Fuzzy Interval Number (FIN) from a population of samples stored in vector x. Hence, a FIN represents a population of samples.

---

[1] The median $median(x)$ of a vector $x = [x_1, x_2,...,x_n]$ of numbers is a number such that half of vector x entries $x_1, x_2,...,x_n$ are less than $median(x)$ and the other half are larger than $median(x)$. For example, $median([1, 3, 7])= 3$; whereas, the $median([-1, 2, 6, 9])$ was computed as $median([-1, 2, 6, 9])= (2+6)/2= 4$.

Algorithm CALFIN computes two vectors, i.e. *pts* and *val*, where vector *val* includes the degrees of (fuzzy) membership of the corresponding real numbers in vector *pts*. Algorithm CALFIN produces a positive FIN with membership function $\mu(x)$ such that $\mu(x)=1$ for exactly one number x.

When CALFIN is applied on a population $x_1, x_2, \ldots, x_n$ of samples drawn independently according to a probabilistic distribution function $p_0(x)$ and a FIN F is constructed then interval $\{x: F(h) \neq 0\}$ constitutes, by definition, an *interval of confidence at level-h* in the following sense: A random number drawn according to $p_0(x)$ is expected to fall (1) *inside* interval $\{x: F(h) \neq 0\}$ with probability $100(1-h)\%$, (2) *below* interval $\{x: F(h) \neq 0\}$ with probability $50h\%$, and (3) *above* interval $\{x: F(h) \neq 0\}$ with probability $50h\%$.

A number of density estimation techniques have been presented in the literature (Parzen 1962; Kim and Mendel 1995; Magdon-Ismail and Atiya 2002). A potentially useful alternative is presented in the following based on a one-one correspondence between *cumulative distribution functions* (CDFs) and FINs.

In the one direction, let x_0 be such that $G(x_0)=0.5$, where $G(x)$ is a CDF. Define the membership function $\mu_F(.)$ of a FIN F such that $\mu_F(x)=2G(x)$ for $x \leq x_0$ furthermore $\mu_F(x)=2[1-G(x)]$ for $x \geq x_0$. In the other direction, let x_0 be such that $\mu_F(x_0)=1$, where $\mu_F(x)$ is the membership function of a FIN F constructed by algorithm CALFIN. Define a CDF $G(x)$ such that $G(x)=0.5\mu_F(x)$ for $x \leq x_0$ furthermore $G(x)=1-0.5\mu_F(x)$ for $x \geq x_0$. Due to the aforementioned 'one-to-one correspondence' between FINs and CDFs it follows that a FIN captures statistics of all orders. The previous analysis also implies that metric $d_K(.,.)$ can be employed for calculating a distance between either two *probability density functions* (*pdf*s) or two CDFs.

In a different context, using a fuzzy set theoretic terminology, a positive FIN F may be interpreted as a nonparametric *possibility distribution* (Zadeh 1978), where the corresponding fuzzy membership function $\mu_F(.)$ is interpreted as a (linguistic) constraint (Zadeh 1999). Note that a statistical interpretation for a FIN does not exclude a linguistic (fuzzy) interpretation. For instance, Dubois and Prade (1986) have presented transformations between probability and possibility measures; moreover Dubois and Prade (1986) present algorithms, similar with algorithm CALFIN, for inducing fuzzy membership functions from statistical interval data.

The application of algorithm CALFIN is demonstrated in Fig.4-9 on a population of samples. In particular, a FIN is computed in Fig.4-9(b2) from a population of 50 samples. The identical figures Fig.4-9(a1) and Fig.4-9(a2) show the corresponding 63 *median* values computed in vector *pts* by algorithm CALFIN. Fig.4-9(b1) shows in a histogram the distribution of the 63 median values in intervals of 400.

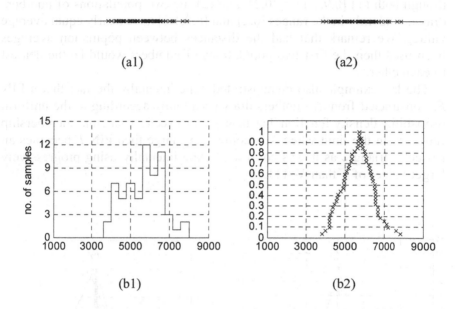

(a1)   (a2)

(b1)   (b2)

Fig.4-9   Calculation of a FIN from a population of samples/measurements.
(a1), (a2) 63 median values computed by algorithm CALFIN.
(b1) A histogram of the 63 median values in steps of 400.
(b2) The 63 median values of vector *pts* have been mapped to the corresponding entries of vector *val* computed by algorithm CALFIN.

Fig.4-10 displays more FINs constructed by algorithm CALFIN. In particular, Fig.4-10(a) displays FIN G constructed from 222 random numbers generated according to the normal pdf N(0,1) with mean 0 and standard deviation 1. Note that the maximum value (1) of FIN G is attained close to the mean 0 of probability density function N(0,1).

Three populations of random numbers were generated randomly according to the uniform pdf in the ranges [0,2], [0.9,1.1], and [0.2, 2.2], respectively. Each population included 111 random numbers. The averages of the three populations of random numbers were, respectively, 1.0467, 1.0030 and 1.1625. Fig.4-10(b) shows the corresponding FINs constructed using algorithm CALFIN. The following distances were calculated between pairs of FINs: $d_K(F_1,F_2)= 0.2843$, $d_K(F_2,F_3)= 0.3480$ and $d_K(F_1,F_3)= 0.1002$. The aforementioned distances d_K confirm what might have been expected by observing Fig.4-10(b), that is the proximity of the two populations of random numbers in the ranges [0,2] and [0.2, 2.2] is larger than the proximity of the two populations of numbers in the ranges [0,2] and [0.9,1.1], even though both (1) [0.9,1.1] $\subseteq$ [0,2], and (2) the two populations of numbers drawn randomly in the ranges [0,2] and [0.9,1.1] have nearly equal average values. We remark that had the distances between population averages been used then the first two populations of numbers would be the nearest to each other.

The last example also demonstrated experimentally the fact that a FIN F, constructed from N numbers drawn randomly according to the uniform probability density function, has nearly an isosceles triangular membership function. In the limit when N becomes very large then FIN F becomes an isosceles triangle as it was confirmed experimentally using progressively larger values of N than N=111.

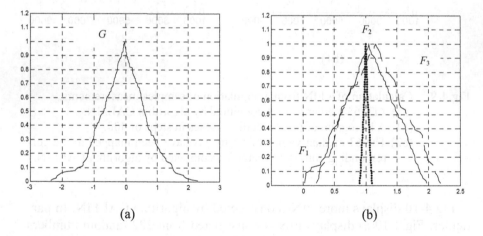

(a)   (b)

Fig.4-10 (a) FIN G in this figure was constructed by applying algorithm CALFIN to 222 random numbers generated according to the normal (Gaussian) probability density function $N(0,1)$ with mean 0 and standard deviation 1.

(b) The three FINs F_1 (solid line), F_2 (dotted line), and F_3 (dashed line) were computed from three populations of samples generated randomly according to the uniform probability distribution with ranges [0, 2], [0.9, 1.1] and [0.2, 2.2], respectively. It was computed $d_K(F_1,F_2)$= 0.2843, $d_K(F_2,F_3)$= 0.3480 and $d_K(F_1,F_3)$= 0.1002.

4.6 Integrable Real Functions

Consider the set $\mathcal{F}$ of non-negative real functions $f\colon \mathrm{R}\to\mathrm{R}_0^+$. The set $\mathcal{F}$ under relation $f \leq g \Leftrightarrow f(x) \leq g(x)$, $x\in\mathrm{R}$ is partially ordered. It turns out that $(\mathcal{F},\leq)$ is a lattice such that for $f,g\in\mathcal{F}$ it is $(f\wedge g)(x)=\min(f(x),g(x))$ and $(f\vee g)(x)=\max(f(x),g(x))$, $x\in\mathrm{R}$. Our interest is, in particular, in the subset $\mathcal{G}$ of integrable functions in $\mathcal{F}$. It turns out that $(\mathcal{G},\leq)$ is a sublattice of $(\mathcal{F},\leq)$.

Real function $v\colon \mathcal{G}\to\mathrm{R}$ given by $v(f)=\int_{-\infty}^{\infty} f(x)dx$ is a *monotone valuation* in lattice $(\mathcal{G},\leq)$. Therefore, function $d(f,g) = v(f\vee g)-v(f\wedge g)$ defines a *pseudometric* in lattice $(\mathcal{G},\leq)$. It follows that condition $d(f,g)=0$ implies a *congruence relation*, which maps $\mathcal{G}$ onto the metric lattice $(\mathcal{G}^*,\leq)$ of the corresponding equivalence classes of $\mathcal{G}$ – For definition of a *congruence relation* see in Appendix A.

In conclusion, $(\mathcal{G}^*,\leq)$ is a metric lattice with metric given by $d(f,g)=v(f\vee g)-v(f\wedge g)$. Moreover, based on *positive valuation* function $v(f)=\int_{-\infty}^{\infty} f(x)dx$, inclusion measures $k(f,g)= v(u)/v(x\vee u)$ and $s(f,g)= v(x\wedge u)/v(x)$ are implied in lattice $(\mathcal{G}^*,\leq)$.

4.7 Linear Metric Spaces

Let (L,d) denote a linear metric space, where L is a (real) linear space and $d\colon L{\times}L\to\mathrm{R}_0^+$ is a metric. A *ball* centered on a vector $c\in L$ with a radius $r{\geq}0$ is defined as set $ball(c,r)= \{x\in L\colon d(x,c)\leq r\}$, where $c\in L$, $r{\geq}0$. The collection B of balls in (L,d) has been studied in various computational contexts. For instance, the VC dimension of the set B of balls has been calculated (Wenocurn and Dudley 1981). This subsection summarizes some results regarding an ordering in B (Kaburlasos et al. 1999).

We write '$ball(k,a) \leq ball(c,r)$' if and only if $ball(k,a) \subseteq ball(c,r)$. It can be shown that the set B of balls is partially ordered. However, poset $(B,\leq)$

is not a lattice because a pair $(ball(c,r), ball(k,a))$ may not have neither a least upper bound nor a greatest lower bound. Nevertheless, it is possible to introduce both a metric and an inclusion measure in poset $(B,\leq)$ by mapping B to a *hidden lattice* $(\Phi,\leq)$ whose underlying lattice L is equipped with a positive valuation function as explained in the following.

Consider the set S of spheres in a linear metric space (L,d), where a sphere with *center* $c\in L$ and *radius* $r\geq 0$ is denoted by $sphere(c,r)$. There exists a bijection (that is an one-to-one map and onto) b_B: $B\leftrightarrow S$ between the set B of balls and the set S of spheres. Hence, a partial ordering relation is implied in S, symbolically $sphere(k,a)\leq sphere(c,r)$, if and only if $ball(k,a)\leq ball(c,r)$. The set Φ of *generalized spheres* is introduced as follows.

Definition 4-14   The set Φ of *generalized spheres* in a linear metric space (L,d) is defined by $\Phi=\{(c,r)\colon c\in L, r\in R\}$.

To a $sphere(c,r)$ there corresponds $ball(c,|r|)$. A partial ordering relation is defined in Φ as follows.

H1.   If $r_1, r_2 \geq 0$ then:
    $sphere(c_1,r_1) \leq sphere(c_2,r_2) \Leftrightarrow ball(c_1,r_1) \leq ball(c_2,r_2)$
H2.   If $r_1, r_2 < 0$ then:
    $sphere(c_1,r_1) \leq sphere(c_2,r_2) \Leftrightarrow ball(c_2,|r_2|) \leq ball(c_1,|r_1|)$
H3.   If $r_1 < 0, r_2 \geq 0$ then:
    $sphere(c_1,r_1) \leq sphere(c_2,r_2) \Leftrightarrow ball(c_1,|r_1|)\cap ball(c_2,r_2)\neq\varnothing$
In all other cases generalized spheres $sphere(c_1,r_1)$ and $sphere(c_2,r_2)$ are *incomparable*, symbolically $sphere(c_1,r_1)\|sphere(c_2,r_2)$.

The partial ordering of *generalized spheres* is illustrated geometrically on the plane in Fig.4-11 in analogy with *generalized intervals* in Fig.4-2.

It turns out that the partially ordered set Φ of generalized spheres is a *hidden lattice* with *underlying lattice* the set R of real numbers and *corresponding map* given by the radius of a generalized sphere. Hence, any strictly increasing function of the radius, e.g. function $v(x)=x$, constitutes a positive valuation in R.

The aforementioned assumptions imply an injective monotone map φ: $B\rightarrow\Phi$ given by $\varphi(ball(c,r))= sphere(c,r)$. In particular, φ implies $ball(k,a)\leq ball(c,r) \Rightarrow \varphi(ball(k,a))\leq\varphi(ball(c,r)) \Rightarrow sphere(k,a)\leq sphere(c,r)$. Therefore both a metric and an inclusion measure in Φ can be used in B. Note that the computation of a metric between two cycles as shown in Fig.4-11(a) and (b) might be of particular interest in various practical applications including RBF networks (Bianchini et al. 1995).

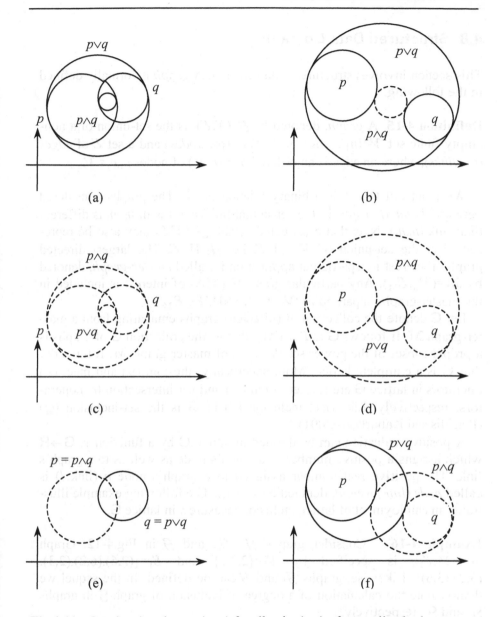

Fig.4-11 *Join* (*p*∨*q*) and *meet* (*p*∧*q*) for all pairs (*p*,*q*) of generalized spheres.
(a) 'Intersecting' positive generalized spheres.
(b) 'Non-intersecting' positive generalized spheres.
(c) 'Intersecting' negative generalized spheres.
(d) 'Non-intersecting' negative generalized spheres.
(e) 'Intersecting' positive and negative generalized spheres.
(f) 'Non-intersecting' positive and negative generalized spheres.
In all other cases *p* and *q* are *incomparable*, symbolically *p*‖*q*.

4.8 Structured Data Domains

This section involves structured data (graphs). A *graph* is formally defined in the following.

Definition 4-15 A *graph*, denoted by $\mathcal{G}=(V,\mathcal{E})$, is the set-union of a nonempty finite set $V=\{n_1,\ldots,n_N\}$ of *vertices* (or, *nodes*) and a set $\mathcal{E}$ of *edges* (or, *links*), where an edge is an ordered pair (n_i,n_j) of nodes $n_i,n_j \in V$.

We point out that $\mathcal{E}$ is a binary relation on V. The graphs considered here are *directed graphs* in the sense that link (n_j,n_i) with $n_j \neq n_i$ is different than link (n_i,n_j). Note that a (directed) graph $\mathcal{G}=(V,\mathcal{E})$ may also be represented as the set-union of V and $\mathcal{E}$, i.e. $\mathcal{G}=V \cup \mathcal{E}$. The largest directed graph of interest in a particular application is called *master-graph*, denoted by $\mathcal{M}=(V_\mathcal{M},\mathcal{E}_\mathcal{M})$. Any particular graph $\mathcal{G}=(V,\mathcal{E})$ of interest is included in the master-graph, in particular '$V \subseteq V_\mathcal{M}$'.AND.'$\mathcal{E} \subseteq \mathcal{E}_\mathcal{M}$'.

Let G denote the collection of (directed) graphs emanating from a master-graph $\mathcal{M}$. It follows $G \subset Pow(\mathcal{M})$; that is, the collection G of graphs is a proper subset of the power set $Pow(\mathcal{M})$ of master-graph $\mathcal{M}$. It turns out that G is a complete lattice. More specifically, the join ($\vee$) and meet ($\wedge$) operators in lattice G are the set-union ($\cup$) and set-intersection ($\cap$) operators, respectively. Likewise, inclusion ($\leq$) in G is the set-inclusion ($\subseteq$) (Petridis and Kaburlasos 2001).

A positive valuation can be defined in lattice G by a function $v: G \to R$ which assigns a positive number to a graph's node as well as to a graph's link. The positive real number assigned to a graph's node n (link l) is called *node* (*link*) *weight*, denoted by w_n (w_l). The following example illustrates an employment of lattice inclusion measure s in lattice G.

Example 4-16   Consider graphs $\mathcal{G}_1$, $\mathcal{G}_2$, and $\mathcal{G}$ in Fig.4-12. Graph $\mathcal{G}_1=(V_1,\mathcal{E}_1)$, is specified by $V_1=\{2,3,6\}$ and $\mathcal{E}_1=\{(2,6),(6,2),(2,3),(3,2),(3,6)\}$. Likewise, graphs $\mathcal{G}_2$ and $\mathcal{G}$ can be defined. In the sequel we demonstrate the calculation of a degree of inclusion of graph $\mathcal{G}$ in graphs $\mathcal{G}_1$ and $\mathcal{G}_2$, respectively.

$$s(\mathcal{G} \leq \mathcal{G}_1) = \frac{v(\mathcal{G} \wedge \mathcal{G}_1)}{v(\mathcal{G})} = \frac{2w_n + 2w_l}{5w_n + 7w_l}, \text{ and}$$

$$s(\mathcal{G} \leq \mathcal{G}_2) = \frac{v(\mathcal{G} \wedge \mathcal{G}_2)}{v(\mathcal{G})} = \frac{2w_n + w_l}{5w_n + 7w_l}.$$

Furthermore, $k(\mathcal{G} \leq \mathcal{G}_1) = \dfrac{v(\mathcal{G}_1)}{v(\mathcal{G} \vee \mathcal{G}_1)} = \dfrac{3w_n + 5w_l}{6w_n + 10w_l}$, and

$$k(\mathcal{G} \leq \mathcal{G}_2) = \frac{v(\mathcal{G}_2)}{v(\mathcal{G} \vee \mathcal{G}_2)} = \frac{4w_n + 4w_l}{7w_n + 10w_l}.$$

Using $w_n = 1.0$, $w_l = 0.1$ it follows $s(\mathcal{G} \leq \mathcal{G}_1) = 0.3860 > 0.3684 = s(\mathcal{G} \leq \mathcal{G}_2)$; whereas $k(\mathcal{G} \leq \mathcal{G}_1) = 0.5 < 0.55 = k(\mathcal{G} \leq \mathcal{G}_2)$.

Hence, given a positive valuation function, graph $\mathcal{G}$ might be included more in graph $\mathcal{G}_1$ (or less) than in graph $\mathcal{G}_2$ depending upon the inclusion measure (s or k) used.

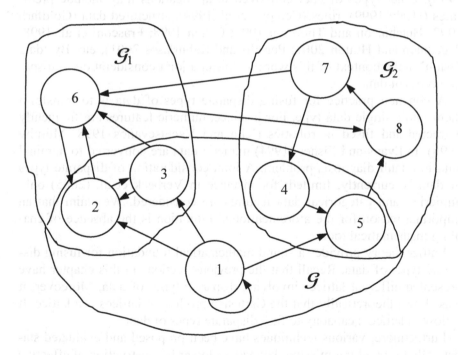

Fig.4-12 Graph $\mathcal{G}$ is included, in a lattice inclusion measure $s(.)$ sense, more in graph $\mathcal{G}_1$ than it is in graph $\mathcal{G}_2$.

4.9 Disparate Data Fusion

Data fusion in the literature typically considers N-dimensional vectors of real numbers stemming from an assortment of sensors /modalities in agricultural, manufacturing, medical, military, transportation, and other applications (Hongyi Li et al. 1995; Jones et al. 1995; Hall and Llinas 1997; Iyengar and Brooks 1997; Varshney 1997; Ben-Yacoub et al. 1999; Dawant and Garbay 1999; Solaiman et al. 1999a, 1999b; Wan and Fraser 1999; de Jesus et al. 2000; Chen and Nakano 2003; Culler et al. 2004). Multi-modal information processing may also be used for person identification (Hong and Jain 1998; Chatzis et al. 1999), information retrieval (Srihari et al. 2000), human interactive robots (Lund 2004).

In general, models use a single data type, i.e. vectors of real numbers (Vidyasagar 1978, 1997; Pindyck and Rubinfeld 1991; Mitchell 1997; Citterio et al. 1999; Vapnik 1999); intervals may also be used (Horiuchi 1998). Other types of data employed in applications may include predicates (Healy 1999), rules (Cercone et al. 1999), structured data (Goldfarb 1992; Hutchinson and Thornton 1996; Cohen 1998; Frasconi et al. 1998; Paccanaro and Hinton 2001; Petridis and Kaburlasos 2001), etc. By 'data fusion' in the context of this work we mean a joint consideration of disparate types of data.

A common practice for fusing disparate types of data is to transform them into a single data type. For instance, numeric features are frequently extracted and fused in robotics (Kim and Vachtsevanos 1998; Murphy 1998). In Evans and Fisher (1994) numeric data are converted to nominal data in a fault diagnosis problem. A joint consideration of disparate types of data is, currently, limited; for instance in Verbeek et al. (2005) only numerical and categorical data features are considered. We claim that an important reason for the aforementioned limitation is the absence of enabling mathematical tools.

Lattice theory provides a sound mathematical foundation for fusing disparate types of data. Recall that the previous sections in this chapter have presented different lattices involving disparate types of data. Moreover, it was shown theoretically that the Cartesian product of lattices is a lattice. It follows a lattice's capacity to fuse disparate types of data.

Furthermore, various techniques have been proposed and evaluated statistically for handling missing data values towards construction of effective prediction models in R^N (Myrtveit et al. 2001). This work has proposed techniques for dealing with both 'missing' and 'don't care' data values in a general (complete) lattice including the unit hypercube in R^N.

5 Knowledge Representation

This chapter outlines knowledge representation practices

Consider the following two definitions proposed by the International Federation for Information Processing in conjunction with the International Computation Centre in 1966 (Holmes 2001).

• **Data**. A representation of facts or ideas in a formalized manner capable of being communicated or manipulated by some process.

• **Information**. In *automatic data processing* the meaning that a human assigns to *data* by means of the known conventions used in its representation.

We remark that people process information, whereas machines process data. This chapter illustrates how lattices may enhance machine data processing towards human information processing. The term 'semantics' is used below as a synonym to 'information' above.

5.1 Data, Semantics, and Knowledge

Reading is difficult for machines because there are complex phenomena in language that are hard to represent. The latter is one example where data representation may be instrumental in applications. For a simpler example consider classification of two tones (sounds). It is understood that the aforementioned tones may be separated easily when represented in the frequency domain. Various data representations are reviewed next.

The most popular data representation uses N-dimensional vectors. Hence, usually, modeling is carried out in R^N based on 'algebraic semantics'; the latter quoted term means sound mathematical techniques (in R^N). The popular practice is to force a (real) vector observation out in a problem in order to enable modeling in R^N. For example, consider the problem of document classification. A popular representation for a document is a 'bag of words', i.e. a N-dimensional vector of word features (Lecun et al. 1998; Drucker et al. 1999; Massey 2003; Anagnostopoulos et al. 2004). In con-

Vassilis G. Kaburlasos: *Knowledge Representation*, Studies in Computational Intelligence (SCI) **27**, 63–66 (2006)
www.springerlink.com

clusion, statistical- and other models are used in R^N (Sahami 1998; Manning and Schütze 1999; Yang 1999; Sebastiani 2002). Various enhancements have been proposed for document classification such as relating documents based on content rather than merely on statistics (Chang and Hsu 1999). Sometimes a priori knowledge is employed including the Wordnet semantic concordance (Green 1999). Another enhancement is to consider vectors of senses or, equivalently, vectors of meanings of words (Kehagias et al. 2003). However, all aforementioned representations are restricted in R^N and, at large, ignore important aspects of text such as structure, word order, syntax, etc.

A straightforward extension considers (fuzzy) sets in R^N; in particular, (fuzzy) numbers and intervals have been employed even with standard languages for knowledge representation including Prolog (Guadarrama et al. 2004). Fuzzy sets may represent linguistic data and accommodate ambiguity in applications towards computing with words (Zadeh 1996). Moreover fuzzy sets have been proposed for computing with perceptions (Zadeh 2004), and a fuzzy membership function is interpreted as a constraint in a constraint satisfaction problem (Zadeh 1999). The aforementioned data representations are based on real numbers one way or another. Selected non-numeric data representations are reviewed in the following.

In bioinformatics, frequently appear models for processing non-numeric data including strings of characters (Sabatti and Lange 2002), graphs (Lubeck et al. 2002), Boolean probabilistic data (Shmulevich et al. 2002), etc. Furthermore, an integration of multiple types of data has been considered in bioinformatics (Moreau et al. 2002). Distributed data representations, holographic reduced, have been considered for cognitive computing (Plate 2003). Helixes have been proposed as a data structure for representing spatiotemporal events (Agouris and Stefanidis 2003). In artificial intelligence a symbolic measurement process was proposed both in theory and in practice (Goldfarb and Deshpande 1997; Goldfarb et al. 2000) resulting in strings of characters. Guglielmo and Rowe (1996) retrieve images represented by descriptive captions, whereas in Jain and Bin Yu (1998) an image is represented by a (block adjacency) graph.

Graphs are popular for structured data representation. For instance, in formal concept analysis (Ganter and Wille 1999) graph-based ontologies may represent structure in a text for learning lattice-ordered concept hierarchies towards improved classification (Bloehdorn et al. 2005). Of particular importance is the employment of (structured) ontologies for Semantic Web applications (Berners-Lee et al. 2001; Berners-Lee 2003). In neural computing, lately, interest has shifted to architectures that process structured data (Sperduti and Starita 1997; Frasconi et al. 1998; Hagenbuchner et al. 2003; Hammer et al. 2004).

Analytical and empirical proofs were presented regarding the equivalence between certain knowledge representations including deterministic/fuzzy finite state automata and recurrent neural networks (Giles et al. 1999). Sometimes data are represented hierarchically (Samet 1988; Pedrycz 1992). Of particular interest is also integration of disparate data (Castano and de Antonellis 2001). However, even though it can be useful, a rigorous treatment of disparate types of data is not currently established due also to the absence of a sound mathematical foundation. Instead, the common practice is to 'expert' transform disparate data to a single data domain, the latter is usually R^N, as was explained in section 2.2.

Data are assigned a meaning or, equivalently, an interpretation, or semantics, by humans. For instance, a positive FIN may be interpreted as a possibility distribution or as a probability distribution. Semantics are often defined differently in different contexts (Harel and Rumpe 2004) thus causing confusion. There are various practices to assign semantics to data. For instance, in robotics, based on work in psychology (Harnad 2002), the meaning of internal symbols may be induced from the data and firmly anchored to real-world representations (Chella et al. 2001). Furthermore, data structures such as ontologies may be assigned semantics by an expert.

When mapping non-numeric data to R^N it turns out that original problem semantics are abandoned. However, useful 'algebraic semantics' exist in R^N to carry out meaningfully a syntactic manipulation of formal symbols including *number crunching* (Wong et al. 1987; Berry et al. 1995, 1999; Lee et al. 1997). It is understood that a transformation to R^N is eligible in practice as long as it produces good results.

In practice many useful data representations are lattice ordered. Moreover, information processing in a lattice data domain is both possible and it may retain, flexibly, original problem semantics as explained below. Conditions under which computing in a general lattice might be preferable than computing (after data transformation) in R^N is a topic of future research. Another reason space R^N cannot retain common sense semantics is an employment of 'energy inspired' objective functions as explained next.

An objective function is a standard tool for optimizing learning. More specifically, Gaussian least-square-error (LSE) estimation provides the basis for a number of estimation theories and techniques (Sorenson 1985) including supervised learning algorithms in the context of computational intelligence (CI). In terms of linear signal processing, LSE minimizes error energy; whereas, for non-linear signal processing, a more appropriate approach might be to constrain the entropy (Erdogmus and Principe 2002). Note that the inspiration for an objective function typically derives from an energy function in physics (Bishop et al. 1998; Kaburlasos and Papadakis 2006). Another energy type function, namely Lyapunov function, is frequently used in studies of stability regarding non-linear systems. Nevertheless, an energy /entropy /Lyapunov function cannot accommodate common sense semantics; the latter can be retained in a lattice data domain as explained below. The notion 'knowledge' is defined next.

Knowledge has been an object of scientific philosophical speculation (Russell 2000). By 'knowledge' here we mean a set $(x_i, f(x_i))$, $i \in \{1,...,n\}$ of samples of a function $f: L \rightarrow K$, where both L and K are sets of data. Note that a sample $x_i \rightarrow f(x_i)$, $i \in \{1,...,n\}$ could be interpreted either as a rule or as a causal relationship (Pearl 2000). A major advantage of rule-based or expert systems is that they can display the rules they use to reach their conclusions. Unfortunately, rule-based systems can solve problems effectively only when they are provided with extensive, high quality, specific knowledge regarding a narrow problem domain. The performance of conventional expert systems is notorious for dropping off sharply when the problem deviates even slightly from the expected problem domain. A major contribution of this book is in presenting an induction of a function $f:$ $L \rightarrow K$ from samples $(x_i, f(x_i))$, $i \in \{1,...,n\}$ regarding lattices L and K such that function f retains a capacity for interpolation and/or extrapolation.

5.2 Tunable Lattice Semantics

It is understood that a syntactic manipulation of symbols alone is insufficient to establish semantics (Searle 1980). However, there are tools in a product lattice $(L, \leq) = (L_1, \leq) \times ... \times (L_N, \leq)$ data domain for introducing and/or tuning semantics including: (1) the lattice order relation, (2) structured data, and (3) positive valuation functions, as explained next.

(1) Different lattice orders can be defined. For instance, for $x = (x_1,...,x_N)$, $y = (y_1,...,y_N) \in L$, first, $x \leq_1 y \Leftrightarrow x_i \leq y_i$, where $x_i, y_i \in L_i$, $i = 1,...,N$ and, second, $x \leq_2 y \Leftrightarrow \exists k \in \{1,...,N\}: x_i = y_i$ for $i < k$, and $x_k \leq y_k$. The latter order $(\leq_2)$ is called *lexicographic*. More order relations can be defined in lattice $(L, \leq)$.

(2) User-defined, structured data relations may express semantics. For instance, Fig.9-1 shows a 'parent-of' relation (Kaburlasos 2003).

(3) A positive valuation (v) may tune semantics by tuning an inclusion measure function such as $k(x \leq u) = v(u)/v(x \vee u)$. Recall that the utility of positive valuation functions is to enable manipulating elements in a lattice of semantics by operating on real numbers. Note that inclusion measure $k(x \leq u)$ quantifies 'how much lattice element u is short of including another lattice element x'; in particular, inclusion measure $k(x \leq u)$ measures the diagonal of u in terms of the diagonal of $x \vee u$. Finally, product lattice $(L, \leq)$ implies an inherently hierarchical, modular, and disparate data domain.

6 The Modeling Problem and its Formulation

Function estimation is the issue here

A *function f*: L→K is defined as a rule which, to each element of a *domain* set L, assigns an element of a *range* set K. A function is a one-to-one relation. The words *function* and *model* are used interchangeably in this work.

6.1 Models in Soft Computing

The above definition for a function is widely employed in practice, where the existence of a function *f*: L→K is often assumed. A popular practical problem is to induce, empirically from training data, an optimal estimate $\hat{f}$ of *f*. The induction of $\hat{f}$ may be cast as a *search* problem, or as a *learning* problem (Vidyasagar 1997; Sutton and Barto 1999). In terms of mathematics the induction of optimal estimate $\hat{f}$ can be formulated as a *function approximation* problem.

Perhaps the best known function approximation problem involves a function *f*: $R^N \rightarrow R^M$ to be approximated optimally, in a least square error (LSE) sense, from data samples $(x_i, y_i) \in R^N \times R^M$, *i*=1,...,*n*. A number of parametric models have been proposed in various application domains including statistical regression models (Pindyck and Rubinfeld 1991), ARMA models (Box and Jenkins 1976), etc.

Cybenko (1989) has shown that a single hidden layer is sufficient to uniformly approximate any continuous function. Moreover, Hornik et al. (1989) showed that a feedforward multiplayer perceptron with two layers of neurons can approximate any piecewise continuous function provided sufficiently many neurons in the hidden layer. The capacities as well as the limits of neural networks for function approximation were also studied by other authors (Cotter 1990; Gori et al. 1998). In all cases a drawback remains, that is neural networks cannot explain their answers.

Vassilis G. Kaburlasos: *The Modeling Problem and its Formulation*, Studies in Computational Intelligence (SCI) **27**, 67–70 (2006)
www.springerlink.com

Fuzzy system models, using (implicitly) mass function $m(x)=1$, often employ parametric *possibility distributions* (i.e. fuzzy sets) as devices for introducing tunable non-linearities (Wang and Mendel 1992a; Castro and Delgado 1996; Dickerson and Kosko 1996; Zeng and Singh 1996; Nauck and Kruse 1999; Runkler and Bezdek 1999a; Pomares et al. 2000; Liu 2002). Moreover, fuzzy systems can explain their answers using rules.

Note that the specifications for a useful function estimate $\hat{f}$ include a capacity for *generalization* that is a capacity to perform well on hitherto unknown data. A necessary condition for generalization is that the training data do not *overfit*, i.e. they do not produce (near) zero training error.

With the proliferation of information technologies non-numeric data have proliferated in applications. For instance, Boolean data, images, symbols, structured data (graphs), frames, (fuzzy) sets, etc. have proliferated in practice. While some of the aforementioned data can be transformed to vector data in a straightforward manner others cannot. Moreover, there might be a need to employ various functions in the aforementioned data domains, e.g. functions for classification, regression, etc. In soft computing, various biologically inspired models such as neural networks (Hornik et al. 1989; Chen and Chen 1995; Frasconi et al. 1998; Scarselli and Tsoi 1998; Citterio et al. 1999; Chuang et al. 2002; González et al. 2002) as well as linguistically inspired models such as fuzzy inference systems (Wang and Mendel 1992a; Kosko 1994; Castro and Delgado 1996; Dickerson and Kosko 1996; Zeng and Singh 1996; Nauck and Kruse 1999; Runkler and Bezdek 1999a, 1999b; Pomares et al. 2000; Mitaim and Kosko 2001; Liu 2002) have been proposed for real function approximation involving non-numeric (fuzzy) data. However, there is a need to broaden the notion of a function in practice.

A simple model is a *classifier* that is a function c: $L \rightarrow K$ taking values in a finite set K of category labels. Often it is $L = R^N$, hence f: $R^N \rightarrow K$. The latter classifier ultimately partitions R^N into a number of mutually disjoint (Voronoi) subregions and a unique label is attached to a subregion. Note that a classifier's general domain L may also include non-numeric data.

A classifier can be the basis for learning other functions including real-valued ones, namely *regression models* (Mitchell 1997). Furthermore note that classification, regression as well as probability density estimation can be described as function estimation problems (Vapnik 1999).

Function approximation may involve uncertainty and/or ambiguity. In a Bayesian context uncertainty is often handled by a one-to-many relation that maps a single element (vector) to a conditional distribution defined in R^N. However, if a conditional distribution is interpreted as a single datum then the aforementioned one-to-many relation becomes a functional (one-to-one) relation.

A more general problem involves learning a many-to-many relation. An example of the latter is mapping one fuzzy set to another one. Likewise, when a fuzzy (sub)set in the universe of discourse is interpreted as a single element, that is an *information granule*, then the aforementioned many-to-many relation becomes a functional (one-to-one) relation.

Another interesting situation arises when a function's domain involves N-tuples of disparate data. Note that there is evidence in the literature that consideration of disparate types of input data can improve accuracy in classification applications (Hong and Jain 1998; Ben-Yacoub et al. 1999; Chatzis et al. 1999; de Jesus et al. 2000; Garg et al. 2003).

In the aforementioned function approximation problems 'heuristics' as well as 'domain-specific' techniques abound. A unifying mathematical framework, for inducing rigorously useful functions in soft computing based on disparate types of data, is currently missing. A proposal is presented next based on lattice theory.

6.2 Model Unification

Consider functions $f: \mathsf{L} \to \mathsf{K}$, where both $(\mathsf{L}, \leq)$ and $(\mathsf{K}, \leq)$ are lattices. A unification of useful models in soft computing is outlined as follows. First, if $\mathsf{L} = \mathsf{R}^N$ and $\mathsf{K} = \mathsf{R}^M$ then it follows a conventional system model. Second, if K is a set of labels then it follows a classifier model. Third, if both L and K equal the lattice of fuzzy numbers then it follows a Mamdani type fuzzy inference system (FIS). Fourth, if L equals the lattice of fuzzy numbers and K is a lattice of functions, e.g. polynomial linear functions, then it follows a Sugeno type FIS. Fifth, if L equals R^N and K is a lattice of conditional distribution functions then it follows a conventional regressor model. Other options also exist involving alternative lattices. In this context the general learning problem is described as follows.

Consider a training data set including samples $(x_i, y_i) \in \mathsf{L} \times \mathsf{K}$, $i = 1, \ldots, n$ of a function $f: \mathsf{L} \to \mathsf{K}$. Positive valuation functions are assumed in both lattices L and K, therefore a metric distance (d) as well as an inclusion measure function (σ) is implied in lattice L as well as in lattice K. Recall that algorithmic decision-making in computational intelligence applications typically boils down to selecting the 'best' among different alternatives based on either a similarity measure or a distance function. In the context of this work, induction of an estimate $\hat{f}$ of function f as well as generalization are driven by either inclusion measure $\sigma(x \leq c)$ or metric $d(x,c)$, where x and c are lattice elements.

An optimal estimate $\hat{f}$ of function f: $L \rightarrow K$ may be induced by mapping (1) the lattice-join $\bigvee_{k \in A \subset \{1,...,n\}} x_k = x'$ of 'adjacent' elements x_k in L to (2) the lattice-join $\bigvee_{k \in A \subset \{1,...,n\}} y_k = y'$ of 'adjacent' image elements y_k in K. Note that if a lattice element x is represented as trivial interval $[x,x]$ then both x' and y' above are lattice intervals interpreted as *information granules*. The aforementioned elements x' and y' could be in a causal relationship (Pearl 2000). In particular, mapping $x' \rightarrow y'$ may be interpreted as a 'rule' induced from the training data. The words *interval* and *cluster* are used interchangeably in this work; in particular, an interval in lattice R^N corresponds to a hyperbox. The induction of a classifier model is discussed next, furthermore a useful notation is introduced.

Let $(L, \leq)$ be a lattice, $(\tau(L), \leq)$ be the corresponding lattice of intervals, and let g: $a(L) \rightarrow K$ be a *category function*, where $a(L)$ is the set of atoms in lattice $(L, \leq)$, and K is a finite set of category labels. The domain of function g can be extended to the lattice $(\tau(L), \leq)$ of intervals as follows: if $\Delta \in \tau(L)$ and $g_0 \in K$ then $g(\Delta) = g_0$ if and only if $g(t_r) = g_0$ for all atoms $t_r \leq \Delta$. Likewise the domain of g can be extended to the collection $\mathcal{F}_{\tau(L)}$ of families of intervals as follows: if $f = \{w_j\}_{j \in J} \in \mathcal{F}_{\tau(L)}$ and $g_0 \in K$ then $g(f) = g_0$ if and only if $g(w_j) = g_0$ for all $j \in J$. Furthermore, if $g(f) = g_0 \in K$ for all families $\{w_j\}_{j \in J} \in \mathcal{F}_{\tau(L)}$ which represent a single class $c = \bigcup_{j \in J} w_j$ then we write $g(c) = g_0$. The training data to a classifier model consist of pairs $(\Delta_i, g(\Delta_i))$, $i = 1,...,n$, where $\Delta_i \in \tau(L)$ and $g(\Delta_i)$ is the category label of Δ_i.

6.3 Optimization

An algorithm applicable in a general lattice cannot employ any optimization technique. However, optimization can be pursued in a general lattice by search algorithms, e.g. genetic algorithms (GAs). More specifically, optimization in the context of this work typically involves an optimal induction of underlying positive valuation function(s) from training data.

7 Algorithms for Clustering, Classification, and Regression

This chapter presents algorithms for inducing a function

A large number of algorithms for clustering, classification, and regression have been presented in the literature (Mitchell 1997; Jain et al. 1999). Many of the aforementioned algorithms have their own unique advantages. However, even though an algorithm may perform well in specific applications, the same algorithm often performs poorly in other applications.

The corresponding algorithms presented in this chapter have two unique characteristics. First, the same algorithm is applicable in disparate data domains. Second, performance can be tuned by a positive valuation function. Both aforementioned characteristics are based on lattice theory.

7.1 Algorithms for Clustering

Two families of algorithms are presented for competitive clustering, namely the FLN family and the grSOM family. First, FLN stands for *Fuzzy Lattice Neurocomputing* (Kaburlasos and Petridis 1997, 2000, 2002; Petridis and Kaburlasos 1998, 2000, 2001; Kaburlasos et al. 1999; Kaburlasos and Kazarlis 2002; Athanasiadis et al. 2003; Cripps et al. 2003; Kaburlasos et al. 2003; Kaburlasos 2003, 2004b; Kehagias et al. 2003) and it originates from Carpenter's fuzzy version of Grossberg's adaptive resonance theory or ART for short (Carpenter and Grossberg 1987a; Carpenter et al. 1991b, 1992). Second, grSOM stands for *granular self-organizing map* (Papadakis et al. 2004; Kaburlasos et al. 2005; Papadakis and Kaburlasos 2005; Kaburlasos and Kehagias 2006b; Kaburlasos and Papadakis 2006) and it originates from Kohonen's self-organizing map paradigm or KSOM for short (Kohonen 1990, 1995).

Both abovementioned families emphasize a neural (parallel) implementation. Nevertheless, the emphasis here is on functionality. Hence, the proposed algorithms do not adhere to specific architectures.

Vassilis G. Kaburlasos: *Algorithms for Clustering, Classification, and Regression*, Studies in Computational Intelligence (SCI) **27**, 71–94 (2006)
www.springerlink.com

7.1.1 FLN models

A FLN model is applicable in a complete lattice. The most popular among the FLN models for clustering is the σ-FLN, which employs an inclusion measure function (σ) as described in the following.

The σ-FLN model

The σ-FLN has been presented with alternative names including FLN (Kaburlasos and Petridis 1997), FLR (Kaburlasos et al. 1997), FLNN (Petridis and Kaburlasos 1998), and σ-FLL (Petridis and Kaburlasos 1999). It was decided to switch to the name σ-FLN (Petridis and Kaburlasos 2001) in order to signify a neural implementation, moreover letter 'σ' indicates that both the training- and the testing- phase are carried out using an inclusion measure (σ). The training (learning) phase is presented next.

Algorithm σ-FLN for learning (training)

1. The first input $\Delta_1 \in \tau(L)$ is memorized (Assume that at any instant t there is a finite number $M(t)$ of classes c_k in the memory, $k=1,...,M(t)$).
2. Present the next input $\Delta \in \tau(L)$ to the 'set' database of classes $c_1,...,c_{M(t)}$.
3. While there exist 'set' classes c_k, for each 'set' class c_k calculate the fuzzy membership value $m_{c_k}(\Delta)$, $k=1,...,M(t)$.
4. Competition among classes c_k, $k=1,...,M(t)$: Winner is the class c_J with the largest membership value $m_{c_k}(\Delta)$. Let w_L denote the interval of winner family $c_J = \bigcup_i w_{J,i}$ with L= $argmax_i \{\sigma(\Delta \leq w_{J,i})\}$. Break possible ties by selecting interval w_L with the smallest diagonal.
5. *Assimilation Condition*: Test whether the diagonal $diag(\Delta \vee w_L)$ is less than a user-defined threshold diagonal size D_{crit}.
6. If the Assimilation Condition is satisfied, then replace w_L in c_J by $\Delta \vee w_L$. Then, apply the technique of *maximal expansions* followed by *simplification* to calculate the quotient $Q(\{w_{J,i}\})$.
7. If the Assimilation Condition fails then 'reset' c_J.
8. If all the families $c_1,...,c_{M(t)}$ are 'reset' then memorize input interval Δ.
9. At the end of learning, labels may be attached to the unlabelled intervals/clusters of a class by majority voting.

Note that the data processed by σ-FLN are elements in the lattice $(\tau(L), \leq)$ of intervals of a lattice $(L, \leq)$. In the unit hypercube, the σ-FLN processes N-dimensional hyperboxes. For convenience, algorithm σ-FLN for training (learning) is described in Fig.7-1 in a flowchart.

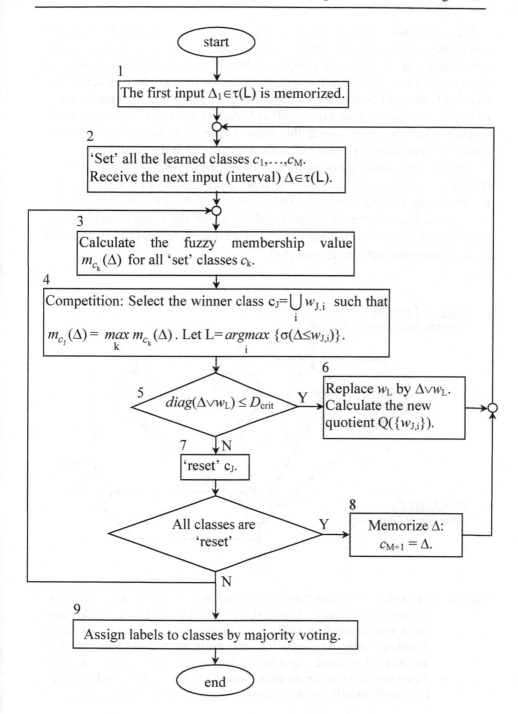

Fig.7-1 Flowchart of algorithm σ-FLN for learning (training).

Fig.7-2 shows a neural architecture for implementing algorithm σ-FLN, where the names and roles of its various subsystems are analogous to the names and roles of the corresponding subsystems in an adaptive resonance theory (ART) model (Petridis and Kaburlasos 1998). More specifically, like ART, the σ-FLN proposes a modifiable recurrent architecture for clustering in two layers. One layer, called *Category Layer* (F_2) is cognitively 'higher' than the other called the *Input Layer* (F_1). Single nodes at the category layer encode patterns of node activities from the input layer.

The category layer consists of L neurons specifying L lattice intervals that define M classes; it is $L \geq M$. The input layer consists of N neurons used for buffering and matching. The two layers are fully interconnected by bi-directional links, where a bi-directional link stores the components of a lattice interval $[x,y]$. The first end, say x, of a lattice interval $[x,y]$ is encoded by its isomorphic lattice element $\theta(x)$.

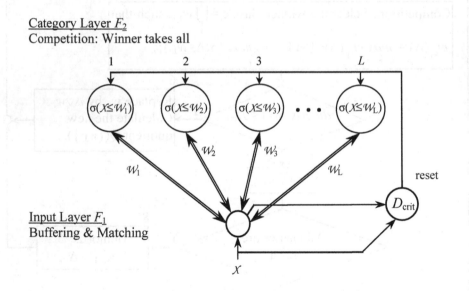

Category Layer F_2
Competition: Winner takes all

Input Layer F_1
Buffering & Matching

Fig.7-2   The σ-FLN neural architecture for clustering. Input $X = (x_1,...,x_N)$ is an element in a product lattice $(L,\leq) = (L_1,\leq) \times ... \times (L_N,\leq)$. A *Category Layer* neuron employs a lattice inclusion measure $\sigma(.,.)$ as its activation function in order to specify by a number in interval $[0,1]$ the degree of inclusion of an input X to a weight lattice element W_k, $k=1,...,L$. *Input Layer* is used to buffer an input datum. A 'reset' node is used for resetting, conditionally, the activity of a node in the *Category Layer*.

Learning proceeds by computing intervals/clusters in a lattice, these are hyperboxes in the unit-hypercube in R^N. A lattice interval keeps increasing until a maximum size criterion (D_{crit}) is met; note that an interval identifies a set of neighboring, in a metric sense, elements in a lattice.

The technique of *maximal expansions*, a *simplification* of intervals (in step 6 above) as well as other techniques are illustrated geometrically on the plane in Appendix C. The objective of the aforementioned techniques is to maximize the degree of inclusion of an interval $\Delta \in \tau(L)$ in a class c by representing class c by its quotient $Q(c)$.

A lattice interval is interpreted as an information granule. Moreover, after potentially attaching labels (in step 9 above), a lattice interval can be interpreted as a rule. In this sense the σ-FLN is an algorithm for inducing rules (knowledge) from the training data.

Criteria for partitioning a set of induced lattice intervals into classes may include: similarity (σ), affinity (d), interval overlapping, etc. It is also possible to consider one class per lattice interval. Furthermore, it is possible to supply, in advance, a set of classes as 'a priori' knowledge.

There is only a single model parameter to tune during learning, that is the user-defined threshold diagonal size D_{crit}, which specifies the maximum size of the lattice intervals/clusters to be learned. Larger or smaller values of D_{crit} effect learning reversely. On the one hand, larger values of D_{crit} result in fewer clusters while the capacity for data discrimination is deteriorated; on the other hand, smaller values of D_{crit} result in more clusters thus improving, up to a point, the capacity for data discrimination.

Regarding the training complexity, it follows that learning by σ-FLN is achieved in one pass through the training data, that is there are no multiple cycles in the algorithm. The worse case training scenario is to keep 'resetting' all L clusters for every input. Hence the training complexity is quadratic $O(n^2)$, where n is the number of data for training.

During the testing (generalization) phase, σ-FLN's learning capacity is disengaged. In the sequel generalization is carried out as follows.

<u>Algorithm σ-FLN for generalization (testing)</u>

1. The next input $\Delta \in \tau(L)$ appears whose category label is to be determined.
2. Calculate the fuzzy membership value $m_{c_k}(\Delta)$ in all classes $k=1,\ldots,M_F$.
3. Competition among classes c_k, $k=1,\ldots,M_F$: Winner is the class c_J with the largest membership value J= $argmax_{k} \ m_{c_k}(\Delta)$.
4. Assign to $\Delta \in \tau(L)$ the label of class c_J.

Advantages of the σ-FLN algorithm include: (1) a capacity to deal with uncertainty in the data in several manners including treatment of 'lattice intervals' these are 'hyperboxes' in the Euclidean unit-hypercube U, (2) an ability for sound, hierarchical, and modular decision making in a lattice data domain based (the decision making) on an inclusion measure σ, and (3) different positive valuation functions. In particular, advantages include the capacity to deal rigorously with: disparate types of data such as numeric and linguistic data, intervals of values, 'missing' and 'don't care' data as explained below.

In order to facilitate an implementation of the σ-FLN by other researchers a list of all required functions during both learning and generalization in the unit hypercube is summarized in the following.

- A lattice interval Δ in the unit-hypercube U corresponds to a hyperbox represented by $\Delta = [a,b] = [(a_1,...,a_N),(b_1,...,b_N)] = [a_1,b_1,...,a_N,b_N]$, where $a_i,b_i \in [0,1]$ and $[a_i,b_i]$ specifies an interval in the ith dimension for $i = 1,...,N$.

- Consider both positive valuation $v_i(x)=x$ and isomorphism $\theta_i(x)=1-x$, $x \in [0,1]$ along the ith dimension for $i=1,...,N$. Then, the positive valuation of a $w = [p,q] = [(p_1,...,p_N),(q_1,...,q_N)] = [p_1,q_1,...,p_N,q_N]$ equals $v(w) = v(\theta(p))+v(q) = v(\theta(p_1,...,p_N))+v(q_1,...,q_N) = v(1-p_1,...,1-p_N)$
$$+v(q_1,...,q_N) = \sum_{i=1}^{N}(1-p_i) + \sum_{i=1}^{N}q_i = N + \sum_{i=1}^{N}(q_i - p_i).$$

- The diagonal $diag(\Delta)$ of box Δ equals $diag(\Delta) = [a_1,b_1,...,a_N,b_N] = $
$$\sum_{i=1}^{N}(b_i - a_i).$$

- The lattice-join $\Delta \vee w$ equals $\Delta \vee w = [a_1,b_1,...,a_N,b_N] \vee [p_1,q_1,...,p_N,q_N] = [a_1 \wedge p_1, b_1 \wedge q_1,...,a_N \vee p_N, b_N \vee q_N]$.

- The degree of inclusion of hyperbox Δ in hyperbox w is calculated as

$$\sigma(\Delta \leq w) = \frac{v(\theta(p)) + v(q)}{v(\theta(a \vee p) + v(b \vee q)} = \frac{N + \sum\limits_{i=1}^{N}(v_i(q_i) - v_i(p_i))}{N + \sum\limits_{i=1}^{N}[v_i(b_i \vee q_i) - v_i(a_i \wedge p_i)]}.$$

- For a concrete family $f = \{w_i\}_{i \in I}$ the degree of inclusion $m_f(\Delta)$ of Δ in f is calculated as $m_f(\Delta) = \max\limits_{i \in I} \{\sigma(\Delta \leq w_i)\}$.

'Missing' data values can be dealt with by replacing them with the least-element; 'don't care' values can be dealt with by replacing them with the greatest-element of the corresponding constituent lattice. For instance, had the constituent lattice been the chain of real numbers I= [0,1] then in place of a missing value 'interval O= [1,0] ' will be used, whereas in place of a don't care value 'interval I= [0,1]' will be used.

From a technical point of view, the σ-FLN is applicable in a complete lattice (L,≤) equipped with (1) a positive valuation function v: L→R, and (2) an isomorphic function θ: L^{∂}→L. Stable learning is carried out rapidly in a single pass through the training data. Typically there is a near 0% error on the training data set. A potential disadvantage of learning by σ-FLN is that the lattice intervals it learns (i.e. the interval total number, size, and location) depend on the order of training data presentation.

The d-FLN model

Replacing, in the σ-FLN clustering algorithm, inclusion measure σ by a distance function d, produces the d-FLN model. Hence, the d-FLN could be regarded as an extension of the conventional radial basis function (RBF) neural networks (Moody and Darken 1989; Chen et al. 1991; Chen and Chen 1995) to a lattice data domain.

7.1.2 grSOM models

A grSOM model is applicable in the non-complete lattice of positive FINs – recall that a positive FIN can be interpreted as a conventional fuzzy number. A model in this family employs metric $d_K(.,.)$. A grSOM model for clustering can be used for classification by attaching labels to computed clusters. Two different models of the grSOM family, namely the *greedy grSOM* and the *incremental grSOM*, are presented next.

The *greedy grSOM* model

The *greedy grSOM* induces N-tuples of FINs from vectors of numbers. In particular, a FIN is computed in a data dimension by algorithm CALFIN. Two different *greedy grSOM* versions have been presented. One version activates the (grid) weights using a *metric* (Papadakis et al. 2004; Papadakis and Kaburlasos 2005; Kaburlasos and Papadakis 2006); the other version activates the (grid) weights using a *fuzzy membership function* (Kaburlasos and Kehagias 2006b). The 'metric' version is presented in the following in two stages, namely Stage-A and Stage-B.

Algorithm *greedy grSOM* for learning (training)

Stage-A: Rule Induction

1. Define the dimensions I and J of a two-dimensional grid I×J of units/neurons. Each unit can store both a N-tuple weight $W_{ij} \in F^N$, and a label K_{ij}, $i=1,...,I$, $j=1,...,J$; the latter indicates the class of the unit.
2. Initialize randomly the weight W_{ij} of each unit from the training data.

   Repeat steps 3 and 4 for a user-defined number *Nepochs* of epochs.

3. For each training input datum $x_r \in F^N$, $r=1,...,n$ do
   (a) Calculate Minkowski metric $d_1(x_r, W_{ij})$, $i=1,...,I$, $j=1,...,J$.
   (b) Competition among I×J grid units: Winner is unit '*p(r)q(r)*' whose weight is the nearest to x_r, i.e. $p(r)q(r) \doteq \arg \min_{i \in \{1,...,I\}, j \in \{1,...,J\}} d_1(x_r, W_{ij})$.
   (c) Assign the training input x_r equally to both the winner unit '*p(r)q(r)*' and to all the units in the neighborhood $N_{p(r)q(r)}(t)$ of the winner.
4. For each unit '*ij*', $i=1,...,I$, $j=1,...,J$ in the grid use algorithm CALFIN to compute the new weight value W_{ij}' based on the data assigned to unit '*ij*' in step 3 of the current epoch.
5. To each unit '*ij*', $i=1,...,I$, $j=1,...,J$ in the grid attach the label of the category, which provided the majority of the input training data to unit '*ij*' during all epochs.

Stage-B: Genetic Optimization

Consider function $f_i(x)= c_{i,1}\tanh((x-a_{i,1})/b_{i,1}) + c_{i,2}\tanh((x-a_{i,2})/b_{i,2}) + c_{i,3}\tanh((x-a_{i,3})/b_{i,3})$, $i=1,...,N$ including 9 parameters a_i, b_i, c_i, $i= 1,2,3$ per a data dimension, where tanh(.) is the *hyperbolic tangent* function.

Calculate, genetically, optimal estimates of functions $f_i(x)$, $i=1,...,N$ that

minimize classification error $E= \sum_{r=1}^{n} \delta(K_r, K_{p(r)q(r)})$, where

$\delta(K_r, K_{p(r)q(r)})=1$ if $K_r=K_{p(r)q(r)}$, $\delta(K_r, K_{p(r)q(r)})=0$ otherwise, and K_r, $K_{p(r)q(r)}$ are class labels; indices '*r*' and '*p(r)q(r)*' are explained in Stage-A.

Note that during training (steps 3 and 4 above) labels are not used. A grid unit with both weight $W_{ij} \in F^N$ and label K_{ij} can be interpreted as a fuzzy rule 'if W_{ij} then K_{ij}'. Therefore, the *greedy grSOM* can be used as a fuzzy neural network for classification (Kaburlasos and Papadakis 2006).

The objective of greedy grSOM in Stage-A above is to identify clusters so as to 'fully cover' the training data by FINs. Hence if, as expected in practice, the training data domain equals the testing data domain then 'full coverage' guarantees that at least one fuzzy rule will be activated for an input. That is why the grSOM algorithm above was named *greedy*: because it expands its fuzzy rule supports so as to cover the training data.

On the one hand, Kohonen's SOM (KSOM) computes reference (weight) vectors $m_i(t)$ in R^N hence KSOM captures, locally, first-order statistics in the training data. On the other hand, *greedy grSOM* above computes a distribution of FINs. Based on the one-to-one correspondence between FINs and probabilistic distribution functions, it follows that *greedy grSOM* captures, locally, all-order statistics in the training data.

Advantages of *greedy grSOM* include: (1) computation, locally, of all-order statistics, (2) induction of descriptive decision-making knowledge (fuzzy rules) from the training data, and (3) generalization beyond rule support. A computational disadvantage is that the *greedy grSOM* needs to 'batch process' the whole training data set in order to compute a new weight (FIN) value $W'_{ij} \in F^N$.

Various authors have proposed genetic algorithm (GA) optimization of the Self-Organizing Maps (Polani 1999). Stage-B of algorithm grSOM here introduces non-linearities in space F^N using a GA. More specifically a different non-linearity is introduced in a different data dimension so as to minimize classification error E= $\sum_{r=1}^{n} \delta(K_r, K_{p(r)q(r)})$. A strictly increasing function is $f(x)= c_1(\tanh(x-a_1)/b_1) + c_2(\tanh(x-a_2)/b_2) + c_3(\tanh(x-a_3)/b_3)$, where 'tanh' is the (saturated) hyperbolic tangent function. Fig.7-3(a) shows function $\tanh(x)$, whereas Fig.7-3(b) plots function $f(x)$ for selected values of the parameters a_i, b_i, c_i, i= 1,2,3.

An individual solution of the GA for the *greedy grSOM* can be represented using 9N parameters, i.e. nine parameters in each of the 'N' data dimensions. Therefore the chromosome of an individual solution consists of 9N genes. Using 16 bits per gene to encode a single parameter value it follows that a chromosome is 144N bits long. The classification performance on the training data set is used as the fitness value of an individual.

The training complexity of Stage-A of the *greedy grSOM* algorithm for training equals O(*Nepochs*IJNn(L+log$n$)), where 'N' is the dimensionality of space R^N, 'n' is the number of data, and 'L' is the number of integration levels from 0 to 1 (Kaburlasos and Papadakis 2006). Stage-B of the grSOM algorithm requires substantial time due to the genetic search. After training the *greedy grSOM* can be used for testing as follows.

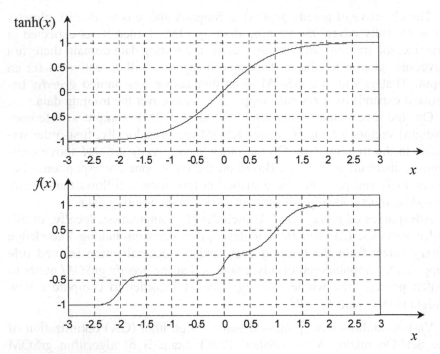

Fig.7-3   (a) The strictly increasing, saturated function tanh(x).
(b) $f(x) = 0.3\tanh((x+2)/0.2)+0.2\tanh(x/0.1)+0.5\tanh((x-1)/0.4)$. The parameter values of a component function 'tanh(.)' specify the location, scale, and the height of the corresponding component function.

Algorithm *greedy grSOM* for generalization (testing)

1. Present an input $x_k \in \mathsf{F}^N$ to the trained *greedy grSOM*.
2. Calculate Minkowski distance $d_1(x_k, W_{ij})$ for all *committed* units W_{ij}, $i=1,\dots,I, j=1,\dots,J$ in the I×J grid.
3. Competition among *committed* units in the I×J grid: Winner is the unit '$p(k)q(k)$' whose weight $W_{p(k)q(k)}$ is nearest to x_k.
4. The class K_k of x_k is defined by $K_k = K_{p(r)q(r)}$.

A 'fuzzy' version of the *greedy grSOM* is produced simply by replacing step 3(b) in stage-A of the *greedy grSOM* algorithm above by: "Competition among I×J grid units: Winner is unit '$p(r)q(r)$' whose weight is included the most in x_r, i.e. $p(r)q(r) \doteq \arg \max_{i\in\{1,\dots,I\},j\in\{1,\dots,J\}} (1/(1+d_1(x_r,W_{ij})))$".

The *incremental grSOM* model

The *incremental grSOM* induces a FIN from a population of real number measurements; then it carries out computations in the cone of positive FINs (Kaburlasos et al. 2005).

Algorithm *incremental grSOM* for learning (training)

1. Define the dimensions I and J of a grid I×J of units/neurons. Each unit can store both a N-dimensional FIN tuple (weight) $W_{ij} \in F^N$, $i \in \{1,...,I\}$, $j \in \{1,...,J\}$ and a label $K_{ij} \in K$, where K is a finite set of category labels. Initially all the units in the I×J grid are *uncommitted*.
2. Memorize the first training data pair $(x_1, K_1) \in F^N \times K$ by randomly committing a unit in the I×J grid.

   Repeat steps 3 through 8 below for a user-defined number $k=1,...,Nepochs$ of epochs.

3. For training datum $(x_r, K_r) \in F^N \times K$, $r=1,...,n$ 'reset' all units in the grid.
4. Calculate Minkowski metric $d_1(x_r, W_{ij})$ between x_r and all *committed* units W_{ij}, $i \in \{1,...,I\}$, $j \in \{1,...,J\}$ in the I×J grid.
5. Competition among the both 'set' and *committed* units in the I×J grid: Winner is the unit '$p(r)q(r)$' whose weight is nearest to x_r, i.e.
$$p(r)q(r) \doteq \arg \min_{i \in \{1,...,I\}, j \in \{1,...,J\}} \{d_1(x_r, W_{ij})\}.$$
6. *Assimilation Condition*: '$d_1(x_r, W_{p(r)q(r)}) \leq d_0$'.AND.'$K_r = K_{p(r)q(r)}$', where d_0 is a user-defined threshold.
7. If the *Assimilation Condition* is satisfied then compute a new value W'_{ij} for the weight W_{ij}, $i \in \{1,...,I\}$, $j \in \{1,...,J\}$ in the neighborhood of the winner '$p(r)q(r)$' using the following convex combination:
$$W'_{ij} = \left[1 - \frac{h(k)}{1 + d_K(W_{p(r)q(r)}, W_{ij})}\right] W_{ij} + \frac{h(k)}{1 + d_K(W_{p(r)q(r)}, W_{ij})} x_r.$$

   Otherwise, if the Assimilation Condition is not satisfied, 'reset' the current winner unit '$p(r)q(r)$' and goto to Step-5 to find a new winner.
8. If all I×J grid units are 'reset' then commit an *uncommitted* unit to memorize the current training datum (x_r, K_r). If no *uncommitted* units exist in the grid then increase the grid dimensions I and/or J by one.

A grSOM algorithm projects a probability distribution from F^N onto a two-dimensional grid I×J. A grid unit (W_{ij}, K_{ij}) is interpreted as a logical rule 'if W_{ij} then K_{ij}' or $W_{ij} \rightarrow K_{ij}$ for short. Hence, a grSOM algorithm can be used as a device for visualizing logical rules.

7.2 Algorithms for Classification

The clustering algorithms above can be used for developing algorithms for classification by supervised clustering. This section illustrates how.

7.2.1 FLN classifiers

Several FLN classifiers have been presented in the literature. The most popular among them is the σ-FLNMAP classifier (Kaburlasos and Petridis 2000; Petridis and Kaburlasos 2000; Kaburlasos and Kazarlis 2002; Athanasiadis et al. 2003; Kaburlasos 2003, 2004b; Kaburlasos et al. 2003; Kehagias et al. 2003) presented in the following.

The σ-FLNMAP classifier

The σ-FLNMAP is a synergy of two σ-FLN modules, namely σ-FLN$_a$ and σ-FLN$_b$ module, interconnected via the MAP field F^{ab}. More specifically, the σ-FLNMAP is a lattice domain extension of the fuzzy-ARTMAP neural network (Carpenter et al. 1992; Anagnostopoulos and Georgiopoulos 2002). The corresponding architecture in shown in Fig.7-4.

During training, a pair $(P,Q=f(P))\in\tau(L)\times\tau(K)$ is presented, where both $(L,\leq)$ and $(K,\leq)$ are lattices. On the one hand, module σ-FLN$_a$ clusters input data P; on the other hand, module σ-FLN$_b$ clusters the corresponding images $Q=f(P)$. The intermediate MAP field F^{ab} associates clusters (intervals) in σ-FLN$_a$ with clusters in σ-FLN$_b$ as explained in the following.

A training datum P activates a lattice interval in layer F_2^a of σ-FLN$_a$ and, ultimately, it activates another interval in layer F_2^b of σ-FLN$_b$ via the MAP field F^{ab}. At the same time, layer F_2^b is activated directly by the corresponding image of P input $Q=f(P)$. If the activated intervals in σ-FLN$_b$ 'match' then both modules σ-FLN$_a$ and σ-FLN$_b$ learn as usual provided that the corresponding join-interval in F_2^a (1) has size smaller than D_a, and (2) is 'uniform', in the sense that all intervals inside it are mapped to the same interval in F_2^b; otherwise, the diagonal D_a is decreased by a small amount ε and the search for a winner interval resumes. If no interval in layer F_2^b matches the current input $f(P)$ adequately then pair $(P,f(P))$ is memorized.

The σ-FLNMAP is an algorithm for learning a function f: $\tau(L) \to \tau(K)$, where both (L,≤) and (K,≤) are lattices. A pair $(P,Q) \in \tau(L) \times \tau(K)$ stored in σ-FLNMAP is interpreted as rule R: 'if P then Q', symbolically R: $P \to Q$, induced empirically from the training data. We remark that $\sigma(x \leq P)$ could be interpreted as the degree of truth of the aforementioned rule R.

The σ-FLNMAP can be used for classification, e.g. it can be used for inducing a function g: $\tau(I^N) \to K$, where I^N is the unit N-dimensional hyper-cube and K is a set of labels. In the latter case we assume $D_b = 0$, that is module σ-FLN$_b$ memorizes all category labels fed to it.

Learning by σ-FLNMAP occurs in one-pass through the training data. Any training datum fed to σ-FLNMAP afterwards will not change the learned clusters/intervals. The latter (intervals) can be interpreted as in-formation granules. During testing, an input P is assigned to a winner in-terval w_J in layer F_2^a of σ-FLN$_a$ only if $diag(P \vee w_J) \leq D_a$; in the latter case the corresponding winner image in F_2^b is assigned to P; otherwise no in-terval of F_2^b is assigned to input P and σ-FLNMAP is called 'undecided'.

The worse case training scenario for σ-FLNMAP is, for each input da-tum, to reset all layer F_2^b nodes followed by resetting all F_2^a nodes. Hence, the training complexity of σ-FLNMAP is $O(n^3)$.

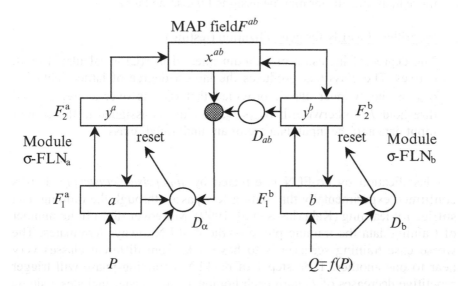

Fig.7-4   The σ-FLNMAP neural network for inducing a function f: $\tau(L) \to \tau(K)$, where both L and K are mathematical lattices.

The dσ-FLN classifier

The letters 'd' and 'σ' show that the training phase and the testing phase are carried out, respectively, based on a distance function (d) and an inclusion measure (σ) (Kaburlasos et al. 1999). The training phase of the algorithm is described next followed by the testing phase.

<u>Algorithm dσ-FLN for learning (training)</u>

1. A 'lower level' supervised clustering is carried out in the training data in the lattice $(\tau(L),\leq)$ of intervals. It is assumed that two training data are in the same cluster if their distance is less than a user-defined distance threshold T_d. A *contradiction* occurs when two data from different classes are put in the same cluster; then a decrease of T_d is triggered followed by a new cycle of step-1. Step-1 keeps repeating until a contradiction-free cycle occurs which provides with *homogeneous clusters*, these are clusters of data from a single class.
2. The 'lower level' supervised clustering concludes by the join operation ($\vee$) of all lattice $(\tau(L),\leq)$ data in the same cluster. Hence, one element in the lattice $(\tau(L),\leq)$ of intervals is defined for each lower level cluster.
3. An 'upper level' clustering is carried out by attaching a label to each lower level cluster. An external teacher attaches the labels. Note that more than one cluster may be assigned the same label.

<u>Algorithm dσ-FLN for generalization (testing)</u>

The degree of inclusion of a datum interval in $\tau(L)$ is calculated in all classes. The class that produces the largest degree of lattice inclusion σ_{max} is the 'winner class', provided that σ_{max} is over a user-defined threshold T_σ; otherwise, if $\sigma_{max}<T_\sigma$, no class is assigned to the current input data and the input data are of an 'undefined' class.

Classification by dσ-FLN is effected by *supervised clustering*. It was confirmed experimentally that only a few passes through the training data suffice for learning (Kaburlasos et al. 1999). Moreover, for a finite number of training data the training phase of the dσ-FLN always terminates. The worse case training scenario is to have data from different classes very near to one another. Then step-1 of dσ-FLN's training-phase will trigger repetitive decreases of T_d until each homogeneous cluster includes a single datum. It follows that the training complexity of dσ-FLN is $O(n^2)$.

7.2.2 Combinatorial FLN (fit) classifiers

Other FLN classifier models include the *combinatorial* FLN models. Learning in the combinatorial FLN classifiers is effected by computing *fits* in the training data using the lattice-join operation. A *fit* is defined in the following (Kaburlasos and Petridis 2000, 2002; Cripps et al. 2003).

Definition 7-1   A *fit T in a set* $\{(\Delta_i, g(\Delta_i))\}_{i \in \{1,\ldots,n\}}$, where $\Delta_i \in \tau(L)$ and $g: a(L) \rightarrow K$ is a category function, is an interval $T \in \tau(L)$, which satisfies:

  T1. $T = \bigvee_{s \in S} \Delta_s$, where $S \subseteq \{1,\ldots,n\}$ such that $g(\Delta_s) = g_0$ for all $s \in S$.

  T2. For all $\Delta_j, j \in \{1,\ldots,n\}$ with $g(\Delta_j) \neq g_0$ it holds $\Delta_j \wedge T = \varnothing$.
       A fit is called *tightest* if, in addition, it satisfies:

  T3. For all Δ_k, $k \in \{1,\ldots,n\}$ with $g(\Delta_k) = g_0$ (other than the Δ_k in (C1)) it
       holds $(T \vee \Delta_k) \wedge \Delta_j \neq \varnothing$ for some $j \in \{1,\ldots,n\}$ with $g(\Delta_j) \neq g_0$.

Condition (T1) requires a fit to be the lattice join of training lattice intervals that all belong to the same category g_0. Condition (T2) requires that there should not be a *contradiction* between a fit and a training datum, where a *contradiction* occurs when two pairs $(T, g(T))$ and $(\Delta, g(\Delta))$ with $g(\Delta) \neq g(T)$ share an atom. Condition (T3) requires that a fit T be *tightest*, in the sense that no other datum of the same category can be accommodated in T without contradicton. The notion *tightest fit* is explained geometrically in Fig.7-5. More specifically, Fig.7-5(a) shows a collection of points (atoms) in the unit square partitioned in two categories labeled by '•' and 'o', respectively; Fig.7-5(b) shows the corresponding tightest fits.

A generic, three-layer neural architecture used to implement a combinatorial FLN classifier is shown in Fig.7.6. A node in a layer, denoted by a rectangle, is a σ-FLN neural network. Two consecutive layers are fully interconnected. There are no interconnections between non-consecutive layers. The nodes in layers F_d and F_f are drawn in solid lines and they constitute the system's long-term memory. The nodes in layer F_h are drawn in dotted lines and they constitute the system's short-term memory.

Layer F_d stores one-by-one all the incoming training data; layer F_f keeps the ever-updated *fits*; layer F_h is active only during learning and it computes intermediate lattice intervals. A double link is used for transmitting both an interval and the corresponding category index from one σ-FLN to another. A single link from the Hypothesis Testing Layer to the Training Data Layer is used for transmitting 'binary acknowledgement' signals.

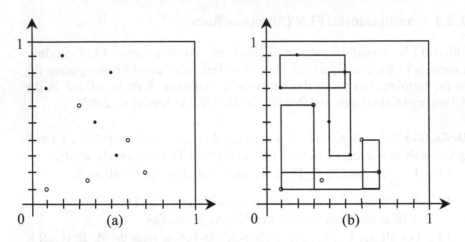

Fig.7-5   (a) Five different atoms (i.e. trivial intervals) are given from each one of two distinct categories labeled, respectively, by '•' and 'o'.
   (b) The collection of the tightest fits (boxes) that correspond to the atoms of Fig.7-5(a).

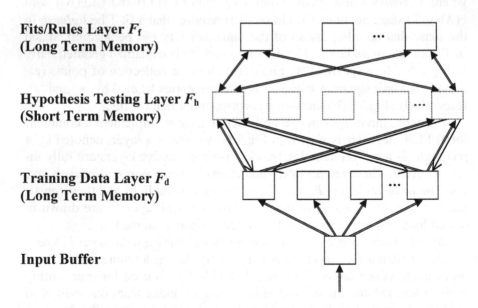

Fits/Rules Layer F_f
(Long Term Memory)

Hypothesis Testing Layer F_h
(Short Term Memory)

Training Data Layer F_d
(Long Term Memory)

Input Buffer

Fig.7-6   The basic three layer architecture for implementing a generic FLN fit classifier. A node, denoted by a rectangle, is a σ-FLN neural network module.

The *FLN with tightest fits*, or FLNtf for short, is presented first inspired from (1) the probably approximately correct (PAC) statistical learning model, in particular the Rectangle Learning Game (Blumer et al. 1989; Kearns and Vazirani 1994), and (2) the min-max neural networks (Simpson 1992, 1993).

Algorithm FLNtf for learning (training)

1. Let a be the next training datum. Add the trivial interval for the training datum a to the set of fits for datum a.
2. Consider the next fit, say I, containing datum a. Initially, the aforementioned set of fits contains only the trivial interval formed from the training datum a.
3. For the next training datum x with $g(x) = g(a)$ compute $I \vee x$.
4. If there is a training datum z such that $g(z) \neq g(a)$ and datum z is inside interval $I \vee x$, symbolically $\sigma(z \leq I \vee x) = 1$, then contradictory datum z implies that interval $I \vee x$ cannot be a fit.
5. Otherwise, if there is no contradictory datum z, then add the interval $I \vee x$ to the fits of datum a.
6. Repeat Steps 3, 4 and 5, in turn, for each datum.
7. Repeat Steps 2, 3, 4, 5 and 6, in turn, for each fit.
8. Repeat Steps 1, 2, 3, 4, 5, 6 and 7, in turn, for each datum.
9. The fits contained in no other fit are tightest fits.

When binary data (i.e. '0' or '1') are used instead of fuzzy data (i.e. in the interval [0,1]) then the FLNtf produces the same results as a standard Karnaugh map simplification algorithm in digital logic. Hence, FLNtf is more general because it may consider any point in the unit hypercube.

In summary, the FLNtf possesses the following advantages: (1) it identifies clusters/rules in a data order independent fashion, (2) it guarantees 0% recognition error on the training data, (3) it guarantees maximization of the degree of inclusion of the testing data in a class due to the employment of the quotient of a class, and (4) it can deal with disparate lattice elements. The FLNtf can justify its answers by inducing descriptive decision-making knowledge (rules) from the data in the Rules Layer F_f (Fig.7-6). A fit is interpreted as a 'rule for classification' induced from the training data.

The performance of FLNtf is not usually the best in practical applications; the latter is attributed to the noise sensitivity of FLNtf. Therefore, a number of alternative FLN fit classifiers, namely FLN first fit (FLNff), FLN ordered tightest fit (FLNotf), FLN selective fit (FLNsf), and FLN max tightest fit (FLNmtf), are presented in the following.

Algorithm FLNff for learning (training)

1. (Initialization) Mark all the training data as 'unused'.
2. Let a be the next 'unused' training datum. Mark a as 'used'. Consider trivial interval I which contains only the training datum a.
3. For the next 'unused' training datum x with $g(x) = g(a)$ compute $I \vee x$.
4. If there is a training datum z such that $g(z) \neq g(a)$ and datum z is inside interval $I \vee x$, symbolically $\sigma(z \leq I \vee x)=1$, then contradictory datum z implies that interval $I \vee x$ cannot be a fit.
5. Otherwise, if there is no contradictory datum z, then mark x as 'used' then replace interval I by interval $I \vee x$.
6. Repeat Steps 3, 4 and 5, in turn, for each 'unused' datum.
7. Repeat Steps 2, 3, 4, 5 and 6, in turn, for each 'unused' datum.

Algorithm FLNotf for learning (training)

1-6. Steps 1-6 are the same as for classifier FLNff above.
7. Repeat Steps 3, 4, 5, and 6, in turn, for each 'used' datum outside interval I.
8. Repeat Steps 2, 3, 4, 5, 6, and 7, in turn, for each 'unused' datum.

Algorithm FLNsf for learning (training)

1-6. Steps 1-6 are the same as for classifier FLNff above.
7. Repeat Steps 3, 4, 5, and 6 with 'used' data x outside interval I such that the corners of interval $I \vee x$ reside either inside interval I or inside the interval which contains x.
8. Repeat Steps 2, 3, 4, 5, 6, and 7, in turn, for each 'unused' datum.

Algorithm FLNmtf for learning (training)

1. Let a be the next training datum. Consider the trivial interval I which contains only the training datum a.
2. For the next training datum x with $g(x) = g(a)$ compute $I \vee x$.
3. If there is a training datum z such that $g(z) \neq g(a)$ and datum z is inside interval $I \vee x$, symbolically $\sigma(z \leq I \vee x)=1$, then contradictory datum z implies that interval $I \vee x$ cannot be a fit.
4. Repeat Steps 2 and 3, in turn, for each datum.
5. Select fit formed in Step 3 with minimal value $0 < \sigma(x \leq I) < 1$ and replace interval I by interval $I \vee x$.
6. Repeat Steps 2, 3, 4, and 5, in turn, while there are fits I and datum x meeting the criteria $0 < \sigma(x \leq I) < 1$.
7. Repeat Steps 1, 2, 3, 4, 5, and 6, in turn, for each datum.

Training of the previous classifiers is incremental, memory-based, and polynomial $O(n^3)$ in the number n of the training data. For algorithm FLNsf note that previously 'used' data x are employed only if join-interval $I \lor x$ corners meet certain containment criteria. Also, the difference between algorithms FLNsf and FLNotf is that FLNsf requires all join-interval $I \lor x$ corners to reside either inside join-interval I or inside the join-interval which contains x, whereas FLNotf poses no restrictions whatsoever regarding the corners of join-interval $I \lor x$. The testing phase for all aforementioned FLN classifiers is the same as explained next.

A generic 'FLN fit' algorithm for generalization (testing)

A datum x is classified to the category whose label is attached to the interval in which x is included the most in an inclusion measure sense.

Fig.7-7 and Fig.7-8 illustrate geometrically on the plane the mechanics of the aforementioned classifiers. In particular, Fig.7-7(a) shows the training data presented in the order a, b, c, d, e, f, g, h (category '•') and 1, 2, 3, 4, 5, 6 (category 'o'). Fig.7-7(b), and Fig.7-8 (a), (b), (c) and (d) illustrate the differences of algorithms FLNff, FLNotf, FLNsf, FLNtf and FLNmtf, respectively. Note that fits in different categories may overlap without contradicting on the training data as illustrated in Fig.7-8 (c) and (d).

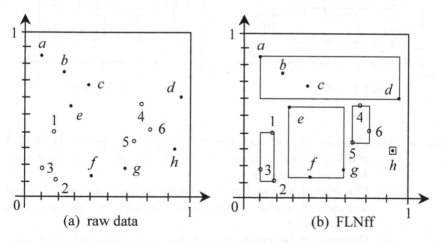

(a) raw data   (b) FLNff

Fig.7-7 (a) Fourteen training data are presented in the order a, b, c, d, e, f, g, h (category '•') and 1, 2, 3, 4, 5, 6 (category 'o').

(b) Classifier FLNff computes two *fits* for category 'o', i.e. $1 \lor 2 \lor 3$ and $4 \lor 5 \lor 6$, and another three *fits* for category '•', i.e. $a \lor b \lor c \lor d$, $e \lor f \lor g$, and h.

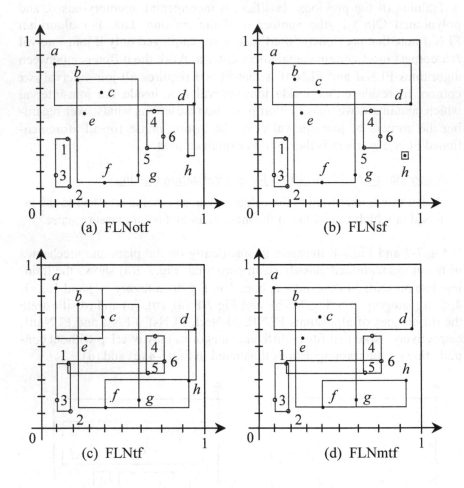

Fig.7-8 (a) Classifier FLNotf computes two *fits* for category 'o', i.e. 1∨2∨3 and 4∨5∨6, and another three *fits* for category '•', i.e. a∨b∨c∨d, b∨c∨e∨f∨g, and d∨h.

(b) Classifier FLNsf computes two *fits* for category 'o', i.e. 1∨2∨3 and 4∨5∨6, and another three *fits* for category '•', i.e. a∨b∨c∨d, c∨e∨f∨g and h.

(c) Classifier FLNtf computes three *fits* for category 'o', i.e. 1∨2∨3, 4∨5∨6 and 1∨5∨6, and another five *fits* for category '•', i.e. a∨b∨c∨d, a∨b∨c∨e, b∨c∨e∨f∨g, f∨g∨h, and d∨h.

(d) Classifier FLNmtf computes three *fits* for category 'o', i.e. 1∨2∨3, 4∨5∨6 and 1∨5∨6, and another three *fits* for category '•', i.e. a∨b∨c∨d, b∨c∨e∨f∨g, and f∨g∨h.

7.2.3 FINkNN classifier

A σ-FLNMAP classifier based on an inclusion measure (σ) suffers a substantial drawback when applied to fuzzy numbers as shown in section 4.5.2; that is, an employment of σ may produce a counter-intuitive equality. The aforementioned drawback can be mended using metric d_K as demonstrated in Example 4-12. Based on metric d_K a k-nearest neighbor classifier, namely FINkNN, can be introduced as explained next.

It is already known that metric function d_K can compute a distance between arbitrary-shaped FINs; the latter can be induced by algorithm CALFIN from populations of samples. Since (F,d_K) is a metric lattice, it follows that $d_p(E,H)= \{d_K(E_1,H_1)^p+\ldots+d_K(E_N,H_N)^p\}^{1/p}$, $p{\ge}1$, where $E= (E_1,\ldots,E_N)$, $H=(H_1,\ldots,H_N){\in}F^N$, is a metric in product lattice F^N.

Let g be a *category* function g: $F^N{\to}K$ which maps a positive N-tuple FIN to an element (a category) of a label set K. Classification can be effected in (F^N,d_K), first, by storing labeled training data $(F_1,g(F_1))$, $\ldots,(F_n,g(F_n))$ and, second, by mapping a new FIN H to a category $g(F){\in}K$ that receives the majority vote among the k Nearest Neighbor N-tuple FIN categories. The aforementioned procedure forms the basis for a (memory-based) k-nearest-neighbor classifier, namely Fuzzy Interval Number k-Nearest Neighbor classifier or FINkNN for short. Classifier FINkNN is usually applied in metric lattice (F^N,d_1) (Petridis and Kaburlasos 2003).

Algorithm FINkNN for generalization (testing)
1. Store all labeled training data pairs $(F_1,g(F_1)),\ldots,(F_n,g(F_n))$, where $F_i{\in}F^N$, $g(F_i){\in}K$ for $i{=}1,\ldots,n$.
2. Classify a new datum $H{\in}F^N$ to the 'majority category' $g(F_{i_j})$, $j{=}1,\ldots,k$, where $i_j{\in}\{1,\ldots,n\}$ are the indices of the 'k' nearest neighbors to datum H using metric $d_1(H,F_i)$, $i{=}1,\ldots,n$.

7.3 Optimized Classifier Ensembles

The intervals (hyperboxes/clusters) computed by σ-FLNMAP, and ultimately σ-FLNMAP's classification performance, depend on the order of presenting the training data. In the aforementioned sense, the σ-FLNMAP is an *unstable* classifier. Hence, the σ-FLNMAP is a promising candidate for majority-voting classification (Schapire 1990; Breiman 1996).

The idea is to train an ensemble of σ-FLNMAP modules, namely *voters*, using a different permutation of the training data per module; finally, a testing datum is classified to the category that receives the majority vote in the ensemble. Hence, the *Voting σ-FLNMAP* classifier emerges (Petridis and Kaburlasos 2000; Kaburlasos 2003; Kaburlasos et al. 2003; Kehagias et al. 2003), which may be regarded as a panel of experts each of who induces its own set of rules from the training data.

The following parameters are considered for the Voting σ-FLNMAP model: (1) the diagonal D_a in module σ-FLN$_a$, (2) the number n_V of σ-FLNMAP voters, and (3) the positive valuation functions employed in each constituent lattice; in particular, note that linear positive valuations $y = w_i x$ correspond to different weight coefficients w_i, $i=1,...,N$ in different data dimensions. In the interest of simplicity, assuming constant weight coefficients $w_i=1$, $i=1,...,N$ above, it follows that the parameters D_a and n_V can be estimated optimally from the training data by exhaustive search according to the following algorithm (Kaburlasos 2003).

<u>Algorithm PAREST for Estimating Model Parameters D_a and n_V</u>

Consider random partitions of the training data *TRN* into 2 sets *TRN*$_k$ and *TST*$_k$, such that *TRN*$_k \cup$*TST*$_k$=*TRN*, *TRN*$_k \cap$*TST*$_k$=∅, $k=1,...,K$. Furthermore consider a user-defined grid of pairs $((D_a)_i,(n_V)_j)$, $i=1,...,I, j=1,...,J$.

Repeat steps 1 to 2 below for $k=1,...,K$.

1. For each pair $((D_a)_i,(n_V)_j)$ train a Voting σ-FLNMAP classifier using the *TRN*$_k$ data. Record the *ensemble classification accuracy* $E_{k,i,j}$, furthermore compute the *average classification accuracy* $A_{k,i,j}$ of individual σ-FLNMAP classifiers in the ensemble.

2. Compute numbers $\delta_{k,i} = \dfrac{1}{J}\sum\limits_{j=1}^{J}(E_{k,i,j} - A_{k,i,j})$.

    Consider vector $\delta_k = (\delta_{k,1}, ..., \delta_{k,I})$.

3. Compute vector $\delta = \dfrac{1}{K}\sum\limits_{k=1}^{K}\delta_k$. Let M be the index of the maximum number in vector δ. An optimal estimate for D_a is $\hat{D}_a = (D_a)_M$.

4. Use $\hat{D}_a$ as well as the training data set *TRN*$_k$, $k=1,...,K$ to train an ensemble with $(n_V)_j$, $j=1,...,J$ voters.

5. Let n_k, $k=1,...,K$ be the smallest number of voters, which maximizes classification accuracy. An optimal estimate for n_V is $\hat{n}_V = \left\lfloor \dfrac{1}{K}\sum\limits_{k=1}^{K}n_k \right\rfloor$.

By *classification accuracy* above is meant the percentage of correct classifications in the testing data TST_k. Moreover, the *floor* function $\lfloor . \rfloor$ by definition equals the largest integer smaller than its real number argument.

In all, algorithm PAREST is computationally intensive. However, it was confirmed in many experiments that Voting σ-FLNMAP may clearly improve the classification accuracy of an individual σ-FLNMAP voter; moreover, it was observed experimentally that classification accuracy was clearly more sensitive in parameter D_a than in parameter n_V. Further improvements can be pursued by estimating optimally the underlying positive valuation functions as explained in the following.

A σ-FLNMAP voter of the Voting σ-FLNMAP classifier can assume a different 'weight function' in each data dimension. The latter practice can introduce useful non-linearities at the expense of longer time required (1) to compute optimal 'weight functions', and (2) to carry out calculations with the computed optimal 'weight functions'.

7.4 Algorithms for Regression

Regression is a term used in statistical pattern recognition for real-valued target function approximation (Michell 1997, section 8.2.3).

Conventional regression may map a real number x_0 to a (probabilistic) cumulative distribution function. The aforementioned mapping is subsumed here in the problem of learning a general function $f: L \rightarrow K$. In particular, for L=R and K=F₊ conventional regression emerges. Furthermore, a use of different lattices L or K may extend conventional regression.

A closed-form algebraic expression $f: L \rightarrow K$ for regression is convenient in practice. However, the development of a closed-form algebraic expression for regression usually takes time, if possible at all.

A useful alternative is to pursue regression 'by classification' using the classification algorithms presented in the previous sections. Note that regression by classification is a common practice in machine learning (Mitchell 1997). A disadvantage of the aforementioned alternative is that, typically, it is used only in interpolation applications. Nevertheless, regression 'by classification' is characterized by short development/tuning times (Papadakis et al. 2005). Moreover, regression 'by classification' is technically feasible due to the dropping costs of both computing power and computer memory.

7.5 Genetic Algorithm Optimization

Genetic Algorithms (GAs) include a set of global optimization algorithms inspired from Darwin's theory of evolution. GAs work by maintaining a population of solutions to a specific problem in an encoded form (genotypes) on which they apply simplified metaphors of principles like: genetic material recombination via genotype mating, genetic mutation of genotypes, survival of the fittest, etc. Through those operations GAs evolve the population of solutions for several generations thus producing gradually fitter individuals (i.e. better solutions) and driving the population towards an optimal region of the search space. GAs have proved their efficiency in difficult optimization problems (Goldberg 1989; Fogel 1999).

In the context of this work, GAs were used for improving performance of a model in two different manners (1) by optimizing the structure of a model, and (2) by optimizing the positive valuation functions in constituent lattices. Note that GA-based *structure optimization* has been pursued by various authors (Ishigami et al. 1995; Shimojima et al. 1995; Polani 1999; Su and Chang 2000; Hartemink et al. 2002; Shmulevich et al. 2002; Leung et al. 2003). However, optimization of positive valuation functions has been neglected. The reason for the latter is that only mass function $m(t)=1$ is considered (implicitly) in the literature. This work has demonstrated how alternative mass functions can introduce useful non-linearities in a lattice data-domain including R^N (Kaburlasos et al. 1997; Kaburlasos and Petridis 1997, 1998, 2000, 2002; Petridis and Kaburlasos 1999; Kaburlasos 2002, 2003, 2004a, 2004b; Kaburlasos and Kehagias 2006a, 2006b). Note that an employment of GAs for estimating optimally positive valuation functions has been very encouraging at a cost of high training complexity (Kaburlasos and Kazarlis 2002; Kaburlasos et al. 2003; Kaburlasos and Papadakis 2006; Papadakis and Kaburlasos 2005).

In this work *simple* GAs have been employed with binary encoding of the solutions, standard crossover and mutation operators, adaptive crossover and mutation rates, roulette wheel parent selection, generational replacement, and elitism. Furthermore, the GA fitness function was an index of the quality of function approximation, e.g. the mean square error, on part of the training data. Typical models, which have used GAs successfully, include the Voting σ-FLNMAP and the grSOM models for both structure- and positive valuation function optimization.

PART III: APPLICATIONS AND COMPARISONS

Part III presents, comparatively, experimental results.
It also describes connections with established paradigms.

8 Numeric Data Applications

This chapter presents numeric data applications

This chapter demonstrates comparatively applications of the previous algorithms in a number of problems involving numeric data.

8.1 Artificial Data Sets

Two artificial data sets were used to demonstrate the capacity of a lattice algorithm for learning.

8.1.1 Nonlinearly Separable Data

An artificial two-dimensional dataset consisting of 112 points was composed to test the ability of the σ-FLN algorithm for clustering non-linearly separable data.

Starting with the cluster above and proceeding to the one underneath in Fig.8-1(a), the order of data presentation was from left to right and from bottom up. A critical threshold diagonal size $D_{crit}= 0.0202$ was employed. The σ-FLN clustered the data into the two clusters using 78 boxes, assigning 39 boxes per cluster (Fig.8-1(b)). Note that a box in Fig.8-1(b) is often a trivial one that is a line segment. This problem was originally presented in Kaburlasos and Petridis (1997).

8.1.2 Nonconvex Data

An artificial two-dimensional dataset consisting of 84 points was composed to test the ability of σ-FLN for clustering a non-convex dataset.

Starting from the top-left corner, the order of the data presentation in Fig.8-2(a) was to the right and downwards. A critical threshold diagonal size $D_{crit}= 0. 0408$ was employed.

Vassilis G. Kaburlasos: *Numeric Data Applications*, Studies in Computational Intelligence (SCI) **27**, 97–122 (2006)
www.springerlink.com

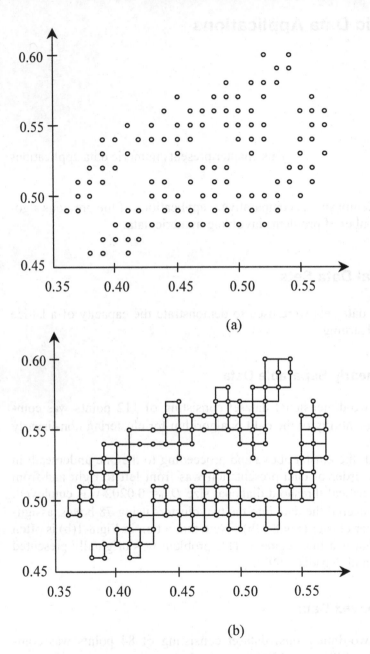

(a)

(b)

Fig.8-1   (a) Non-linearly separable two-dimensional categories.
(b) Separation was possible by σ-FLN clustering. Thirty-nine overlapping boxes identified each one of two categories.

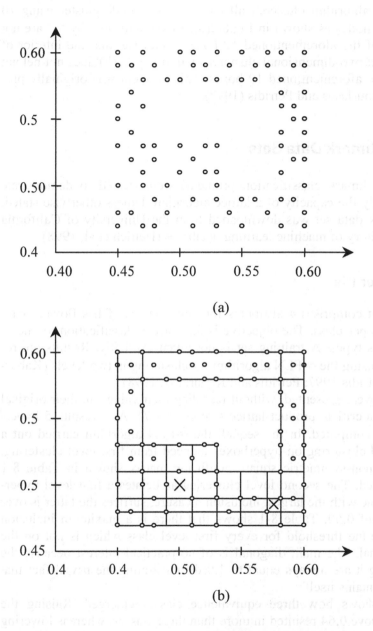

Fig.8-2   (a) A non-convex (hollow) two-dimensional category.
(b) The identification of this non-convex category was possible by
the σ-FLN using 10 overlapping boxes. Boxes indicated by an
'×' are not inside any of the 10 boxes.

The σ-FLN algorithm clustered all the data in a single cluster using 10 overlapping nodes as shown in Fig.8-2(b). Boxes indicated by 'x' are not inside any of the aforementioned 10 boxes. Likewise, that the interior of the identified two-dimensional cluster shown in Fig.8.2(b) does not belong to any of the aforementioned 10 boxes. This problem was originally presented in Kaburlasos and Petridis (1997).

8.2 Benchmark Data Sets

Several benchmark classification problems were treated to demonstrate comparatively the capacity of a lattice algorithm. Unless otherwise stated, a benchmark data set was downloaded from the University of California Irvine repository of machine learning databases (Hettich et al. 1998).

8.2.1 Fisher Iris

This data set comprises 4 attributes for three classes of Iris flowers with fifty flowers per class. The objective is the correct classification to one of the three Iris types. A training set is not given explicitly. Results are reported next using the σ-FLN algorithm for clustering in two levels (Kaburlasos and Petridis 1997; Petridis and Kaburlasos 1998).

The data were presented, without rescaling their values, in their original order using a critical product-lattice size D_{crit}= 1.71. As a result 15 hyperboxes were computed. In the sequel, the σ-FLN algorithm carried out a second level clustering on hyperboxes induced from first level clustering. Hence, the non-symmetric square confusion matrix shown in Table 8-1 was computed. The second level clustering associates a first level hyperbox to the one with the largest inclusion measure, suffices the latter is over a threshold of 0.60. Table 8-1 shows that there is a maximum inclusion value above the threshold for every first level class which is not on the main diagonal. The main diagonal is of no practical interest because the values along it are always equal to 1.0 demonstrating the trivial fact that any class contains itself.

Fig.8-3 shows how three equivalence classes emerged. Raising the threshold above 0.64 resulted in more than three classes, whereas lowering the threshold to less than 0.57 resulted in fewer than three classes. In conclusion, without the presence of a teacher but using exclusively σ-FLN for clustering, the Fisher's Iris data were clustered in three classes with only three (3) data misclassifactions as summarized in Table 8-2.

TABLE 8-1   This non-symmetric confusion matrix shows the inclusion measure value of a hyperbox, induced from the Iris data by the σ-FLN algorithm, into another hyperbox. The largest inclusion measure value (outside the main diagonal) in each row is underlined and associates a hyperbox with another one.

hyperbox	1	2	3	4	5	6	7	8	9	10	11	12	13	14	15
1	1.0	<u>.67</u>	.42	.64	.66	.38	.59	.54	.45	.36	.30	.43	.25	.27	.29
2	.67	1.0	.52	.61	.61	.45	.65	<u>.68</u>	.57	.43	.35	.54	.28	.29	.32
3	.54	.68	1.0	.58	.51	.71	.57	<u>.82</u>	.72	.66	.58	.81	.45	.40	.44
4	.64	.61	.44	1.0	.58	.40	<u>.65</u>	.56	.43	.36	.31	.42	.26	.24	.26
5	<u>.93</u>	.86	.55	.82	1.0	.50	.82	.74	.60	.48	.40	.57	.34	.36	.39
6	.53	.63	.76	.56	.50	1.0	.55	<u>.83</u>	.74	.74	.67	.80	.51	.46	.49
7	.85	<u>.93</u>	.62	<u>.93</u>	.83	.56	1.0	.81	.65	.53	.44	.63	.37	.36	.39
8	.76	<u>.93</u>	.89	.75	.76	.81	.78	1.0	.85	.72	.58	.83	.46	.44	.48
9	.65	.81	.79	.62	.61	.76	.65	<u>.90</u>	1.0	.71	.55	.83	.44	.44	.49
10	.50	.61	.71	.50	.48	.74	.52	.74	.70	1.0	.71	<u>.90</u>	.54	.52	.59
11	.40	.46	.59	.42	.38	.63	.41	.56	.51	.67	1.0	<u>.74</u>	.65	.55	.60
12	.61	.78	.87	.60	.58	.80	.63	.87	.85	<u>.92</u>	.80	1.0	.59	.53	.60
13	.25	.28	.34	.26	.24	.37	.26	.33	.31	.38	.49	.41	1.0	<u>.68</u>	.64
14	.27	.29	.31	.24	.26	.33	.25	.32	.31	.37	.42	.37	.68	1.0	<u>.72</u>
15	.41	.44	.47	.37	.39	.49	.38	.48	.47	.56	.61	.56	.90	<u>.95</u>	1.0

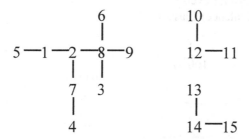

Fig.8-3   The second level of σ-FLN clustering, partitions the first level's hyperboxes into three equivalence classes.

Table 8-3 shows the corresponding best results obtained using the min-max neural network (Simpson 1993). Besides σ-FLN's superior classification results, it is also important to note that the σ-FLN tracked down 3 classes after two levels of clustering, whereas the min-max neural network identified a total of 14 classes in one stage of clustering which were then assigned to the 3 Iris classes by an external teacher.

Note that a two-stage clustering has been proposed in neural computing by other authors (Wang et al. 2003), where two different neural networks were used for processing N-dimensional points. Nevertheless, the second level clustering by the σ-FLN here involves *information granules* (i.e. N-dimensional hyperboxes) instead of N-dimensional points.

Additional experimental results are shown in Table 8-4. The results in the first two lines of Table 8-4 were produced in the context of this work using the σ-FLNMAP algorithm and, respectively, the leave-1-out and leave-25%-out training paradigms. The (marginally) better performance of the σ-FLNMAP algorithm compared to the σ-FLN was attributed to the fact that σ-FLNMAP uses 'labeled data' during training, whereas σ-FLN attaches labels only after clustering. Results by various SOM versions are from Kaburlasos and Papadakis (2006). The remaining of the results in Table 8-4 are from Petridis and Kaburlasos (1998).

TABLE 8-2   2-level σ-FLN clustering on of the Iris Benchmark

Number of Clustering Levels = 2
Number of (Equivalence) Classes = 3

class	1	2	3
1	100 %	-	-
2	-	94 %	6 %
3	-	-	100 %

TABLE 8-3   Min-Max Neural Clustering on the Iris Benchmark

Number of Clustering Levels = 1
Number of Classes = 14

class	1	2	3
1	100 %	-	-
2	-	88 %	12 %
3	-	10 %	90 %

TABLE 8-4 Recognition results for the Fisher Iris benchmark data.

Method	% Right on Testing
σ-FLNMAP for classification, leave-1-out	99.30
σ-FLNMAP for classification, leave-25%-out	99.10 (*)
2-level σ-FLN for clustering	98.00
greedy grSOM, GA optimized, leave-34%-out	97.60 (*)
SLF	95.84
greedy grSOM, leave-34%-out	95.60 (*)
σ-FLN for clustering, leave-25%-out	95.04 (*)
Back Propagation	94.67
σ-FLN for clustering, leave-1-out	94.00
Min-max neural net for clustering	92.67
CMOC	92.00
Kohonen SOM	91.00 (*)

(*) Average in 10 random experiments

8.2.2 Wine - Glass - Ionosphere - Sonar

Whenever a training data set is not defined explicitly below, a training set was defined either by leaving 1 datum out or by leaving randomly 25% of the data out. For leave-1-out, the experiment was repeated so as to leave, in turn, all data out. For leave-25%-out, typically ten random experiments were carried out and the average performance was reported.

Wine

The data are 13-dimensional vectors that correspond to various wine ingredients. A total number of 178 data vectors are given, distributed by 59, 71, and 48 in three wine types. The objective is classification to one among three wine types. A training set is not given explicitly.

Results with the σ-FLNMAP were produced in this work. Results by the *greedy grSOM* are from Kaburlasos and Papadakis (2006). The other results in Table 8-5 are from Petridis and Kaburlasos (1998) for training data sets of comparable size.

Glass

The data are 9-dimensional vectors that specify various chemical elements in two types of glasses, namely float-processed glass and non-float processed window glass. A total number of 87 vectors are given for float-processed glass; moreover 76 vectors are given for non-float processed window glass. The objective is the classification of a glass in one of the two classes. No training set is given explicitly.

Results with *Voting σ-FLNMAP* are from Kaburlasos et al. (2003); note that the average number of voters in the ensemble was 7.0 with a standard deviation of 3.29 in 10 experiments. Results with the σ-FLNMAP were produced in this work. The other results in Table 8-6 are from Petridis and Kaburlasos (1998) for training data sets of comparable size.

Ionosphere

The objective is classification of radar return signals in one of two classes based on a vector of 34 attributes. More specifically, the classes are 'good' radar returns, which show evidence of some type of structure in the iono-sphere, and 'bad' radar returns that do not. A total number of 200 instances are for training, whereas 151 instances are for testing.

Results with the σ-FLNMAP were produced in this work. The other results in Table 8-7 are from Petridis and Kaburlasos (1998) for training data sets of comparable size.

Sonar

The data are 60-dimensional vectors of numbers in the range 0.0 to 1.0 that correspond to sonar signals bounced off either a metal cylinder or a cylindrical rock. More specifically, 104 vectors are for training (including 49 'Mine' and 55 'Rock' vectors) and another 104 are for testing (including 62 'Mine' and 42 'Rock' vectors). The objective was the classification of sonar testing signals to one of the two classes: 'Mine' or 'Rock'.

Results with the σ-FLNMAP were produced in this work. The other results in Table 8-8 are from Petridis and Kaburlasos (1998) for training data sets of comparable size.

TABLE 8-5   Recognition results for the Wine benchmark data set.

Method	% Right on Testing
greedy grSOM, leave-10%-out	100.00 (*)
σ-FLNMAP for classification, leave-1-out	100.00
Rprop	100.00
RDA	100.00
σ-FLNMAP for classification, leave-25%-out	99.70 (*)
QDA	99.40
greedy grSOM, GA optimized, leave-34%-out	99.30 (*)
LDA	98.90
σ-FLN for clustering, leave-25%-out	96.50 (*)
1NN	96.10
FLR for clustering, leave-1-out	96.10
CLUSTER	84.09

(*) Average in 10 random experiments

TABLE 8-6 Recognition results for the Glass benchmark data set.

Method	% Right on Testing
σ-FLNMAP for classification, leave-1-out	99.30
σ-FLNMAP for classification, leave-25%-out	98.80 (*)
Voting σ-FLNMAP (10-fold cross validation)	96.87
Rprop	95.23
FASTCLUS	89.25
Nearest Neighbor	82.82
Beagle	82.20
σ-FLN for clustering, leave-25%-out	81.84 (*)
σ-FLN for clustering, leave-1-out	80.98
Discriminant Analysis	73.62

(*) Average in 10 random experiments

Table 8-7 Recognition results for the Ionosphere benchmark data set.

Method	% Right on Testing
σ-FLNMAP for classification	98.01
IB3	96.70
Rprop	96.23
σ-FLN for clustering	96.00
Backpropagation	96.00
C4	94.00
Multiresolution Algorithm	92.87
Nearest Neighbor	92.10
Non-linear Perceptron	92.00
Linear Perceptron	90.70

Table 8-8 Recognition results for the Sonar benchmark data set.

Method	% Right on Testing
σ-FLNMAP for classification	96.15
σ-FLN for clustering	94.23
Fuzzy ARAM	94.20
K-Nearest Neighbor	91.60
BackProp: Angle-Dep. [1]	90.40
Nearest Neighbor	82.70
BackProp: Angle-Ind. [2]	77.10

(1) Angle Dependent data ordering. (2) Angle Independent data ordering.

8.2.3 Vowel

This dataset includes 10-dimensional vectors of the linear predictive coefficients regarding eleven steady state vowels of British English. The training data set includes 528 vectors, whereas the testing data set includes 462 vectors. The objective is classification of the testing data to their corresponding classes; the latter are the eleven steady state vowels.

The σ-FLN was used as described in Petridis and Kaburlasos (1999). Details of σ-FLN's application are illustrated in Table 8-9. Four different types of data normalization were used as indicated in a row in Table 8-9. For each normalization type eight different ranges Range-1,...,Range-8 of the diagonal size D_{crit} were tested. A cell of Table 8-9 displays both the maximum (M) and minimum (m) '% correct' on the testing data set.

In cases (c1)-(c4) positive valuation functions $v(x_1,...,x_{10})=x_1+...+x_{10}$ were employed. Cases (c5)-(c7) employed the same normalizing intervals as cases (c2)-(c4) above, nevertheless the positive valuation functions were scaled by the corresponding standard deviations; more specifically we used positive valuation function $v_{new}(x_1,...,x_{10})=\sigma_1 x_1+...+\sigma_{10} x_{10}$, where σ_i, $i=1,...,10$ is the standard deviation of the corresponding vector entry in the training data set.

An employment of the aforementioned positive valuation function was motivated by the hypothesis that entries clustered more closely around their means convey a lesser discriminatory power than other, more dispersed entries. The experimental testing confirmed the aforementioned hypothesis. In particular, in case (c7) and Range-6 the σ-FLN achieved a maximum classification accuracy of 60.17% on the testing data; moreover 195 hyperboxes were computed.

As expected theoretically, the total number of hyperboxes computed in the experiments was a decreasing function of diagonal size D_{crit}. For instance the numbers of hyperboxes computed for sizes D_{crit}= 0.005, 0.055, 0.105, 0.155 and normalization type (c7) were 527, 344, 217, and 162, respectively. Table 8-10 summarizes selected classification results by different methods (Petridis and Kaburlasos 1999).

The *Voting σ-FLNMAP* produced the best result (67%) than any other FLN model in this problem. Estimates $\hat{D}_a$ and $\hat{n}_V$ were induced from the training data, in particular $\hat{D}_a$ =0.75 and $\hat{n}_V$ =442 as explained in Kaburlasos (2003). The average number of hyperboxes was 104.10 with standard deviation 6.20. Note that the classification accuracy has been fairly stable with respect to both parameters D_a and n_V.

TABLE 8-9   Seven combinations of data normalization and positive valuations (c1)-(c7) were used with the σ-FLNMAP for eight different ranges of D_{crit}. Rows (c5)-(c7), marked with an asterisk, used different positive valuations than rows (c2)-(c4). A Table cell shows both the max (M) and the min (m) '% correct' on the testing data.

		Type of data normalization						
		(c1)	(c2)	(c3)	(c4)	(c5)	(c6)	(c7)
Ranges of D_{crit}		Full	[-6,6]	[-8,8]	[-10,10]	[-6,6]*	[-8,8]*	[-10,10]*
Range-1	M	50.65	55.84	55.84	55.84	57.79	57.79	57.79
(0-0.018)	m	50.65	55.84	55.84	55.84	57.79	57.79	57.79
Range-2	M	50.65	55.84	56.49	56.49	57.79	58.44	58.44
(0.02-0.038)	m	50.65	55.63	55.63	55.41	57.79	57.79	57.58
Range-3	M	50.65	56.28	56.06	58.66	58.44	58.23	58.44
(0.04-0.058)	m	50.65	55.63	54.33	54.55	57.79	56.06	55.63
Range-4	M	50.65	56.28	58.87	58.01	57.58	58.66	58.44
(0.06-0.078)	m	50.65	54.55	56.71	54.98	55.84	57.14	56.49
Range-5	M	50.65	58.87	56.49	56.49	58.23	58.44	58.87
(0.08-0.098)	m	50.65	57.58	54.11	53.03	57.36	55.63	56.49
Range-6	M	50.87	58.44	56.49	57.36	58.66	58.66	60.17
(0.10-0.118)	m	50.65	55.63	52.81	55.19	57.14	56.28	57.14
Range-7	M	51.95	55.84	56.93	56.06	58.01	59.09	58.87
(0.12-0.138)	m	51.08	52.81	54.76	52.60	56.06	56.71	56.06
Range-8	M	52.81	56.28	57.58	53.46	59.09	59.52	57.79
(0.14-0.158)	m	52.38	53.03	55.84	51.06	56.71	58.44	54.55

Table 8-10   Performance of various methods, including the σ-FLN algorithm for clustering, in classifying eleven vowels.

Method	No. Hidden Units	% Correct
Voting σ-FLNMAP	104	67
3-D Growing Cell Structures	154	67
Gaussian ARTMAP (5 voters)	273	63
σ-FLN for clustering	195	60.17
Gaussian ARTMAP (w/o voting)	55	59
Nearest Neighbor	-	56.27
Gaussian Node Network	528	54.54
Radial Basis Function	528	53.46
Fuzzy ARTMAP (5 voters)	279	53
Multi-layer Perceptron	88	50.64
Modified Karneva Model	528	50.00
Single-layer Perceptron	-	33.33

8.2.4 Cleveland Heart

This benchmark includes 303 14-dimensional vectors. There are a few 'missing' data. The objective is to diagnose the presence of heart disease. The severity of a heart disease is denoted by an integer between 0 (no presence) and 4. Past experiments have only tried to distinguish between absence (value 0) from presence (values 1,2,3,4) of heart disease; the latter is referred to as '2-categories problem'. The '5-categories problem' is the classification problem that considers all 5 categories.

Table 8-11 summarizes the classification results by different methods reported in the literature for the 2-categories problem (Kaburlasos and Petridis 2000). The best result in Table 8-11 was produced by the *greedy grSOM* (Kaburlasos and Papadakis 2006). Table 8-12 details results by the FLNtf algorithm (Kaburlasos and Petridis 2000). In order to provide a good basis for comparison, 250 data were randomly 'kept-in' for training.

The SOM algorithms were evaluated using the one-sided 'matched pairs' statistical t test with df=9 degrees of freedom. It was found that the testing classification accuracy of both *greedy grSOM* variants was clearly better than the accuracy of *crisp SOM*; moreover, the performance of the *GA optimized greedy grSOM* was clearly better than the corresponding performance of the *greedy grSOM* (Kaburlasos and Papadakis 2006).

A 10-fold cross-validation was further carried out (Table 8-12) by partitioning the first 300 data of this benchmark into 10 consecutive parts of 30 data each. This experiment was repeated 10 times, such that each one of the aforementioned ten parts was employed for testing. The average classification accuracy of the testing data was 79.34% with a standard deviation 5.84; the average of the corresponding number of rules/boxes was 60 with standard deviation 3.52. Note that the performance of the FLNtf for the 2-categories problem is comparable to the best of other methods shown in Table 8-11. The 10-fold cross-validation experiment was also repeated for the 5-categories problem. The average testing classification accuracy in 10 experiments was 57.34 % with standard deviation 12.15; the corresponding average number of rules/boxes was 95.90 with standard deviation 3.41. Again, a larger number of rules/boxes resulted in the 5-categories problem.

Using a simple ensemble of σ-FLNMAP classifiers, as detailed in Kaburlasos (2003), the average classification accuracy error was 9.0% (Table 8-13). Furthermore, using a genetically (GA) optimized ensemble of σ-FLNMAP classifiers, based on probabilistic distribution functions as described in Kaburlasos et al. (2003), the average classification error in 10 experiments was reduced to 5.5%. In the latter case there corresponded an average number of 6.1 voters with standard deviation 1.45. The remaining of the entries in Table 8-13 are from Opitz and Maclin (1999).

Table 8-14 shows results in both the '2-category problem' and the '5-category problem' using various *FLN fit* classifiers in a series of experiments (Cripps et al. 2003). Note that the same 100 randomly generated data sets were used in both the '2-category' and the '5-category' problems. The clear improvement over FLNtf is attributed to the computation of *fits* smaller than *tightest fits*. Table 8-14 also shows that classifiers FLNff, FLNotf and FLNsf clearly produce a smaller number of rules than FLNtf.

Table 8-11   Performance of various methods from the literature in classifying the 2-categories Cleveland's Heart benchmark data set.

Method	Classification Accuracy (%)
greedy grSOM, GA optimized	83.20 (*)
greedy grSOM	79.90 (*)
Probability Analysis	79
Conceptual Clustering (CLASSIT)	78.9
ARTMAP-IC (10 voters)	78
Discriminant Analysis	77
Instance Based Prediction (Ntgrowth)	77
Kohonen *SOM*	75.20 (*)
Instance Based Prediction (C4)	74.8
Fuzzy ARTMAP (10 voters)	74
KNN (10 neighbors)	67

(*) Average in 10 random experiments

Table 8-12   Performance statistics regarding the FLNtf in classifying the 2-categories Cleveland's Heart benchmark data set.

Method	% classification accuracy		number of rules	
	average	stdv	average	stdv
FLNtf, 2-categories 10-fold Cross-Validation (*)	79.34	5.84	60.00	3.52
FLNtf, 2-categories Keep-250-in (**)	77.88	4.58	53.80	4.58
FLNtf, 5-categories 10-fold Cross-Validation (*)	57.34	12.15	95.90	3.41
FLNtf, 5-categories Keep-250-in (**)	56.74	7.23	86.81	4.67

(*) 10 random experiments   (**) 100 random experiments

Table 8-13   Classification error by various ensemble methods in the 2-categories Cleveland Heart benchmark data set.

Method	Classification Error (%)
Voting σ-FLNMAP (GA optimized)	5.5
Voting σ-FLNMAP (simple)	9.0
Bagging Backpropagation	17.0
Arcing Backpropagation	20.7
Boosting Backpropagation	21.1
Bagging C4.5	19.5
Arcing C4.5	21.5
Boosting C4.5	20.8

Table 8-14   Accuracy of classifiers FLNtf, FLNff, FLNotf, FLNmtf, and FLNsf in the 2-categories Cleveland Heart benchmark data set.

	FLN classifier	Statistics of % classification accuracy		Statistics of the number of rules	
		average	stdv	average	stdv
2-Categories Problem	FLNotf	89.51	3.60	34.79	6.58
	FLNsf	89.51	3.60	34.80	7.12
	FLNff	89.13	3.92	34.71	7.22
	FLNmtf	87.38	4.21	36.55	4.33
	FLNtf	77.88	4.58	53.80	4.58
5-Categories Problem	FLNff	66.74	4.96	53.47	10.3
	FLNotf	66.60	5.52	51.18	11.0
	FLNsf	66.49	5.59	49.47	9.8
	FLNmtf	65.38	5.0	61.24	7.0
	FLNtf	56.74	7.23	86.81	4.67

8.2.5 Credit Approval

We considered two data sets for classification. The first one, namely 'Australian Credit Approval' contains data regarding credit card applications. It includes 690 instances with 14 attributes per instance plus one class attribute, i.e. credit card approval /disapproval; hence, the number of classes is 2. The second data set, namely 'German Credit Database', contains data regarding good or bad human creditors. It includes 1000 instances with 24 attributes per instance plus one class attribute; again, the number of classes is 2. For both datasets, the first 90% of the instances were used for training and the last 10% for testing as in Quinlan (1996).

A few *Voting σ-FLNMAP* models were employed using different positive valuation functions (Kaburlasos and Kazarlis 2002; Kaburlasos et al. 2003). A model's parameters included (1) the number n_V of σ-FLNMAP voters, and (2) the weights w_i, $i=1...14$ of a positive valuation function $v= w_1 v_1 + ... + w_{14} v_{14}$, where an individual positive valuation v_i, $i=1...14$ in a constituent lattice $l= [0,1]$ was equal to either x or the corresponding cumulative distribution function (CDF) induced from the training data.

We employed a simple GA that used binary encoding of solutions, standard crossover and mutation operators (for the recombination and perturbation of solution genotypes), roulette wheel parent selection (for selecting parent genotypes for mating), generational replacement (total replacement of the parent population with the offspring population), and elitism. The fitness function of the GA was the classification accuracy of the *Voting σ-FLNMAP* model on the testing data. The population size was 20 genotypes per generation randomly generated at the beginning. The GA was left to evolve for 100 generations.

Classification results by various methods are shown in Table 8-15. The term 'non-weighted' means $w_1=...=w_{14}=1$. The results in the first two lines were produced using CDFs. The GA-computed weights w_i, $i=1,...,14$, which produced the best result in line 1 of Table 8-15, are shown in Table 8-16. Lines 3 and 4 in Table 8-15 show comparatively the corresponding classification results using a linear underlying positive valuation function $v_i(x)= w_i x$, $i=1,...,14$ in each constituent lattice $l= [0,1]$. Line 5 is from Quinlan (1996).

The results in Table 8-15 demonstrate the importance of configurable (genetically optimized) weights for the positive valuation functions of the constituent lattices. Moreover, Table 8-15 demonstrates an improved capacity for classification using weighted CDFs. It should also be pointed

out that the classification accuracies achieved by individual σ-FLNMAP
voters in an ensemble were clearly less than the classification accuracy of
the ensemble. For instance for 'CDF, weighted' classification, i.e. line 1 of
Table 8-15, the classification accuracies of the five σ-FLNMAP voters involved in the ensemble have been 82.60, 86.95, 88.40, 81.15 and 78.26%,
respectively. Note also that the corresponding average number of hyper-
boxes/rules computed in the ensemble has been 67.4 hyperboxes with
standard deviation 2.88.

For the 'German Credit Database' only one set of experiments was carried out using the *Voting σ-FLNMAP* model with positive valuation function $v(x_1,..., x_{14})= w_1x_1+...+w_{14}x_{14}$. The corresponding classification result
was 88% (Kaburlasos and Kazarlis 2002); for comparison note that a classification result of 71.6% has been reported in the literature using the C4.5
Rel.8 algorithm (Quinlan 1996).

Table 8-15   Performance of various methods in classifying the Australian
Credit Approval benchmark data set.

Method	Classification Accuracy (%)
Voting σ-FLNMAP (CDF, weighted)	94.2
Voting σ-FLNMAP (CDF, non-weighted)	76.8
Voting σ-FLNMAP (linear, weighted)	91.3
Voting σ-FLNMAP (linear, non-weighted)	84.0
C4.5 Rel.8	85.3

Table 8-16   GA-computed weights for 14 constituent lattices l= [0,1] in the
Australian Credit Approval classification problem.

Constituent Lattice	Weight	Constituent Lattice	Weight
l_1	0.28	l_8	0.83
l_2	0.44	l_9	0.10
l_3	0.59	l_{10}	0.71
l_4	0.45	l_{11}	0.71
l_5	0.62	l_{12}	0.83
l_6	0.06	l_{13}	0.40
l_7	0.20	l_{14}	0.18

8.2.6 Brown Corpus

The principal objective of this work was to test whether *senses*, i.e. *meanings* of words, give better text classification results than *words* (Kehagias et al. 2003). The *Brown Corpus* collection of documents was used which is distributed along with the *Wordnet*; the latter is an electronic thesaurus organized around the distinction between words and senses.

The Brown Corpus collection is a *semantic concordance* that is a combination of documents and a thesaurus. The documents are combined such that every substantive word in a document is linked to its appropriate sense in the thesaurus. The Brown Corpus semantic concordance makes use of 352 out of the 500 Brown Corpus documents, which are classified into fifteen categories. Linguists have manually performed *semantic tagging*, i.e. annotation of the 352 texts with WordNet senses.

In the context of this work each document in the Brown Corpus was represented by a vector of either words or senses. The *vocabulary* in a classification problem, involving a set of documents, was defined to be the set of N_w words $w_1, \ldots, w_{Nw}$ (or, the set of N_s sense $s_1, \ldots, s_{Ns}$) which appear in at least one document. A few different document representations are presented in the following for 'words'. The same representations were used for 'senses' as well. A document d was represented by either

(1) a *relative frequency* representation vector $\mathbf{d}=[d_1, \ldots, d_n, \ldots, d_{Nw}]$, where

$$d_n = \frac{\text{no. of times the } n^{\text{th}} \text{ word } w_n \text{ appears in document d}}{\text{total no. of words in document d}}, \text{ or}$$

(2) a *normalized frequency* representation vector $\mathbf{d}=[d_1, \ldots, d_n, \ldots, d_{Nw}]$ whose component d_n was defined as

$$d_n = \frac{\text{no. of times word } w_n \text{ appears in document d}}{\text{maximum no. of times word } w_n \text{ appears in any document}}.$$

Moreover, document d was represented by a *Boolean vector* $\mathbf{d}=[d_1, \ldots, d_n, \ldots, d_{Nw}]$ where d_n is either 1 or 0 when, respectively, the nth word w_n appears or does-not appear in document d.

A number of classification problems were treated. In particular, either *all the 15 categories* of the Brown Corpus with 352 documents, or *only 3 categories* (i.e. categories 1, 2 and 10) with 100 documents were considered. For each one of the aforementioned two problems either (1) the *nouns, verbs, adjectives* and *adverbs*, or (2) only the *nouns* and *verbs* were considered in a document. For all previous 2×2=4 combinations of problems, vectors of either *words* or *senses* were produced from the documents.

Three different classification algorithms were employed, namely *k-Nearest Neighbor* (kNN), *Naïve Bayes Classifiers* (NBC), and the *Voting σ-FLNMAP*; the kNN was employed with both a *relative frequency* and a *Boolean* representation. The NBC was employed with the *frequency* representation of a document. Each time, an algorithm was applied for several values of its parameter vector on ten different random partitions of a data set such that 2/3 of the data were used for training and the remaining 1/3 for testing. The classification results are summarized in Table 8-17. It follows that 'senses' performed better than 'words' in 36 out of 40 cases. Only in 3 cases the performance of senses was over 5 percentage points than the performance of words. Therefore it is concluded that, in this problem, the use of senses improves classification accuracy only marginally.

Among the three classification algorithms the *Voting σ-FLNMAP* performed best whereas NBC performed worst. The kNN gave fairly good classification results especially for the *Boolean* representation. The good performance of *Voting σ-FLNMAP* was attributed to both the effectiveness of inclusion measure σ and the model's capacity for generalization (Kehagias et al. 2003). Moreover, the performance of the *Voting σ-FLNMAP* remained stable for a wide range of values of the corresponding vigilance parameter (ρ_a) and it dropped sharply as ρ_a approached 1 as shown in Fig.8-4 and Fig. 8-6 for the 15- and the 3- categories problem, respectively (For details regarding ρ_a see in section 10.1.2). Fig.8-5 demonstrates that the number of hyperboxes increases exponentially as ρ_a approaches 1. Fig.8-7 demonstrates that for selected values of the vigilance parameter (ρ_a) an ensemble of σ-FLNMAP voters can improve stably the performance of an individual σ-FLNMAP voter in the ensemble.

Table 8-17	The average % classification accuracy in 10 experiments for various combinations of (1) parts of speech, (2) no. of categories, (3) a classification algorithm, and (4) a document representation. The code-words 'n', 'v', 'adj' and 'adv' stand, respectively, for 'noun', 'verb', 'adjective' and 'adverb'.

Parts of speech	no. of categ	kNN (Boolean)		kNN (R. Freq.)		NBC (Freq.)		*Voting σ-FLNMAP* (Boolean)		*Voting σ-FLNMAP* (N. Freq.)	
		w	s	w	s	w	s	w	s	w	s
n, adj,	3	80	76	76	80	53	56	79	81	80	82
v, adv	15	45	46	42	46	39	38	43	46	44	48
n, v	3	80	75	75	80	55	61	80	81	80	82
	15	45	47	41	46	41	38	41	47	44	47

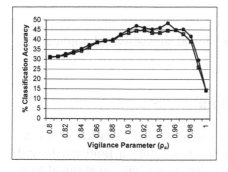

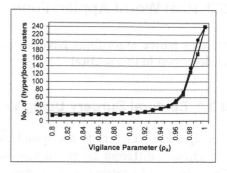

Fig.8-4
Average classification accuracy of *words* (□) and *senses* (•) versus the vigilance parameter (ρ_a) for 10 random training/testing data sets and the 15 categories of Brown Corpus. Senses have resulted in a marginally better classification accuracy.

Fig.8-5
Average *minimum* (□) and *maximum* (•) number of (hyper)boxes/clusters computed by *Voting σ-FLNMAP* versus the vigilance parameter (ρ_a) for 10 random training/testing data sets and the 15 categories of Brown Corpus. The number of boxes increases exponentially as ρ_a approaches 1; for $\rho_a=1$ the number of (hyper)boxes computed equals the number of training data.

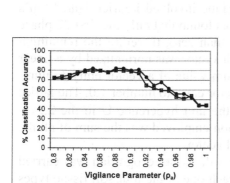

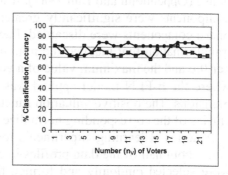

Fig.8-6
Average classification accuracy of *words* (□) and *senses* (•) versus the vigilance parameter (ρ_a) for 10 random training/testing data sets and 3 categories of Brown Corpus. Senses have resulted in a marginally better classification accuracy.

Fig.8-7
Classification of σ-FLNMAPs (□) and *Voting σ-FLNMAP* (•) with an increasing number n_V of voters versus the number n_V of voters for $\rho_a=0.89$. The *Voting σ-FLNMAP* resulted in both stability and an improvement in classification accuracy.

8.3 Real World Applications

Various lattice algorithms have been employed in real-world applications as described in this section.

8.3.1 Epidural Surgery Using Vectors

The epidural puncture to the spinal cavity involves the insertion of a needle through different layers of visco-elastic tissue into a fluid cavity and the delivery of anesthesia. An automated electro-mechanical surgical tool, that is a *mechatronics tool*, has been envisaged in order to control penetration through soft tissues in the epidural puncture (Petridis and Kaburlasos 2000). From an information processing point of view, the capacity to automate the epidural puncture boils down to the capacity for recognizing the type of soft tissues being penetrated.

Twenty-nine Raman spectra profiles were available that corresponded to four soft tissues encountered in the epidural surgical procedure. In particular, 6 data corresponded to *connective* tissue, 9 data to *muscle* tissue, 6 to *skin* tissue, and another 8 data corresponded to *fat* tissue. Typical Raman spectra profiles for various soft tissues are shown in Fig.8-8.

An efficient feature extraction technique involved Fourier Transform's phase (Oppenheim and Lim 1981). It was found that only the first 22 phase components were significantly greater than zero, therefore the remaining phase components were discarded. Each vector was normalized by a linear transformation that mapped the minimum value of a vector entry to number 0.0 and the maximum value of a vector entry to number 1.0. The lattice where the σ-FLN model applied was the unit hypercube U in the 22 dimensions. The positive valuation function employed was the sum of all the entries of the corresponding normalized vector.

A series of learning experiments for soft tissue recognition was carried out. About half of the data profiles in each one of the four soft tissue types were selected randomly and formed the training data set. The remaining data were the testing data set. Every time, and as a result of training, four different clusters were computed corresponding to each one of the 'connective', 'muscle', 'skin', and 'fat' tissues. A teacher attached labels to learned 22-dimensional hyperboxes. Then the testing data were applied. No misclassifications were recorded here. This problem was originally presented in Petridis and Kaburlasos (1999).

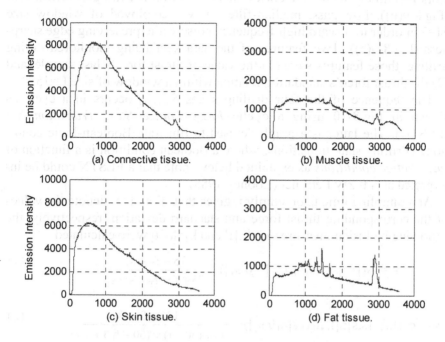

Fig.8-8   Raman spectra profiles that correspond to four soft tissues encountered in the epidural surgical procedure. A Raman profile shows laser emission intensities (y-axis) versus wavenumber offsets away from the driving frequency at 633 nm (x-axis).

8.3.2 Orthopedics Bone Drilling

Minimal access surgical drilling of long bones was pursued in orthopedics (Kaburlasos et al. 1997). In particular, the problem was to monitor the advancement of drilling through a long bone and to identify incipient breakthrough in order to contain the drilling bit upon breakthrough.

Fresh swine long bones were used experimentally. More specifically, 24 drilling profiles were used partitioned in groups of eight profiles for various combinations of feed rate (in mm/min) and rotational speed (in RPM). For example, a thrust force drilling profile is shown in Fig.8-9(a), where the two lobes correspond to the two cortical walls of a long bone.

The objective was to identify four drilling states: (S_a) the drilling bit is outside the cortex, (S_b) the bit is entering a cortical wall, (S_c) the bit is working entirely within the cortex, and (S_d) the bit is exiting a cortical wall (Fig.8-9(b)). Low-pass, median-filtering was employed of window size M=3 in order to discard high frequency noise while preserving edge sharpness (Fig.8-9(c)). Two features of the data kept being monitored during drilling, those features were (1) the value of thrust force's last sample, and (2) the thrust force's standard deviation within a window of size L=15.

The sequence of the four drilling states S_a...S_d occurs in a cycle as shown in Fig.8-10 using a cyclic *Fuzzy Lattice State Transition Net* (FLSTN). The latter is a graph devised to illustrate Boolean-logic conditioned drilling state transitions, where a Boolean condition is a function of *fuzzy lattice conditions* as explained below; note that a FLSTN could be interpreted as a fuzzy Petri net (Looney 1988).

At a sampling time t_i, or simply i, given that 'f' and 's' denote the values of the corresponding thrust force and standard deviation, respectively, the following inclusion measures $k_1=k_1[i]$ and $k_2=k_2[i]$ were calculated

$$k_1 = \sigma([f,f]\times[s,s] \leq [0,f_T]\times[0,s_T]) = \frac{75 + f_T + s_T}{75 + f \vee f_T + s \vee s_T} \tag{1}$$

$$k_2 = \sigma([f,f]\times[s,s] \leq [f_L,f_{MAX}]\times[0,s_S]) = \frac{135 - f_L + s_S}{75 + (60 - f) \vee (60 - f_L) + s \vee s_S} \tag{2}$$

where subscripts 'T', 'S', and 'L' stand for 'tiny', 'small', and 'large', respectively. The four Boolean conditions B_a...B_d were computed as follows.

B_a : ($k_1[i] \leq 0.95$).AND.($k_2[i] > 0.75$),
B_b : ($k_1[i] \leq 0.79$).AND.($k_2[i] > 0.99$),
B_c : ($k_1[i] > 0.79$).AND.($k_2[i] \leq 0.99$), and
B_d : ($k_1[i] > 0.95$).AND.($k_2[i] \leq 0.80$),

where the thresholds 0.75, 0.80, 0.95, 0.99 were user-defined by trial-and-error. The aforementioned scheme was interpreted as a simple *fuzzy lattice reasoning* (FLR) scheme (Kaburlasos et al. 1997).

The cyclic FLSTN was applied on the 24 available drilling profiles. In all profiles the four states S_a through S_d, including the final bone breakthrough, were captured pointedly. The vertical dashdotted lines shown in Fig.8-9(d) were drawn automatically to separate two successive states among S_a...S_d. Note that the transition from state S_d to the state S_a marks the conclusion of a cortical wall's drilling.

The FLR, comparatively with alternative detection algorithms, has demonstrated a better capacity to support drilling of *fractured* bones. Another merit of FLR includes a capacity for handling *granules* of sensory data so as to compensate for uncertainty in the measurements.

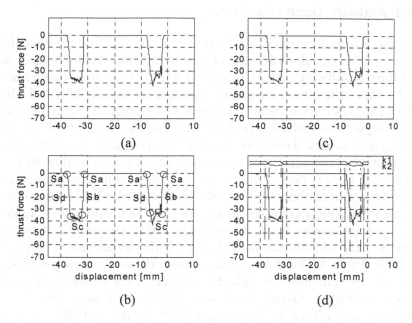

Fig.8-9 (a) A long bone drilling profile through both cortical walls.
(b) The four states S_a, S_b, S_c, S_d for each cortical wall drilling. Transitions between successive states are indicated by a small circle.
(c) Low-pass, median-filtered drilling profile of a long bone.
(d) Above a drilling profile are plotted inclusion measures k_1 and k_2.

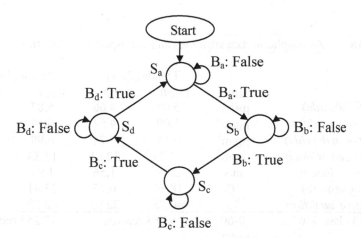

Fig.8-10 A Fuzzy Lattice State Transition Net (FLSTN) for drilling through a long bone's cortical wall. Two encirclements $S_a \rightarrow S_b \rightarrow S_c \rightarrow S_d$ of the FLSTN correspond to crossing both cortical walls of a long bone.

8.3.3 Ambient Ozone Estimation

Air quality is frequently assessed 'on line' by humans; computationally intensive mathematical models are also used 'off-line'. A multi-agent decision-support software system, namely O_3RTAA, was developed for assisting humans in real-time air quality assessment (Athanasiadis and Mitkas 2004). The O_3RTAA system is equipped with predictive (software) agents, which implement various classifiers including the σ-FLNMAP, to fill in missing data values. This work demonstrates prediction of *ozone concentration* by classification, based on meteorological and air quality data.

The data were collected from a meteorological station in the district of Valencia, Spain as described in Athanasiadis et al. (2003). In summary, several meteorological attributes and air-pollutants values were recorded on a quarter-hourly basis during year 2001. Originally, there were approximately 35,000 records, with seven attributes per record plus a class attribute (Table 8-18). After removing records with missing values, the data set was split into two subsets: one subset for training and another subset for testing. Ozone level was characterized as either 'low' or 'medium' for values in the ranges 0-60 $\mu g/m^3$ and 60-100 $\mu g/m^3$, respectively. The prediction accuracies of three classification algorithms are shown in Table 8-19.

Various backpropagation neural networks were employed. A best performance of 82.15% on the testing data was obtained by a neural network with 11 hidden neurons, linear transfer functions (for the hidden

Table 8-18   Atmospheric data attributes and corresponding statistics

Attribute	Units	Training Data mean	Training Data stdv	Testing Data mean	Testing Data stdv
SO_2 (*Sulfur dioxide*)	$\mu g/m^3$	5.08	4.66	5.87	5.84
NO (*Nitrogen oxide*)	$\mu g/m^3$	5.09	4.54	6.49	7.55
NO_2 (*Nitrogen dioxide*)	$\mu g/m^3$	9.55	7.74	6.98	5.80
NO_x (*Nitrogen oxides*))	$\mu g/m^3$	17.09	12.61	15.83	13.97
VEL (*Wind velocity*)	m/s	2.19	1.56	1.91	1.25
TEM (*Temperature*)	°C	18.34	6.77	23.41	7.60
HR (*Relative humidity*)	%	60.23	22.93	82.59	17.54
ozone (O_3) class 'low'	0-60 $\mu g/m^3$	6,265 records		12,255 records	
ozone (O_3) class 'medium'	60-100 $\mu g/m^3$	4,761 records		5,138 records	

layer neurons), sigmoid transfer functions (for the output layer neurons), and the resilient backpropagation algorithm. The C4.5 algorithm for classification produced a decision tree whose nodes specified inequalities for the values of environmental attributes and the tree leaves specified an output class. The best performance on the testing data set was 73.75% for a pruned decision tree including 66 leaves. The σ-FLNMAP produced a best testing data classification performance of 83.24% for $\rho= 0.59$. Due to the inclusion measure function $\sigma(u{\le}w)= v(w)/v(u{\vee}w)$, a datum $x= (x_1,...,x_{13})$ can always be assigned to a category even if x is outside all hyperboxes.

The classification results by σ–FLNMAP were only marginally better than those by Backpropagation, and they clearly outperformed the classification results by C4.5. Additional advantages of the σ–FLNMAP classifier included faster training in one pass through the data. Also, only three rules/hypercubes were induced from the training data (Table 8-20). Note that rules (represented as hyperboxes) are easily understood by humans, whereas the answers of a backpropagation neural network are not easily deciphered.

Table 8-19   Confusion Matrices for three classifiers

	Backpropagation		C4.5		σ-FLNMAP	
	no. data classified as		no data classified as		data classified as	
no. data in class	'low'	'med'	'low'	'med'	'low'	'med'
'low'	9,905	2,351	8,487	3,769	11,244	1,012
'med'	754	4,384	798	4,340	1,904	3,234
overall success:	82.15%		73.75%		83.24%	

Table 8-20   Rules induced from the training data by the σ-FLNMAP classifier

Rule	IF	THEN
R1	$(SO_2{\in}[3,87])$ and $(NO{\in}[2,74])$ and $(NO_2{\in}[4,57])$ and $(NO_x{\in}[6,151])$ and $(VEL{\in}[0.1,9.4])$ and $(TEM{\in}[4,28.6])$and $(HR{\in}[9,99])$	Ozone is 'low'
R2	$(SO_2{\in}[3,47])$ and $(NO{\in}[2,24])$ and $(NO_2{\in}[4,36])$ and $(NO_x{\in}[6,54])$ and $(VEL{\in}[0.1,11.1])$ and $(TEM{\in}[5,35])$ and $(HR{\in}[8,99])$	Ozone is 'med'
R3	$(SO_2{\in}[3,52])$ and $(NO{\in}[2,89])$ and $(NO_2{\in}[4,65])$ and $(NO_x{\in}[6,176])$ and $(VEL{\in}[0.1,7.5])$ and $(TEM{\in}[9,35])$ and $(HR{\in}[24,99])$	Ozone is 'low'

8.4 Discussion of the Results

The application of several lattice algorithms in clustering, classification, and regression problems involving numeric data has shown that the aforementioned algorithms can be highly effective. The presentation of experimental results here was brief. For further application details the interested reader may refer to the cited publications.

A key idea behind the proposed algorithms is the induction of *information granules* from the data. A critical parameter is the *size* of induced information granules. In particular, regarding the computation of hyperboxes in R^N, an information granule corresponds to a *hyperbox*; moreover, the corresponding size is a hyperbox *diagonal*. An optimum information granule size can be estimated by a search algorithm, e.g. a steepest-descent- or a genetic algorithm.

Classifier *Voting σ-FLNMAP* was computationally intensive. Nevertheless, when applied, it usually produced very good classification results.

Where applicable, a 'missing' datum in a complete lattice $(L,\leq)$ was replaced by the corresponding minimum interval [I,O]; moreover a 'don't care' datum was replaced by the corresponding maximum interval [O,I].

It was confirmed experimentally that the technique of *maximal expansions* does not improve significantly classification performance, moreover it introduces significant computational overhead. Therefore the technique of maximal expansions was often omitted.

Mass functions were not employed to their full potential in this work. An optimal estimation of mass functions is a topic for future research.

9 Nonnumeric Data Applications

This chapter presents non-numeric data applications

This chapter demonstrates comparatively applications of the previous algorithms in a number of problems including non-numeric data.

9.1 Benchmark Data Sets

Several benchmark classification problems were treated to demonstrate comparatively the capacity of a lattice algorithm. Unless otherwise stated, a benchmark data set was downloaded from the University of California Irvine repository of machine learning databases (Hettich et al. 1998).

9.1.1 Shuttle

The Shuttle benchmark is a tiny data set including 15 data vectors. The data consist of six *totally ordered* nominal attributes (which determine conditions for space shuttle landing) plus a class label per data vector (which determines whether landing should be automatic or manual). Six and nine data vectors correspond to landing 'by manual control' and 'by automatic control', respectively. No training set is given explicitly. About 29% of the attribute values are 'don't care'.

We carried out experiments with the FLNtf algorithm considering one *constituent lattice* per *totally ordered* nominal attribute. The FLNtf was employed in the leave-1-out mode; that is, one of the 15 data vectors was left out for testing, and the remaining 14 data vectors were used for training. The experiment was repeated leaving, in turn, all data out for testing (Kaburlasos and Petridis 2000).

Several different linear positive valuations were selected heuristically and the best results were obtained for the linear coefficients $(c_1,c_2,c_3,c_4,c_5,c_6)= (1,2,1,1,1,1)$; in the latter experiment 1 and 2 data have been misclassified from the classes 'automatic landing' and 'manual land-

Vassilis G. Kaburlasos: *Nonnumeric Data Applications*, Studies in Computational Intelligence (SCI) **27**, 123–140 (2006)
www.springerlink.com
© Springer-Verlag Berlin Heidelberg 2006

ing', respectively. In different experiments, the number of tightest fits (rules) calculated for class 'automatic control' was between 2 and 4, whereas the corresponding number of tightest fits (rules) for class 'manual control' was between 3 and 5.

In conclusion, the FLNtf has demonstrated a capacity to generalize successfully on 12 out of 15 data vectors with nominal attributes; in other words 80% of new and hitherto unknown *totally ordered* nominal data were classified correctly.

9.1.2 Abalone

The Abalone benchmark includes gender data as well as physical measurements regarding marine snails, namely abalones. There exist 4177 records of data including 3133 records for training and 1044 records for testing. Each data record has 9 entries. The first 8 entries are one gender plus seven physical (numeric) measurements and they are used as inputs; more specifically 'I', 'M', 'F' stand for Infant, Male, and Female, respectively. The ninth entry indicates one of three age groups and it is used as an output. The objective is to learn predicting the age of an abalone.

Other researchers have arbitrarily mapped the gender of an abalone to R^3; then, they fed a 10-dimensional vector to a neural network as explained in Petridis and Kaburlasos (1999). In the context of this work the lattice ontology in Fig.9-1 was used for representing the *parent-of* relation, where the symbol 'G' was inserted to render the corresponding constituent lattice a complete one; hence, the first input was encoded as a single input to a σ-FLN(MAP). Fig.9-1 also shows the values of the employed positive valuation function. The employed isomorphic function θ has been $\theta(I)=G$, $\theta(F)=F$, $\theta(M)=M$, and $\theta(G)=I$. The remaining 7 (numeric) Abalone input entries were encoded in the lattice '7-dimensional unit-hypercube'.

Algorithm σ-FLNMAP was employed in a series of experiments (Kaburlasos 2003). The training data set was partitioned into a pair of sets (TRN_k,TST_k), $k=1,\dots,10$. A grid of pairs (D_a,n_V) was considered for $D_a=4$, 3.99, 3.98,..., 0.05 and $n_V=2,3,\dots,500$. Optimal estimates $\hat{D}_a=4.32$ and $\hat{n}_V=36$ were calculated, and a 68.29% classification accuracy resulted in. The other results in Table 9-1 are from Kaburlasos and Petridis (2002).

Only a single pass through the training data sufficed for stable learning with the σ-FLN. Then a teacher attached labels to the learned intervals (clusters) by majority voting. A category included a number of intervals. The total number of clusters was a decreasing function of the threshold size D_{crit}. For instance, sizes $D_{crit}=0.500$, 0.600, and 0.700 yielded 346,

256, and 192 clusters, respectively. The average classification correct rate on the testing data has been 60.65% with standard deviation 1.24. The performance peaked to 65.32% for threshold size $D_{crit}=0.636$; moreover 223 clusters were induced. We remark that the best classification results were recorded for the lattice ordering and the positive valuation function shown in Fig.9-1. For alternative lattice orderings and/or alternative positive valuations the performance deteriorated up to 3 percentage points.

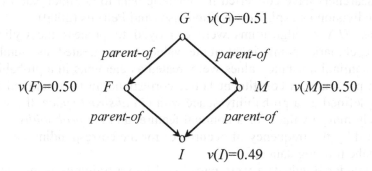

$$G \quad v(G)=0.51$$

parent-of   *parent-of*

$v(F)=0.50 \quad F \qquad\qquad M \quad v(M)=0.50$

parent-of   *parent-of*

$$I \quad v(I)=0.49$$

Fig.9-1   The three symbols 'I', 'F', 'M' in the first input entry of the Abalone benchmark were considered to be lattice ordered. Symbol 'G' was introduced to make lattice L a *complete* one. The values of the employed positive valuation function $v(.)$ are also shown.

Table 9-1   Performance of various methods in classifying the Abalone benchmark data. Due to its treatment of the Abalone data as lattice elements, a σ-FLN(MAP) algorithm reduced the total number of inputs.

Method	% Correct
σ-FLNMAP (36 voters)	68.29
Cascade-Correlation (5 hidden nodes)	65.61
σ-FLN	65.32
Back-Propagation	64.00
Nearest Neighbor (k=5)	62.46
Cascade-Correlation (no hidden nodes)	61.40
C4.5	59.20
Dystal	55.00
Linear Discriminant Analysis	32.57

9.1.3 Cylinder Bands

The Cylinder Bands benchmark data set consists of 540 data records including 20 numeric plus 20 nominal attributes per data vector. The *class* attribute is included in the nominal data and it specifies one of two classes those are 'band' and 'noband' classes. In all, there exist 228 and 312 data records for the 'band' and 'noband' classes, respectively. About 4.75% of the overall data attributes are 'missing'. No training set is given explicitly. Other researchers have converted the numeric data to nominal data by interval subdivision as explained in Kaburlasos and Petridis (2000).

Various *FLN fit* algorithms were employed to process the Cylinder Bands benchmark. Numeric attribute values were treated as numbers whereas nominal attribute values were treated as elements in a probability space. In particular, a constituent lattice corresponding to a nominal feature was defined as a probability space with (1) *abstract space*, the set of all (finitely many) values for a nominal feature, and (2) *probability measure* defined by the frequency of occurrence for the corresponding attribute values in the training data set.

The Cylinder Bands data were partitioned into a training-testing pair of data sets in several different ways. Positive valuation function $v=v_1+\ldots+v_N$ was employed. Furthermore, *complement coding* was used to represent a data vector $(x_1,\ldots,x_N)$ as vector $(1-x_1,x_1,\ldots,1-x_N,x_N)$. The result by both 'Apos' and FLNtf in Table 9-2 are from Kaburlasos and Petridis (2000). The other results in Table 9-2 are from Cripps et al. (2003) for training/testing data sets of comparable size. Only for FLNtf, was the computed clusters/rules independent of the order presenting the training data. In all cases a tightest fit was interpreted as a 'rule'.

Table 9-2   Performance of various methods on the Cylinder Bands benchmark. The last column shows the number of induced rules.

Method	% Correct	Testing data misclassifications (band + noband)	no. of rules
FLNotf	85.56	17 + 9 = 26	11
FLNff	80.56	27 + 8 = 35	10
FLNsf	80.56	27 + 8 = 35	10
FLNtf	78.88	29 + 9 = 38	10
FLNmtf	78.33	25 + 14 = 39	11
Apos	63.64	-	-

9.1.4 Forest Covertype

A much larger data set was considered in this section in order to address the issue of a lattice algorithm's scalability and applicability to real world problems. The Forest Covertype benchmark contains 581,012 data records including 11,340 records for training, 3,780 for validation, and 565,892 for testing. This data set involves jointly 12 numeric and nominal attributes per data record. The first 10 attributes are numeric variables, whereas the last 2 attributes are Boolean strings of zeros and ones. The 'class' of a data record is denoted by an integer from 1 to 7, and specifies a forest cover type. There are no missing attribute values.

Table 9-3 summarizes classification results by various methods. Results by backpropagation, FLNtf, and linear discriminant analysis are from Kaburlasos and Petridis (2000); the remaining results in Table 9-3 are from Cripps et al. (2003). A *FLN fit* treated the numeric attributes as numbers, whereas for each nominal attribute the corresponding Boolean lattice of zeros and ones has been considered. Positive valuation function $v=v_1+...+v_N$ was employed. Furthermore, *complement coding* was used to represent a data vector $(x_1,...,x_N)$ as vector $(1-x_1,x_1,...,1-x_N,x_N)$.

The experiments have demonstrated that a *tightest fit* FLN algorithm produces poorer classification results than another *FLN fit* algorithm due to the effects of noise/outliers in the data; that is the computation of tightest fits, in the presence of noise/outliers in the training data, deteriorates the capacity for generalization. All three FLNff, FLNotf, and FLNsf have resulted in considerably fewer rules than FLNtf. Number 6600 (within parentheses) in the first line of Table 9-3 indicates the total number of connection weights in the corresponding backpropagation neural network.

Table 9-3   Performance of various methods on the Forest Covertype benchmark data set. The last column shows the number of induced rules.

Method	% Correct	no. of rules
Backpropagation	70	(6600)
FLNff	68.25	654
FLNmtf	66.27	1566
FLNotf	66.13	516
FLNsf	66.10	503
FLNtf	62.58	3684
Linear discriminant analysis	58	-

9.1.5 Reuters

Documents were drawn from the Reuters-21578 benchmark collection, and the following three problems were treated (Petridis and Kaburlasos 2001).
• **Problem 1**: 148 documents were drawn from categories: 'colombia', 'singapore', and 'taiwan' from the *PLACES set of categories*.
• **Problem 2**: 143 documents were drawn from categories: 'nasdaq', and 'nyse' from the *EXCHANGES set of categories* of Reuters-21578.
• **Problem 3**: 169 documents were drawn from categories: 'opec', and 'worldbank' from the *ORGS set of categories* of Reuters-21578.

Approximately 2/3 of the documents in a problem were used for training and the remaining 1/3 for testing. The objective was to learn predicting the correct category of a document in the testing set.

Two different problems were treated. First, clustering of graphs, emanated from a Thesaurus of the English language synonyms for computing *hyperwords* (*hw*) towards dimensionality reduction based on semantics. Second, computing hyperboxes in the unit hypercube. The σ-FLN was used in the first problem, and the σ-FLNMAP in the second problem.

Various dimensionality reduction techniques are known in the literature based on: document frequency thresholding, information gain, mutual information, term strength, and χ^2 (Drucker et al. 1999; Quinlan 1992). In addition, 'word stemming' and a 'stop list' have been used in order to further reduce the dimensionality of the feature space (Sahami 1998).

In the context of this work the MOBY Thesaurus of synonyms was used including 30,260 records with more than 2.5 million synonyms. In a series of preprocessing steps the MOBY Thesaurus was reduced to 11,769 records and 628,242 synonyms. A *master-graph* $\mathcal{M}$ with 11,769 nodes, or vertices, emerged such that a link (n_i,n_j) was assumed in $\mathcal{M}$ if and only if the term corresponding to node n_j is a synonym of the term in node n_i. The objective of σ-FLN was to learn clusters of words, namely *hyperwords*, which retain a similar meaning.

A conjunctive version of σ-FLN with two vigilance parameters ρ_1 and ρ_2 was employed for learning hyperwords (Petridis and Kaburlasos 2001) – For details regarding ρ see in section 10.1.2. Inclusion measure $s(x,u)=v(x \wedge u)/v(x)$ gave better results than inclusion measure $k(x,u)= v(u)/v(x \vee u)$. The σ-FLN with parameters $\rho_1=0.001$ and $\rho_2=0.1$ computed 2,601 hyperwords. Table 9-4 displays a list of selected hyperwords.

A document was represented in two different ways: first, by a *vocabulary of stemmed words* and, second, by a *vocabulary of hyperwords*. An ensemble of nine σ-FLNMAPs was employed with vigilance parameter ρ=0.90. Table 9-5 summarizes the results. 'Words' resulted in an average of 13, 12.55, and 15.11 hyperboxes, respectively, in problems 1, 2, and 3; whereas 'hyperwords' resulted in an average of 21.77, 17.22, and 27.55 hyperboxes in problems 1, 2, and 3, respectively. The larger number of hyperboxes for hyperwords was attributed to the smaller dimension of the corresponding vectors of features. This work has presented experimental evidence that the employment of words favors document classification problems involving shorter documents, whereas the employment of hyperwords favors document problems involving longer documents.

TABLE 9-4   Hyperwords (hw), these are clusters of semantically related words, have been calculated by σ-FLN on the reduced MOBY thesaurus.

Hyperword	Words included in a hyperword
hw#74	housekeeper caretaker jailer keeper warden guardian
hw#157	autocrat despot dictator oppressor tyrant
hw#405	considerably greatly highly exceedingly awfully
hw#410	dynamite decapitate disassemble unmake incinerate dismantle devastate vandalize vaporize pulverize devour demolish pillage slaughter shatter butcher batter atomize maul dissolve assault consume disintegrate overwhelm ravage spoil wreck kill confound destroy
hw#1883	inexact inaccurate imprecise
hw#1925	intention ambition inspiration motive aspiration

TABLE 9-5   The σ-FLNMAP was employed in the unit N-dimensional hypercube with parameter ρ=0.90. The average number of hyperboxes computed by an ensemble of nine σ-FLNMAPs is also shown.

Prob lem no.	no. cat- egories	Document length statistics		Classification performance using words		hyperwords	
		ave	stdv	ave no. boxes	% success	ave no. boxes	% success
1	3	71.79	51.59	13	77.58	21.77	81.03
2	2	37.58	34.47	12.55	95.74	17.22	91.48
3	2	95.77	51.11	15.11	94.82	27.55	96.55

9.2 Real World Applications

Various lattice algorithms have been employed in real-world applications including non-numeric data as described in this section.

9.2.1 Epidural Surgery Using Waveforms

The problem as well as the data, i.e. Raman spectra profiles of soft tissues, have been described in section 8.3.1, where a data preprocessing step was carried out for numeric feature extraction. This section explains how soft tissue recognition can be achieved without numeric feature extraction.

A Raman spectrum was treated here as a single datum in a fuzzy lattice of integrable functions. Experiments were carried out using the *Voting σ-FLNMAP* algorithm. A training data set and a testing data set were defined randomly including 16 and 13 Raman spectra profiles, respectively. The training data set was partitioned into a pair of sets (TRN_k, TST_k), $k=1,...,5$. A grid of pairs (ρ_a, n_V) was considered for $\rho_a = 0.70, 0.65, 0.60,..., 0.20$ and $n_V = 2,3,...,30$. Note that all data sets included Raman spectra profiles representative of all four tissues. Estimates $\hat{\rho}_a = 0.5$ and $\hat{n}_V = 7$ were calculated. In conclusion, 100% classification accuracy on the corresponding testing data set of 13 Raman spectra profiles resulted in (Kaburlasos 2003).

Fig.9-2 shows the performance of individual σ-FLNMAP voters versus an ensemble of a progressively increasing number of up to 30 voters using equal training and testing vigilance parameters $\rho_a = 0.5$. Note that ensemble classification performance has been better than an individual voter's performance (Petridis and Kaburlasos 2000).

9.2.2 Stapedotomy Surgery

Stapedotomy is a corrective surgical procedure whose objective is to overcome *otosclerosis*, the latter is excessive bone growth within the mid-ear cavity which typically leads to hearing impairment. A stapedotomy is typically carried out by expert drilling a hole through the base of the stapes bone. Hearing is restored by inserting a piston prosthesis through the footplate of the stapes and attaching it suitably (Fig.9-3). The specific technical problem here is estimation of the stapes bone thickness from drilling data.

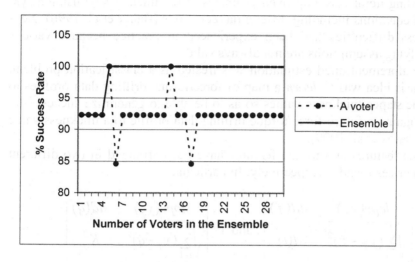

Fig.9-2 Individual vs ensemble performance of σ-FLNMAP using equal training and testing vigilance parameters ρ_a= 0.5.

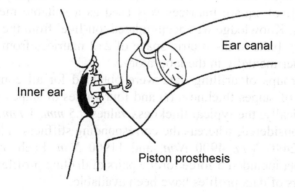

Fig.9-3 Piston prosthesis in the inner ear.

Estimation of stapes thickness (t) has been pursued by other authors by calculating iteratively improved estimates of the thickness (t) using a system of equations including differential equations (Baker et al. 1996). Nevertheless difficulties had been experienced in practice because various simplifying assumptions are not always valid.

The aforementioned estimation was treated as a classification problem. The basic idea was to *learn* a map of force/torque drilling data profiles to specific stapes thickness values so as to be able to generalize successfully by assigning a future force/torque drilling data profile to a thickness value (Kaburlasos et al. 1999).

Force features and torque features have been arranged in two different 2×2 matrices F and T respectively. In particular

$$
\text{F=} \begin{bmatrix} slope(y_f) & std(f) \\ \sqrt{\sum_{i-1}^{l}(y_f - f)^2} & d(l) \end{bmatrix}, \text{ and T=} \begin{bmatrix} slope(y_q) & std(q) \\ \sqrt{\sum_{i-1}^{l}(y_q - q)^2} & K^2 \end{bmatrix}
$$

where l is the number of force (f) and torque (q) data, y_f and y_q are the least square linear regressors of f and q respectively, $slope(.)$ is the slope of its linear argument, $std(.)$ is the standard deviation of its argument, $d(l)$ is the current displacement of the drilling tool, and K is the stiffness parameter of the stapes.

The datum considered for 'learning' was a 2×2 matrix M so as MF=T. The *dσ-FLN* algorithm for competitive clustering was employed. The spectral norm $\|.\|_S$ of square matrices was used as a suitable metric between 2×2 matrices. Knowledge was accumulated 'on-line' from the training data by computing balls in the metric space of 2×2 matrices from drilling data obtained experimentally in the laboratory.

Twelve groups of drilling data were obtained for all combinations of three values of stapes thickness (t) and four values of stapes stiffness (K). More specifically, the typical thickness values 0.5 *mm*, 1 *mm*, and 1.5 *mm* have been considered, whereas the corresponding stiffness values had been 1030 *N/m*, 2440 *N/m*, 4900 *N/m*, and 11900 *N/m*. Each one of the 12 groups of data included 5 force/torque pairs of drilling profiles, hence a total of 60 pairs of data profiles have been available.

The dσ-FLN had to learn twelve classes, where a class was characterized by two labels: (1) a stapes thickness, and (2) a stapes stiffness. In the sequel dσ-FLN's capacity for generalization was tested. Since the stapes stiffness (K) is measurable, the drilling data of a stapes had to be classified to one of the three classes which corresponded to stapes thickness 0.5 *mm*, 1.0 *mm*, and 1.5 *mm*.

Six learning experiments were carried out using the dσ-FLN algorithm employing one, two, or all the force/torque profiles from each of the 12 classes of drilling profiles in two, three, and one experiments, respectively.

Fig.9-4 shows typical results for the 'on line' classification of stapes thickness using the dσ-FLN algorithm. In a sub-figure of Fig.9-4, the dotted line (thickness 0.5 *mm*) leads the way, followed by the dashed line (thickness 1.0 *mm*), the latter is followed by the solid line (thickness 1.5 *mm*). Nevertheless the 'winner class' at the time of stapes bone breakthrough, the aforementioned time is denoted in Fig.9-4 by a vertical dash-dotted line, assigns its label (thickness) to the thickness of the stapes currently being drilled.

In all experiments only one misclassification was recorded. In particular, a misclassification occurred when one training datum per class was employed; then one pair of profiles was misclassified in class '0.5 *mm*' instead of being classified to the correct class '1.0 *mm*'. The aforementioned misclassification was attributed to the small number of data used for training. By increasing the number of training data performance improved to 100% correct classification.

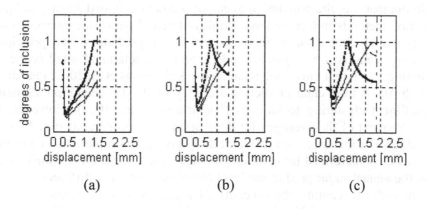

        (a)   (b)   (c)

Fig.9-4   Degrees of inclusion to classes: 0.5 *mm* (dotted line), 1.0 *mm* (dashed line), and 1.5 *mm* (solid line). The maximum inclusion measure at the time of stapes-breakthrough, denoted in a figure by a vertical dash-dotted line, determines the 'winner' and consequently it determines the thickness of the stapes being drilled.

9.2.3 Prediction of Industrial Sugar Production

Sugar is produced in Greece from an annual (in farm practicing) plant, namely *Beta Vulgaris L* or simply *sugar-beet*. The production is organized around factories located, as well, at Larisa, Platy, and Serres. An early season accurate prediction of sugar production is critical for safeguarding the continuous operation of a sugar production factory since interruptions are very costly. Prediction of sugar production is possible on the basis of measurements including both *production* and *meteorological* variables.

Early in a year, a set of *pilot fields* is defined for data sampling. Samples of ten *production variables* (Table 9-6) and eight *meteorological variables* (Table 9-7) were available in this work for eleven years from 1989 to 1999 from three agricultural districts, namely Larisa, Platy, and Serres. Sugar production was calculated as the product POL*RW.

The production variables were sampled every 20 days, whereas the meteorological variables were sampled daily. *Production* and *meteorological* variables are jointly called here *input variables*. The term *population of measurements* is used to denote either (1) production variable samples obtained every 20 days from a pilot field, or (2) meteorological variable samples obtained daily during 20 days.

A population of measurements has been typically represented by a single number, i.e. the population average; moreover, annual sugar yield prediction methods in Greece have employed neural networks, interpolation-, polynomial-, linear autoregression- and neural-predictors, first principles 'energy conservation' system models, and computational intelligence including intelligent clustering techniques (Kaburlasos et al. 2002; Petridis et al. 2001). A best sugar prediction accuracy of 5% has been reported using intelligent clustering techniques (Kaburlasos et al. 2002). However, an employment of the average of a population of measurements may be misleading. For instance, two different daily precipitation patterns in a month may be characterized by identical average values, nevertheless their effect on the annual sugar production level might be drastically different.

In order to capture the diversity of a distribution of measurements, a FIN (Fuzzy Interval Number), computed by algorithm CALFIN, was employed here to represent a population of measurements. Then, in line with the common practice of agriculturalists, prediction of sugar production was pursued by classification in one of the classes 'good', 'medium' or 'poor'. More specifically, the FINkNN classifier was employed with the leave-one-out paradigm for prediction-by-classification.

TABLE 9-6 Production variables used for Prediction of Sugar Production.

	Production Variable Name	Unit
1	Average Root Weight	g
2	POL (: percentage of sugar in fresh root weight)	-
3	α-amino-Nitrogen (α-N)	meq/100 g root
4	Potassium (K)	meq/100 g root
5	Sodium (Na)	meq/100 g root
6	Leaf Area Index (LAI) (: leaf-area per field-area ratio)	-
7	TOP (: plant top weight)	kg/1000 m^2
8	Roots Weight (RW)	kg/1000 m^2
9	Nitrogen-test (N-test) (: NO$_3$-N content in pedioles)	mg.kg^{-1}
10	Planting Date	-

TABLE 9-7 Meteorological variables used for Prediction of Sugar Production.

	Meteorological Variable Name	Unit
1	Average (daily) Temperature	°C
2	Maximum (daily) temperature	°C
3	minimum (daily) Temperature	°C
4	Relative Humidity	-
5	Wind Speed	miles/h
6	Daily Precipitation	mm
7	Daily Evaporation	mm
8	Sunlight	h/day

A subset of input variables was selected based on an optimization of an *objective/fitness* function. The aforementioned optimization was carried out using, first, a genetic algorithm (GA), second, a GA with local search and, third, human expertise. Two types of distances were considered between two populations of measurements: (1) the metric d_K between FINs, and (2) the L1-distance between the average values of two populations of measurements. Experimental results are summarized in Table 9-8 (Petridis and Kaburlasos 2003).

The first six lines in Table 9-8 show the results of applying a (FIN)kNN classifier for different combinations of selected input variables and distances between populations of measurements. Line 7 shows that selection 'medium' each year resulted in error rates 5.22%, 3.44%, and 5.54% for the Larisa, Platy, and Serres factories, respectively. Line 8 shows the average errors when a year was assigned randomly to one of 'good', 'medium', 'poor'. Finally, line 9 in Table 9-8 shows the minimum prediction error that could be obtained by classification. In conclusion, the best experimental results were obtained for the combination of d_K distances (between FINs) with expert-selected input variables.

Table 9-9 displays experimental results using four different prediction methods including intelligent-clustering techniques and a first-principles model (Kaburlasos et al. 2002), Bayesian Combined Predictor (BCP) (Petridis et al. 2001), the FINkNN classifier (Petridis and Kaburlasos 2003). Note that the 2% error reported in Table 9-9 for classifier FINkNN is the average over three agricultural districts and eleven years. Only classifier FINkNN employed FINs.

TABLE 9-8   Average % prediction error rates using various methods for three factories of Hellenic Sugar Industry, Greece.

	Prediction Method	Larisa	Platy	Serres
1	FINkNN	1.11	2.26	2.74
	(with expert input variable selection)			
2	L1-distances kNN	2.05	2.87	3.17
	(with expert input variable selection)			
3	FINkNN	4.11	3.12	3.81
	(with GA local search input variable selection)			
4	L1-distances kNN	3.89	4.61	4.58
	(with GA local search input variable selection)			
5	FINkNN	4.85	3.39	3.69
	(with GA input variable selection)			
6	L1-distances kNN	5.59	4.05	3.74
	(with GA input variable selection)			
7	'medium' selection	5.22	3.44	5.54
8	random prediction	8.56	4.27	6.62
9	minimum prediction error	1.11	1.44	1.46

TABLE 9-9   Approximate % sugar production prediction error for various prediction methods.

Prediction Method	Approximate % Prediction Error
Classifier FINkNN	2.0
Intelligent clustering techniques	5.0
Bayesian Combined Predictors (BCP)	6.0
First principles model	15.0

9.2.4 Fertilizer Industry Modeling

Industrial fertilizer is produced by spraying Ammonium Nitrate (AN) solution on small particles inside a rotating *pan granulator* mill. Hence, *fertilizer granules* are produced which are then batch-stored. High quality fertilizer specifications require the diameter of at least 95% of the granules to be in the range 2-5 mm. Fertilizer quality can be controlled by pan granulator *operating variables* including the ones shown in Table 9-10.

Two different models were developed in the context of this work. First, a model d: $R^N \rightarrow R$, where $d(\mathbf{x})$ is the average diameter of produced fertilizer granules and $\mathbf{x}$ is a vector of pan granulator operating parameters (Papadakis et al. 2004). Second, a model q: $F^N \rightarrow K$, where function $q(\mathbf{x})$ takes on linguistic values in the set {'good', 'above-average', 'below-average'', 'bad'} and $\mathbf{x}$ is a N-tuple of FINs including geometric and other fertilizer granule features (Kaburlasos et al. 2005). Model development based on 'first principles' was phased out due to the inherent complexity of the industrial process. Instead, model induction by classification was pursued as explained in the following.

First, a function d: $R^N \rightarrow R$ was induced from data using the *greedy grSOM* classifier. In a data preprocessing step eight of the operating variables shown in Table 9-10 could be ignored (Papadakis et al. 2005).

Table 9-10   Operating variables available for modeling the operation of the pan granulator in the Phosphoric Fertilizers Industry, Greece.

	Operating Variable Name	Unit
1	AN Melt Flow	m³/h
2	Recycled Fertilizer	T/h
3	AN Melt Temperature	°C
4	AN Melt Pressure	bar
5	Granulation Temperature	°C
6	Pan Inclination	degrees
7	Pan Rotation Speed	Hz
8	Nozzle Vertical Distance	rings
9	Nozzle Distance from the pan	cm
10	Scraper Speed	Hz
11	Spraying Angle	lines
12	Coarse Screen Vibration	%
13	Fine Screen Vibration	%
14	$Mg(NO_3)_2$ Supply	%

Four different fertilizer granule diameters (classes) were considered, namely 1.5, 2.5, 3.5, and 4.5 mm. There were available 360 input-output pairs of data; the first 240 data were used for training a *greedy grSOM* model with a 4×4 grid of units. For comparison, a conventional KSOM was also trained. The best, the worst, and the average classification rates over the ten runs are shown in Table 9-11.

Second, a function q: $F^N \rightarrow K$ was induced using the grSOM classifier. Geometrical features were measured using histogram thresholding followed by *fuzzy-mathematical-morphology* image processing techniques (Fig.9-5). Hence, the perimeter of each fertilizer granule was computed as well as an index of circularity (CInd) in the closed interval [0,1] for each granule. Apart from geometrical features, additional measurements included the weight of each fertilizer granule (Kaburlasos et al. 2005).

Table 9-11   Classification rates on the industrial problem for classifiers *greedy grSOM* and KSOM.

Experimental	greedy grSOM		KSOM	
Outcome	Training	Testing	Training	Testing
Best	91.7%	87.5%	85.4%	83.3%
Worst	75.0%	58.3%	66.6%	68.3%
Average	85.2%	83.7%	82.1%	80.2%

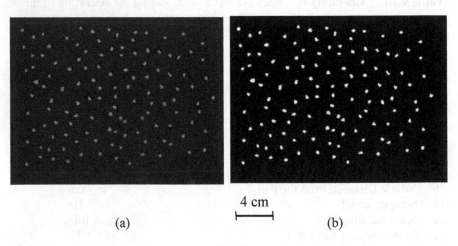

4 cm
⊢———⊣

(a)   (b)

Fig.9-5   Fertilizer granules under white light on black background.
(a) Original grayscale digital image.
(b) Binary digital image after histogram thresholding followed by enhancement using *fuzzy-mathematical-morphology* techniques.

Fig.9-6 shows three FINs Fd, Fc, and Fw for fertilizer granule *diameter*, *circularity*, and *weight*, respectively. The corresponding histograms are shown in Fig. 9-6 (d), (e), and (f), respectively. Note that a histogram may not be a convex function, whereas a FIN is always convex by construction.

Classifier grSOM was employed with a I×J=4×4 grid of units. Forty population triplets were available including ten triplets from each one of the fertilizer categories 'good', 'above-average', 'below-average', and 'bad'. A random permutation of 24 population triplets was employed for training. Algorithm grSOM resulted in a 100% percentage of success. For comparison, a conventional KSOM resulted in a 93.75% percentage of success. An advantage of grSOM in this preliminary work is that it induces descriptive decision-making knowledge (i.e. rules) from the training data, whereas the KSOM does not produce any rules.

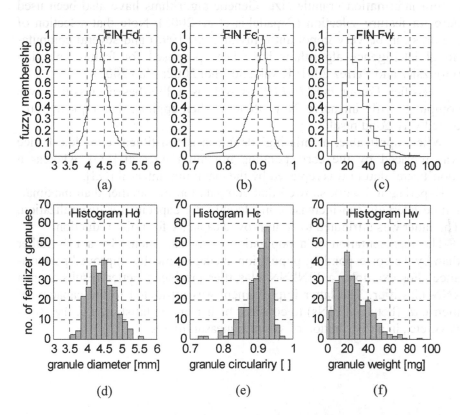

Fig.9-6 Three FINs (a) Fd, (b) Fc, and (c) Fw for, respectively, fertilizer granule *diameter*, *circularity*, and *weight*, computed from a fertilizer CaN27 sample including 254 granules. The corresponding histograms are shown in figures (d), (e), and (f), respectively.

9.3 Discussion of the Results

The application of several lattice algorithms for clustering, classification, and regression, including non-numeric data, showed that the aforementioned algorithms can be very effective. Apart from good experimental results, an additional advantage of the proposed algorithms is 'knowledge acquisition' by inducing a set of rules from the training data. The presentation of experimental results here was brief. For further application details the interested reader may refer to cited publications.

A key feature of the proposed algorithms is an induction of *information granules*; the latter correspond to product-lattice intervals. A critical parameter is the *size* of the induced information granules. A search algorithm, e.g. a steepest-descent- or a genetic algorithm, can estimate an optimum information granule size. Genetic algorithms have also been used here for feature selection (Papadakis et al. 2005). Note that selection of features can be very important (Huynh et al. 1998) and several computational intelligence algorithms have been proposed for feature selection (Gonzalez and Perrez 2001; Kwak and Choi 2002; Mao 2002).

Classifier *Voting σ-FLNMAP*, as well as a FIN-based classifier, was computationally intensive. Nevertheless, when applied, then typically produced very good results.

Where applicable, a 'missing' datum in a constituent complete lattice was replaced by the corresponding minimum interval [I,O], whereas a 'don't care' datum was replaced by the maximum interval [O,I].

Experimental work showed that computation of 'smaller than maximal' lattice intervals can increase considerably the capacity for generalization. The latter was attributed to an improved capacity for noise reduction.

The representation of a population of measurements by a FIN rather than by the corresponding population average typically improved performance. For example, a FINkNN classifier produced better results than a kNN classifier. The latter improvement was attributed to the capacity of metric d_K (between FINs) to consider higher order statistics, e.g. the *skewness*, etc., in a distribution of samples/measurements.

10 Connections with Established Paradigms

This chapter unifies rigorously analysis of established paradigms

Lattice theory is employed either explicitly or implicitly in a number of modeling paradigms. This chapter illustrates how. Tools for further improvements are also presented.

10.1 Adaptive Resonance Theory

Adaptive resonance theory, or ART, evolved from studies of the brain and mind during the 1960s. In particular, first-order, non-linear, differential equations with time-varying coefficients were used to describe interactions among neural cells. Inspired from mathematics and psychology a theory of *embedding fields* was proposed (Grossberg 1969). In conclusion, ART emerged (Grossberg 1976a, 1976b). The original work on ART has inspired a lasting research activity as described in the following.

10.1.1 A review of ART models

ART1 grew out of analysis of *competitive* neural models with influential contributions to the latter models also by von der Malsburg (1973), Amari and Takeuchi (1978), Bienenstock et al. (1982), and Kohonen (1995). In particular, ART1 achieves self-organization of stable recognition binary codes in response to arbitrary binary input patterns (Carpenter and Grossberg 1987a). ART1 addresses the *Stability-Plasticity* dilemma; that is ART1 remains both plastic, so as to be able to learn significant new events, and stable, so as to prevent a degradation of learned codes. Perception and cognition are states of *resonant activity*. Only resonant states enter consciousness and drive adaptive changes. *Adaptive resonance* occurs when feed forward (bottom-up) computations, which correspond to stimuli from the outside world, and feedback (top-down) expectations, are consonant. Theoretical discussions regarding ART are presented in Levine (2000).

Vassilis G. Kaburlasos: *Connections with Established Paradigms*, Studies in Computational Intelligence (SCI) **27**, 141–172 (2006)
www.springerlink.com

Although self-organization of binary patterns is useful in many applications, other applications involve analog patterns. The ART2 model was introduced by splitting the input F1 layer of ART1 into separate sublevels, which carry out a normalization of analog input patterns (Carpenter and Grossberg 1987b). A faster implementation of ART2 is ART 2-A (Carpenter et al. 1991a). Different authors have proposed the *incremental communication* method to reduce the communication and computation costs of ART2 (Chen et al. 2005). Moreover an activation of multilevel category structures, the latter may admit various interpretations, was proposed at the output F2 layer of ART2 (Davenport and Titus 2004).

Perhaps the most popular extension of ART1 is *fuzzy-ART*, which was been introduced for self-organizing both binary and analog patterns in the unit-hypercube (Carpenter et al. 1991b). Note that in practice, typically, fuzzy-ART's *fast learning* mode is used, where arithmetic operations are carried out by computing the 'min' of real numbers. The corresponding model for supervised learning is *fuzzy-ARTMAP* (Carpenter et al. 1992).

Various properties of fuzzy-ART(MAP) have been studied (Georgiopoulos et al. 1994, 1996, 1999; Huang et al. 1995, Dagher et al. 1999) and improvements have been demonstrated (Cano Izquierdo et al. 2001; Gómez Sánchez et al. 2002; Parrado-Hernández et al. 2003; Castro et al. 2005). A number of authors have presented several fuzzy-ARTMAP variants (Williamson 1996, 2001; Healy and Caudell 1997, 1998; Tan 1997; Lavoie et al. 1999; Baraldi and Alpaydin 2002). Moreover, a number of hardware implementations have been proposed as explained next.

A hybrid optoelectronic implementation of ART1 was proposed in Caudell (1992), where parallel computations are relegated to free-space optics while serial operations are performed in VLSI electronics. An alternative hardware implementation of ART1 has been shown (Serrano-Gotarredona and Linares-Barranco 1997). An optoelectronic implementation of the fuzzy-ARTMAP neural model was presented in Blume and Esener (1995).

The focus of ART gradually shifted, at large, from modeling the mind towards information processing and data mining applications of scale (Koufakou et al. 2001; Parsons and Carpenter 2003; Castro et al. 2005). Moreover, researchers in other application domains have employed ART. For instance Lim and Harrison (1997) used fuzzy-ARTMAP for estimating a probability distribution. The FALCON-ART algorithm (Lin and Lin 1997) has employed ART synergistically with backpropagation for structure/parameter learning in automatic control applications. ART learning has also inspired the development of general incremental learning algorithms (Polikar et al. 2001). Moreover, ART has inspired cognitive neural network model extensions (Healy 1999; Healy and Caudell 2004).

The operation of fuzzy-ARTMAP can be described by computations of hyperboxes in the lattice 'unit hypercube' (Kaburlasos and Petridis 2000). In this context, algorithm σ-FLNMAP was introduced as an extension of fuzzy-ARTMAP to a lattice data domain. The latter (extension) has implied novel tools, which may potentially improve fuzzy-ARTMAP's performance in the unit hypercube as explained in the following.

10.1.2 Granular extension and enhancements

Fuzzy-ART employs *mathematical lattices*, implicitly. Consider the following sentence from the caption of Fig.6(b) in Carpenter et al. (1991b): 'During fast learning, R_J expands to $R_J \oplus a$, the smallest rectangle that includes R_J and a, ...'. We remark that the aforementioned 'smallest rectangle' is, by definition, the *lattice join* of R_J and a. A *mathematical lattice* is also employed implicitly by FALCON-ART in the technique of 'rule annihilation' to delete unnecessary or redundant rules (Lin and Lin 1997). In particular, 'rule similarity' is computed on a dimension-by-dimension comparison in order to calculate heuristically the degree of inclusion of a hyperbox into another one.

The domain of fuzzy-ART is, quite restrictively, the unit hypercube. Model σ-FLN is a straightforward extension of fuzzy-ART to a lattice data domain involving, potentially, disparate types of data. Both fuzzy-ART and σ-FLN can be implemented on a two layer architecture (Fig.7-2), where the *Category Layer* F_2 accommodates learned codes/intervals/ hyperboxes. The objective of σ-FLN is to define sets of elements in a complete lattice by the set-union of crisp lattice intervals; the latter correspond, in particular, to hyperboxes in the unit hypercube. Several tools and techniques of fuzzy-ART can be enhanced as explained in the following.

Fuzzy-ART's *complement coding* technique corresponds to a specific isomorphic function θ, i.e. $\theta(x)=1-x$, $x \in [0,1]$. Apparently, there are infinitely many isomorphic functions θ(.), where a function θ(.) corresponds to a *coding* technique. Likewise, there are infinitely many positive valuations $v(.)$ in a data dimension in the unit hypercube.

In order to compare meaningfully σ-FLN and fuzzy-ART in the following we assume both $\theta(x)=1-x$ and $v(x)=x$. It follows that a N-dimensional hyperbox $w= [p,q]= [(p_1,...,p_N),(q_1,...,q_N)]= [p_1,q_1,...,p_N,q_N]$ in the unit hypercube has positive valuation $v(w)= v(\theta(p))+v(q)= v(\theta(p_1,...,p_N)) +$

$$v(q_1,...,q_N)= v(1-p_1,...,1-p_N) + v(q_1,...,q_N)= \sum_{i=1}^{N}(1-p_i)+\sum_{i=1}^{N}q_i .$$

Instrumental to both learning and decision-making by fuzzy-ART are: (1) the *Choice (Weber) Function* $|I \wedge w|/(a+|w|)$, and (2) the *Match Function* $(|I \wedge w|/|I|) \geq \rho_{ART}$, where $\rho_{ART} \in [0,1]$ is a user-defined constant namely *vigilance parameter*. It is shown in the following that, under assumptions (1) trivial inputs (atoms), and (2) fuzzy-ART's *fast learning* mode of operation, both aforementioned functions can be implemented by σ-FLN's inclusion measure σ.

Operator '$\wedge$' is used by fuzzy-ART as the *min* operator in the totally ordered set of real numbers (Carpenter et al. 1991b). The role of (the very small positive) parameter a in the denominator of the *Choice Function* is to break ties in order to select the F_2 node with the smallest diagonal (Kaburlasos 1992). The same result can be attained by the following *double test*: (after calculating numbers $|I \wedge w|/|w|$ for all nodes in layer F_2, first, select the node with largest activation and second, break ties by selecting the F_2 node with the smallest diagonal. Recall that a double test is employed explicitly by σ-FLN. Inclusion measure σ also implies a lattice-theoretic interpretation; in particular, model σ-FLN identifies the F_2 node w that maximizes $\sigma(x \leq w) = v(w)/v(x \vee w)$.

Regarding fuzzy-ART's *Match Function* note that fuzzy-ART accepts (rejects) a winner node when the ratio $|I \wedge w|/|I|$ is larger (smaller) than ART's vigilance parameter ρ_{ART}. It is shown next that fuzzy-ART's *complement coding* implies an implicit comparison of the winner node's diagonal to a threshold diagonal. Using fuzzy-ART's notation, a code w norm equals $|w| = \sum_{i=1}^{N}(w_i + w_i^c)$; in particular for a trivial hyperbox input I it is $|I| = N$ (Carpenter et al. 1991b). Recall that the diagonal of a code in the context of this work is computed as $diag(w) = \sum_{i=1}^{N}((1-w_i^c) - w_i)$, hence it relates to its norm as $diag(w) = N - |w|$. In particular, for a trivial input (atom) I it follows $|I| = N \Leftrightarrow diag(I) = 0$. Therefore fuzzy-ART's match criterion becomes $(|I \wedge w|/|I|) \geq \rho_{ART} \Rightarrow (N - diag(I \wedge w))/N) \geq \rho_{ART} \Rightarrow diag(I \wedge w) \leq N(1-\rho_{ART})$. In words, fuzzy-ART's code $I \wedge w$ is accepted when its diagonal is less than or equal to a threshold $D_0 = N(1-\rho_{ART})$; otherwise, *reset* is triggered and the search for a new winner resumes.

The *Assimilation Condition* of σ-FLN requires an explicit comparison with a user-defined diagonal threshold. Assuming complement coding $\theta(x) = 1-x$ and trivial inputs (atoms), an implicit employment of a vigilance parameter ($\rho_{\sigma FLN}$) is implied by σ-FLN as illustrated in the following. A relation is shown between a code's positive valuation and its diagonal as follows: $diag(w)= \sum_{i=1}^{N}[w_{2i} - (1 - w_{2i-1})]= v(w)\text{-}N$. Since point inputs (atoms)

x are assumed, it follows $v(x)= N$. Model σ-FLN refines an existing code w to $x\vee w$ by an input x if and only if the diagonal of $x\vee w$ is less than or equal to a user-defined diagonal threshold D_{crit}, that is $diag(x\vee w) \leq D_{crit} \Rightarrow$ $N/(diag(x\vee w)+N) \geq N/(D_{crit}+N) = \rho_{\sigma FLN}$. In conclusion, the critical diagonal D_{crit} of model σ–FLN equals $D_{crit}= N(1\text{-}\rho_{\sigma FLN})/\rho_{\sigma FLN}$, where $\rho_{\sigma FLN} \in [0.5,1]$ is the (implicit) *vigilance parameter* of model σ–FLN. Note also that ratio $N/(diag(x\vee w)+N)$ is the degree of inclusion $\sigma(w\leq x)$ of winner code w to input atom x (Kaburlasos and Petridis 2000). It is remarkable that the previous analysis holds in a general lattice.

To recapitulate, given atom input $x= [a,a]$ to σ-FLN, a competition in layer F_2 takes place and it computes the node which *includes* input x 'most', that is $\sigma(x\leq w)$ is maximum. Consequently the winner is accepted if and only if it is *included* inside input x more than an implicit vigilance parameter $\rho_{\sigma FLN}$, i.e. $\sigma(w\leq x) \geq \rho_{\sigma FLN}$. Such a 'double role' of the lattice inclusion measure σ signifies a deep lattice-theoretic interpretation for both *winner choice* and *winner match* for both σ-FLN and fuzzy-ART. It is also possible to attach *Occam razor* semantics to inclusion measure $\sigma(x\leq w)= v(w)/v(x\vee w)$ as explained next. Let $w_1,...,w_L$ be hyperboxes (rule antecedents) competing over input hyperbox x_0, i.e. the largest $\sigma(x_0\leq w_i)$, $i=1,...,L$ is sought. Winner w_J, among hyperboxes $w_1,...,w_L$, is the one whose diagonal size needs to be modified 'comparatively, the least' so as to include x_0 (Kaburlasos 2003). In the aforementioned sense, the winner box w_J correspond to the simplest among hypotheses $x\vee w_i$, $i=1,...,L$, which fit the data. The latter is *Occam razor* semantics (Mitchell 1997).

An *extended choice (Weber) function*, namely $v(x\wedge w)/v(w)$, has also been considered for employment by σ-FLN instead of inclusion measure $\sigma(x\leq w)= v(w)/v(x\vee w)$. Nevertheless, the aforementioned function suffers a serious drawback, which appears when the lattice *meet* of two lattice elements is the least lattice element O as shown by the following simple example in a probabilistic context. Consider the lattice-meet (intersection) $A\cap B$ of the two sets $A= \{a,b,c\}$ and $B= \{d,e\}$ in the power set of $X= \{a,b,c,d,e,f\}$. It follows $A\cap B= \varnothing$. Hence, in the latter example, an employment of the aforementioned *extended choice (Weber) function* $v(A\cap B)/v(B)$ only concludes that the sets A and B are disjoint but it does not quantify affinity for the two sets. On the other hand, σ-FLN's conventional inclusion measure $\sigma(x\leq w)= v(w)/v(x\vee w)$ quantifies by a single number and with respect to a concrete positive valuation 'how much lattice element *u* is short of including another lattice element *x'*. Note that the aforementioned drawback does not arise in the Euclidean space because the lattice *meet* ($\wedge$) and the lattice *join* ($\vee$) of two numbers is their minimum and their maximum, respectively; for instance, $0.3 \wedge 0.8 = 0.3$ and $0.3 \vee 0.8 = 0.8$. Hence, due to fuzzy-ART's restriction in the unit hypercube the aforementioned inherent drawback of ART's choice (Weber) function does not appear in a conventional ART model application.

The reader is cautioned that due to the aforementioned modifications of ART's basic equations, including the replacement of both of ART's choice (Weber) and match functions by an inclusion measure function σ, the learning behavior of σ-FLN in the unit hypercube is not expected to be identical to fuzzy-ART's learning behavior. Nevertheless, model σ-FLN retains several fuzzy-ART features as explained next.

The same way as ART, retraining σ-FLN by another data set (1) does not 'wash away' previous learning and (2) it retains rare training data. Moreover, the vigilance parameter regulates σ-FLN's *granularity of learning*; the latter is the number of clusters in the *Category Layer*. In particular, on the one hand, a large value of ρ within interval $[0,1]$ implies more clusters in *Category Layer* F_2; on the other hand, as ρ decreases, fewer clusters are learned. A single pass through the data suffices for stable learning by both fuzzy-ART and σ-FLN in the sense that additional passes through the data will not alter the arrangement of the learned hyperboxes. Moreover, for both fuzzy-ART and σ-FLN, the families of learned intervals/hyperboxes depend on the order of presenting the training data.

Unique advantages of σ-FLN compared to fuzzy-ART include: First, model σ-FLN is *granular*, whereas fuzzy-ART is not. In particular, σ-FLN can process inputs both points and hyperboxes, whereas fuzzy-ART processes solely points. Second, model σ-FLN is *tunable*, whereas fuzzy-ART is not. In particular, we may calibrate σ-FLN's behavior by the underlying positive valuation function $v(x)$. Whereas, fuzzy-ART employs (implicitly and solely) the same positive valuation function $v(x) = x$. Similar arguments extend to isomorphic function $\theta(x) = 1\text{-}x$. In particular, the employment of a different isomorphic function than $\theta(x) = 1\text{-}x$ implies a different coding technique than fuzzy-ART's *complement coding* technique. Third, σ-FLN is more *versatile* than fuzzy-ART in the sense that σ-FLN can process lattice elements in addition to processing real numbers. In other words, fuzzy-ART is applicable solely in the unit hypercube whereas σ-FLN is applicable to a lattice data domain including fuzzy-ART's domain. In addition, σ-FLN can jointly process disparate types of data due to its applicability to the product of disparate constituent lattices. The latter also accounts for σ-FLN's *modular* and *hierarchical* capacities. Fourth, σ-FLN can employ the technique of *maximal expansions* in order to maximize the degree of inclusion (σ) of an input to a class. Fifth, σ-FLN can cope with both 'missing' and 'don't care' data values in a complete lattice.

From an information processing (and a data mining) point of view both fuzzy-ART(MAP) and σ-FLN(MAP) can be used, quite restrictively, only for function interpolation but not for function extrapolation. However, on the one hand, fuzzy-ART(MAP) remains an instrument for lower-level (physiological) data processing in R^N; on the other hand, σ-FLN(MAP) can rigorously cope with mathematical lattice-ordered objects of perception towards higher-level, cognitive modeling.

10.2 Hyperbox-based Models

Since the work in this monograph has originated from computation of hyperboxes in the unit hypercube as explained in chapter 1, hyperbox-based models will be given special attention in this section.

Hyperbox-based models are *granular*, in the sense that they can cope with both points and neighborhoods of points. From a practical standpoint such a feature can be advantageous in cases we need to compensate for uncertainty of measurements by feeding to a decision making system a neighborhood of values as defined by a N-dimensional hyperbox.

A number of models that compute hyperboxes in R^N have been presented in the literature. For instance, *fuzzy adaptive resonance theory* (ART) neural networks and variants learn by computing hyperboxes in R^N as detailed above. Furthermore, rectangular data sets have been used for improving query performance on spatial databases (An et al. 2003). A hierarchical decomposition in hyperrectangles was also used for database analysis (Faloutsos et al. 1997). In a machine-learning context, the class of axis-parallel rectangles was shown to be efficiently *probably approximately correct* (PAC) learnable (Blumer et al. 1989; Kearns and Vazirani 1994; Long and Tan 1998). Other machine learning models have also considered hyperboxes (Samet 1988; Salzberg 1991; Wettschereck and Dietterich 1995; Dietterich et al. 1997). Nevertheless, the aforementioned models fail to employ lattice theory explicitly for their benefit; potential improvements are shown next.

Salzberg (1991) as well as Wettschereck and Dietterich (1995) define a distance function $d_S(x,[a,b])$ between a point x and an interval $[a,b]$ as follows

$$d_S(x,[a,b]) = \begin{cases} a-x, & x \le a \\ 0, & a \le x \le b \\ x-b, & x \ge b \end{cases}$$

Specific $d_S(.,.)$ distances between points and boxes are computed in Fig.10-1. It turns out $d_S(x_1,u)=0=d_S(x_2,u)$, $d_S(x_1,x_2)=0.8$. However, the latter results are counter-intuitive because they violate the 'common sense' *triangle inequality* $d_S(x_1,x_2) \le d_S(x_1,u)+d_S(u,x_2)$. Here we adhere to Fréchet's original definition for a *metric* shown in Appendix A. Unfortunately, not all *metrics* produce common sense results as demonstrated next.

A known metric between convex sets is the *Hausdorf* metric $d_H(.,.)$ (Diamond and Kloeden 1994). It is known that the distance $d_H(.,.)$ between two intervals $[a,b]$ and $[c,d]$ on the real line equals $d_H([a,b],[c,d])=$ max$\{|a-c|,|b-d|\}$ (Heilpern 1997; Grzegorzewski 2004). Hence, the corresponding L1 distance between boxes in Fig.10-1 equals
$d_H(v,w) = d_H([0.2,0.3],[0.7,0.8]) + d_H([0.6,0.7],[0.2,0.3]) = 0.5 + 0.4 = 0.9$
$d_H(u,w) = d_H([0.2,0.7],[0.7,0.8]) + d_H([0.4,0.7],[0.2,0.3]) = 0.5 + 0.4 = 0.9$

However, it is reasonable to expect $d(v,w) > d(u,w)$ in Fig.10-1 by inspection. Hence, even though the Hausdorf distance $d_H(.,.)$ is *metric*, it produces counter-intuitive results.

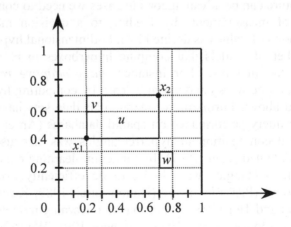

Fig.10-1 Various distances were computed between boxes v, u, w and points x_1, x_2. First, using distance $d_S(.,.)$ it follows $d_S(x_1,u)=0=d_S(x_2,u)$ and $d_S(x_1,x_2)=0.8$. Second, using the Hausdorf metric $d_H(.,.)$ it follows $d_H(v,w)=0.9=d_H(u,w)$. Third, using metric $d_1(.,.)$ it follows $1.8=d_1(v,w)>d_1(u,w)=1.2$; moreover, $d_1(x_1,u)=0.8=d_1(x_2,u)$ and $d_1(x_1,x_2)=1.6$.

A metric distance $d_1(.,.)$ between two boxes in Fig.10-1 is computed next with $v(x)=x$ and $\theta(x)=1-x$ in a constituent lattice $I=[0,1]$. It turns out $d_1([a,b],[c,d]) = \max\{1-a,1-c\} + \max\{b,d\} - \min\{1-a,1-c\} - \min\{b,d\}$.

Hence, the L1 distance $d_1(.,.)$ between two boxes equals
$d_1(v,w) = d_1([0.2,0.3],[0.7,0.8]) + d_1([0.6,0.7],[0.2,0.3]) = 1.8$, and
$d_1(u,w) = d_1([0.2,0.7],[0.7,0.8]) + d_1([0.4,0.7],[0.2,0.3]) = 1.2$

Note that $d_1(v,w) > d_1(u,w)$ as expected intuitively in Fig.10-1 by inspection. Moreover, it is $d_1(x_1,x_2)=1.6$, $d_1(x_1,u)=0.8$ and $d_1(x_2,u)=0.8$. Hence, the 'common sense' triangle-inequality $d_1(x_1,x_2) \leq d_1(x_1,u)+d_1(x_2,u)$ is satisfied. In conclusion, metric $d_1(.,.)$ produces intuitive results.

A specific family of hyperbox-based learning models, namely min-max neural networks, is discussed in the following.

10.2.1 Min-max neural networks

Min-max neural networks, were introduced in Simpson (1992, 1993) and further enhanced by several authors (Gabrys and Bargiela 2000; Rizzi et al. 2002). Note that *extension neural networks* also operate based on similar principles (Wang 2005). All aforementioned neural networks are another example where lattices are employed implicitly as explained next.

The words *min* and *max* in the term 'min-max (neural networks)' directly refer to the lattice *meet* and *join*, respectively. More specifically, a hyperbox is defined in R^N by two extreme points, namely *min* and *max* points. It turns out that point *min* is the lattice-meet of all the points in a hyperbox, whereas point *max* is the corresponding lattice-join.

Min-max neural networks operate similarly to their ART counterparts; however, instead of being biologically motivated, min-max neural networks are computational techniques whose objective is to define sets of points in R^N by hyperboxes.

Model σ-FLN is a straightforward generalization of the min-max neural networks to a lattice data domain. An additional advantage of σ-FLN is its *granularity*; in particular, σ-FLN can process inputs both atoms and intervals, whereas a min-max neural network typically processes solely atoms. Furthermore, model σ-FLN is *tunable*, whereas a min-max neural network is not; in particular, it is possible to calibrate model σ-FLN by an underlying positive valuation function $v(x)$, whereas a min-max neural network employs (implicitly and solely) always the same positive valuation function $v(x)= x$. In addition, σ-FLN can jointly process disparate types of data due to its applicability to the product of disparate constituent lattices.

10.3 Self-Organizing Maps

Kohonen's self-organizing map (KSOM) is a biologically (neural) inspired information-processing paradigm for clustering and visualization of multi-dimensional nonlinear relations (Kohonen 1990, 1995; Rauber et al. 2002).

KSOM variants have been presented, often in information processing applications (Kohonen et al. 1996, 2000). Typically, KSOM implements a non-linear projection of a probability density function $p(x)$ from a high di-mensional data space R^N onto a two-dimensional grid of units/neurons such that topology is preserved. The weights of a neuron are updated by

$$m_i(t+1) = m_i(t) + h(t)[x(t)-m_i(t)], \quad 0<h(t)<1,$$

where $m_i(t) \in R^N$ is a reference (weight) vector, $x(t) \in R^N$ is an input (stimulation) vector, and $h(t)$ is a *neighborhood function*. Note that the above equation can be written as convex combination

$$m_i(t+1) = [1-h(t)]m_i(t) + h(t)x(t), \quad 0<h(t)<1.$$

KSOM variants were presented for learning non-vectorial data using domain-specific distance functions (Kohonen and Somervuo 1998, 2002; Hagenbuchner et al. 2003; Cottrell et al. 2004; Seo and Obermayer 2004; Somervuo 2004). Other authors have proposed KSOM extensions using non-Euclidean metrics (Ritter 1999; Peltonen et al. 2004). Yet, different authors have proposed using *weighting factors* $\lambda_i \geq 0$, i=1,...,n in different input data dimensions in order to improve KSOM's classification perform-ance (Hammer and Villmann 2002). Using the terminology of this book, an aforementioned *factor*, corresponds to mass function $\mu_i(t) = \lambda_i$. Hence, a constant factor λ_i attaches the same weight of significance to all numbers in a data dimension; whereas, a general mass function $\mu_i(t)$ may attach a different weight to a different number along a data dimension.

Another KSOM extension, namely *self-organizing mixture model* (SOMM) was reported based on mixture densities (Verbeek et al. 2005). An advantage of SOMM is a well-defined objective function. However, a SOMM does not consider alternative divergence (distance) functions. Fur-thermore, a SOMM is applicable solely on crisp data. A different KSOM extension using mixture models is the generative topographic mapping (GTM), that is an extension of the latent (hidden) variable framework to allow non-linear transformations (Bishop et al. 1998). Operation of GTM is based on a constrained mixture of Gaussians whose parameters are op-timized by maximum likelihood. Note that objective functions, which are inspired from energy functions in physics, cannot accommodate linguistic data such as fuzzy sets and conventional intervals. Furthermore, a mixture model typically assumes (a priori) parametric probability density func-tions, e.g. Gaussian functions, etc.

Conventional KSOM extensions cannot cope with ambiguity (Krishnapuram and Keller 1993). Additional disadvantages include the lack of sound optimization and convergence strategies (Tsao et al. 1994). In response, a number of 'KSOM type' fuzzy c-means algorithms were proposed (Pal et al. 1993; Tsao et al. 1994; Karayiannis and Bezdek 1997). The latter fuzzy algorithms process crisp data; more specifically, they process vectors of numbers. Fuzzy KSOM extensions were proposed for processing linguistic data represented by simplified 3-dimensional vectors (Mitra and Pal 1996). The latter extensions are simple fuzzy inference systems (FIS) whose rule consequents are category labels. A different fuzzy extension of KSOM was reported for implementing a fuzzy inference system (FIS) in a function $f: R^N \rightarrow R$ approximation problem, where solely triangular fuzzy membership functions, singleton number consequents, non-fuzzy inputs and, finally, constant mass functions $m(t)=1$ were considered (Vuorimaa 1994).

Sound extensions of KSOM from R^N to F_+^N, where F_+ is the metric lattice of positive fuzzy interval numbers (FINs), were proposed based on: (1) algorithm CALFIN, for computing a FIN from a population of real number samples in a data dimension, (2) a Minkowski metric in F_+^N, and (3) convex combinations of FINs. In particular, the name grSOM was used to denote KSOM-inspired architectures, which process FINs. Two types of KSOM extensions were presented, namely *greedy-grSOM* (Papadakis et al. 2004; Papadakis and Kaburlasos 2005; Kaburlasos and Papadakis 2006; Kaburlasos and Kehagias 2006b) and *incremental-grSOM* (Kaburlasos et al. 2005). Their basic difference is that incremental-grSOM uses convex combinations of FINs, whereas greedy-grSOM does not.

Section 7.1.2 presented a greedy-grSOM model, which can induce a distribution of FINs from the training data such that a FIN specifies a data cluster; moreover, a FIN was interpreted statistically as an *information granule*. Learning by grSOM may be driven either by a tunable metric distance $d_1(.,.)$ or by a tunable fuzzy membership function $1/(1+d_1(.,.))$. Moreover, an integrable *mass* function $m(x)$ can tune non-linearly metric $d_1(.,.)$ by attaching a weight of significance to a real number 'x' in a data dimension. A genetic algorithm (GA) can compute optimally a mass function $m(x)$. The induced FINs may be interpreted as antecedents (IF parts) of linguistic (fuzzy) rules; the corresponding rule consequents (THEN parts) may be category labels in a classification problem. A grSOM model in the context of this work can pursue linguistic (fuzzy) rule induction towards classification rather than data visualization. Hence, a grSOM model can be used for granular data mining (Hirota and Pedrycz 1999).

The objectives of conventional KSOM typically include dimensionality reduction and visualization. However, the principal objective of a grSOM model is extraction of descriptive decision-making knowledge (fuzzy rules) from the training data. From a rigorous mathematical point of view a grSOM model treats the Euclidean space R^N differently than a KSOM model. More specifically, the latter treats R^N as a real linear space; whereas a grSOM model treats R^N as the Cartesian product of N totally-ordered lattices R.

10.4 Various Neural Networks

Since the resurgence of neural networks in the mid-1980s (Rumelhart et al. 1986) a number of models have been presented. Perhaps the most popular neural network models are multilayer perceptrons. Interesting connections are shown in the following.

10.4.1 Multilayer perceptrons

Conventional multilayer perceptrons typically employ sigmoid transfer functions (Uykan 2003). Despite the success of conventional multilayer perceptrons, especially in pattern recognition applications, their capacity has been questioned (Gori and Scarselli 1998). A novel proposal follows.

A sigmoid neuron transfer function corresponds to a 'bell-shaped' mass function (Fig.10-3(c)). The latter function emphasizes a neighborhood of real number values. Alternative mass functions may be employed emphasizing more than one neighborhoods. Hence, different positive valuation functions may emerge as 'enhanced' transfer functions for the neurons of multilayer perceptrons (Fig.7-3(b)). Extensive computational experiments as well as their validation are topics for future research.

10.4.2 Nonnumeric neural networks

Connectionist models typically employ 'flat' (vector) data. However, alternative data representations may arise including complex numbers (Hirose 2003), symbols (Omlin and Giles 1996b; Tan 1997), rules (Ishibuchi et al. 1993; Omlin and Giles 1996a; Healy and Caudell 1997), etc.

Non-numeric data are typically preprocessed for 'vector feature extraction' (Lecun et al. 1998). For instance, a problem of grammatical inference was treated by processing strings of real numbers 0s and 1s (Omlin and

Giles 1996a). Further neural processing may proceed on a multilayer perceptron (MLP) architecture (Gori and Scarselli 1998; Scarselli and Tsoi 1998) whose well-known drawbacks include: (1) long training time, (2) erosion of previous knowledge when new training data are presented, and (3) no justification is given for the network's answers.

Experience has shown that terms in first-order logic, blocks in document processing, patterns in structural and syntactic pattern recognition are entities which are best represented as graph structures and they cannot be easily dealt with vector-based architectures (Sperduti and Starita 1997). A number of neural computing architectures have been presented for processing various types of graph (structured) data (Hinton 1990; Frasconi et al. 1998; Hammer et al. 2004; Carozza and Rampone 2005). Furthermore connectionist architectures, including the Recursive Auto-Associative Memory (RAAM) (Pollack 1990), have been developed for processing distributed (non-vector) data representations. More recent techniques include holographic reduced representations (Plate 2003). Nevertheless, the aforementioned architectures cannot accommodate ambiguity.

Fuzzy neural networks (FNNs) were proposed for dealing with ambiguity by processing *linguistic* (fuzzy) data. In particular, a linguistic datum is represented by a fuzzy number defined on the real number R universe of discourse. A FNN combines the 'bottom-up' neural capacity for massively parallel learning and generalization with the 'top-down' capacity of fuzzy logic for human-like reasoning based on uncertain or ill-defined data. Several FNNs have been presented as explained next.

The FNN in Pal and Mitra (1992) suggests fuzzy interpretations to the multilayer perceptron (MLP) and it deals with inputs in numerical, linguistic, and set form by representing N-dimensional patterns as 3N-dimensional vectors. The FNN in Mitra and Pal (1995) builds on Pal and Mitra (1992), nevertheless the emphasis is on an 'expert' connectionist model; moreover the network generates a measure of certainty expressing confidence in its decisions. The knowledge-based network in Mitra et al. (1997) builds on both Mitra and Pal (1995) and Pal and Mitra (1992), and it improves performance by encoding initial knowledge in terms of class *a priori* probabilities.

FNNs have been proposed for integrating human knowledge and numerical data (Ishibuchi et al. 1993, 1995), furthermore the latter FNNs operate by propagating fuzzy numbers through a neural network. Inference in a fuzzy logic-based neural network (FLBN) may be carried out by neuron-implemented fuzzy AND/OR logic rules, moreover antecedents can be weighted by a possibility measure (Liu and Yan 1997). Self-constructed fuzzy inference networks have been studied by several authors (Jang 1993;

Jang and Sun 1995; Jang et al. 1997; Juang and Lin 1998). It turns out that FNNs employ product lattice R^N implicitly. In particular, FNNs typically involve linguistic variables such as 'small', 'large', 'fast', 'old' defined over the *totally ordered* real number universe of discourse R.

Fuzzy lattice neurocomputing (FLN) is a fundamentally different approach to neurocomputing for dealing with non-numeric data without converting them to numeric data. The basic idea is to apply FLN in a lattice data domain for inducing rules involving lattice intervals. In particular, model σ-FLN(MAP) can be interpreted as a FNN (fuzzy neural network) and has demonstrated a capacity to process rigorously vectors, graphs, fuzzy numbers, and other types of data (Kaburlasos and Petridis 1997, 2000, 2002; Petridis and Kaburlasos 1998, 2001; Kehagias et al. 2003).

10.5 Fuzzy Inference Systems

A fuzzy inference system, or FIS for short, includes a knowledge base of fuzzy rules 'if A_i then C_i', symbolically $A_i \rightarrow C_i$, $i=1,\ldots,L$. Antecedent A_i is typically a conjunction of N fuzzy statements involving N fuzzy numbers, moreover consequent C_i may be either a fuzzy statement or an algebraic expression; the former is employed by a Mamdani type FIS based on expert knowledge (Mamdani and Assilian 1975), whereas the latter is employed by a Takagi-Sugeno-Kang (TSK) type FIS based on input-output measurements (Takagi and Sugeno 1985; Sugeno and Kang 1988).

Based typically on fuzzy logic, a FIS input vector $x \in R^N$ activates in parallel rules in the knowledge-base by a *fuzzification* procedure; next, an *inference mechanism* produces the consequents of activated rules; the partial results are combined; finally, a real number vector is produced by a *defuzzification* procedure. Fig.10-2 shows a Mamdani type FIS, involving triangular fuzzy membership functions in L fuzzy rules R1,...,RL. The antecedent (IF part) of a rule is the conjunction of N fuzzy statements, whereas the consequent (THEN part) of a rule is a single fuzzy statement. Various fuzzy number shapes /inference mechanisms/(de)fuzzification procedures have been proposed (De Baets and Kerre 1993; Pedrycz 1994; Runkler 1997; Saade and Diab 2000; Mitaim and Kosko 2001).

A FIS is used as a device for implementing a function $f: R^N \rightarrow K$, where K may be either discrete or continuous. A popular function using a discrete K is a classifier (Ishibuchi et al. 1999; Setnes and Babuska 1999; Mitra and Hayashi 2000), where K is a set of class labels. The Mamdani type FIS in Fig.10-2 implements a function $f: R^N \rightarrow R$ (Nozaki et al. 1997). More generally, a FIS can implement a function $f: R^N \rightarrow R^M$.

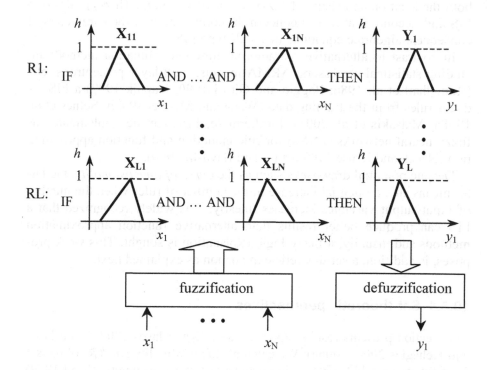

Fig.10-2 A Mamdani type Fuzzy Inference System (FIS) with N inputs $x_1,...,x_N$, one output y_1, and L fuzzy rules R1,...,RL. The above FIS, including both a fuzzification and a defuzzification procedure, implements a function $f: \mathbb{R}^N \rightarrow \mathbb{R}$.

FISs are useful for modeling and control especially in uncertain /ambiguous environments (Elkan 1994; Jang and Sun 1995; Isermann 1998; Passino and Yurkovich 1998; Fuessel and Isermann 2000). FIS were also combined with neural networks in neurocontrol applications (Werbos 1993). Note that various algorithms have been developed for inducing a function $f: \mathbb{R}^N \rightarrow \mathbb{R}^M$ from n pairs $(x_1,y_1),(x_2,y_2),...,(x_n,y_n)$ of training data vectors. It turns out that the design of a FIS typically boils down to a parameter optimization problem with constraints, which seeks minimization of the LSE $\sqrt{\sum_{i=1}^{n}\left\|f(x_i)-y_i\right\|^2}$ (Wang and Mendel 1992a; Kosko 1994; Castro and Delgado 1996; Dickerson and Kosko 1996; Zeng and Singh 1996, 1997, 2003; Mitaim and Kosko 2001; Liu 2002). In particular, the design of a FIS concerns, first, an estimation of parameters which specify

both the location and shape of fuzzy sets involved in the (fuzzy) rules of a
FIS and, second, it may also concern the computation of the parameters of
consequent algebraic equations in a TSK type FIS.

In contrast to alternative 'parametric' function estimation methods including statistical regressors, ARMA models, multilayer perceptrons, etc.
(Rumelhart et al. 1986; Poggio and Girosi 1990; Vapnik 1999) a FIS induces rules from the training data (Wang and Mendel 1992b; Setnes et al.
1998a; Matsakis et al. 2001). Furthermore, FISs can be implemented as
fuzzy neural networks (FNNs) for rule induction and function approximation (Nauck and Kruse 1999; Mitra and Hayashi 2000).

Despite potential drawbacks such as the *curse of dimensionality*, the latter means an exponential increase in the number of rules when the number
of input/output variables increases linearly, it is widely recognized that a
FIS can produce better results than alternative function approximation
methods and, usually, a fuzzy logic explanation is sought. This work proposes, in addition, a set-theoretic explanation as explained next.

10.5.1 Set-theoretic perspectives

This section presents results reported lately (Kaburlasos 2002; Kaburlasos
and Kehagias 2004, 2006a). We calculate the cardinality $card(\mathfrak{F})$ of the set
$\mathfrak{F}$ of functions $f: \mathbf{R}^N \rightarrow \mathbf{R}^M$. Using standard cardinal arithmetic (Stoll 1979)
it follows

$$card(\mathfrak{F}) = \aleph_1^{\aleph_1} = (2^{\aleph_0})^{\aleph_1} = 2^{\aleph_0 \aleph_1} = 2^{\aleph_1} = \aleph_2 > \aleph_1,$$

where $\aleph_1$ denotes the *cardinality* of the set $\mathbf{R}$ of real numbers. Unfortunately a general function f_0 in $\mathfrak{F}$ is practically useless because it lacks a capacity for generalization. More specifically, knowledge of function f_0 values $f_0(x_1),...,f_0(x_n)$ at points $x_1,...,x_n$ cannot give any information regarding
the value of function f_0 at a different point $x_{n+1} \neq x_i$, $i=1,...,n$.

Consider now a parametric family of models (characterized by a capacity for generalization), e.g. polynomials, ARMA models, statistical regressors, radial basis function (RBF) networks, multilayer perceptrons, etc.
Due to the finite number p of parameters involved in a parametric family
of models it follows that the cardinality of any of the aforementioned families equals $\aleph_1^p = (2^{\aleph_0})^p = 2^{\aleph_0 p} = 2^{\aleph_0} = \aleph_1$.

It might be thought that $\aleph_1$ is an adequately large number of models to
choose a 'good' model from in an application. Unfortunately this is not the
case. Consider, for instance, the family of polynomials, which includes $\aleph_1$
models. It is well known that a polynomial may not approximate usefully a

set $(x_1,y_1),...,(x_n,y_n)$ of training data due to *overfitting*; hence a different family of models might be sought, e.g. a ARMA model, a multi-layer perceptron, etc. In the aforementioned sense the cardinality $\aleph_1$ (of a family of models) may be *inherently restrictive*. It might be interesting to note that a closure of functions implementable by a multiplayer perceptron has been studied by other authors (Gori et al. 1998).

What about the cardinality of the set of all FISs? To compute this, let us first compute the cardinality of the set F_n of fuzzy numbers. The next proposition shows the non-obvious result that there are as many fuzzy numbers as there are real numbers.

Proposition 10-1 It is $card(F_n)= \aleph_1$, where $\aleph_1$ is the cardinality of the set R of real numbers.

The proof of Proposition 10.1 is given in Appendix B.

Recall that the set F_+ of positive FINs includes (fuzzy) numbers and intervals. It turns out that $card(F_n) = card(F_+) = card(F) = \aleph_1$.

Now consider, Mamdani type FISs: the rules in a Mamdani type FIS can be interpreted as samples of a function m: $F^N \rightarrow F^M$. Using standard cardinal arithmetic (Stoll 1979) it follows that the cardinality of the set M of Mamdani type FISs equals $card(M)= \aleph_1^{\aleph_1} = \aleph_2 > \aleph_1$. Likewise, the rules in a TSK type FIS can be interpreted as samples of a function s: $F^N \rightarrow P_p$, where P_p is a family of parametric models (e.g. polynomial linear models) with p parameters. It follows that the cardinality of the set S of TSK type FISs also equals $card(S)= \aleph_1^{(2^{\aleph_0})^p} = \aleph_1^{2^{p\aleph_0}} = \aleph_1^{2^{\aleph_0}} = \aleph_1^{\aleph_1} = \aleph_2$. Furthermore, a FIS (of either Mamdani or TSK type) has a capacity for local generalization due to the non-trivial (interval) support of the fuzzy numbers involved in FIS antecedents; in other words, an input vector $x= (x_1,...,x_n)$ within the support of a fuzzy rule activates the aforementioned rule.

In conclusion, a FIS (of either Mamdani or TSK type) can implement, in principle, many more functions than competing families of functions, furthermore a FIS is endowed with a capacity for (local) generalization. Hence the class of FISs is preferable to both the general class $\mathfrak{F}$ of functions (which lacks a capacity for generalization) and parametric families of models (which have a smaller cardinality). It is understood that the aforementioned advantage of the family of FIS models is theoretical rather than practical. However, note that a valid theoretical bound may be useful as a bound to our practical expectations. Additional advantages of FIS include an alleviation of the 'curse of dimensionality' problem, a potential for improved computer memory utilization, and a capacity for granular computing (Kaburlasos 2004a; Kaburlasos and Kehagias 2004, 2006a, 2006b).

Proposition 10.1 also suggests an interesting proposal regarding the preferable fuzzy number (membership function) shape. Note that a variety of such shapes have been proposed in the literature including triangular, trapezoidal, polynomial, bell-shaped, etc. (De Baets and Kerre 1993; Pedrycz 1994; Ishibuchi et al. 1995; Runkler and Bezdek 1999a; Mitaim and Kosko 2001). Any of the aforementioned shapes is described by a finite number p of parameters. For instance, a triangular membership function is described using $p=3$ parameters; hence, there exist $\aleph_1^3 = \aleph_1$ fuzzy numbers of triangular shape. Likewise, there exist $\aleph_1^p = \aleph_1$ fuzzy numbers of any particular parametric shape. Moreover, since the number of different parametric shapes (e.g. triangular, Gaussian, trapezoidal, etc.) is finite, it follows that we have a set of $\aleph_1$ parametric fuzzy numbers altogether. Therefore, using any of the aforementioned families we can generate $\aleph_2$ functions $f\colon F^N \rightarrow F^M$, each function characterized by a (local) capacity for generalization. Hence, Proposition 10-1 implies that any membership function shape enables a FIS to implement, in principle, $\aleph_2$ different functions. In practice triangular membership functions are frequently preferable due to their convenient representation using only $p=3$ parameters.

The 'set-theoretic' analysis above has introduced novel FIS perspectives. 'Lattice theoretic' analysis below proposes novel tools for improving conventional FIS design based on 'RBF network principles'. A short summary of RBF networks is presented next.

10.5.2 Radial basis function networks

RBF (radial basis function) networks were proposed in multivariate interpolation problems in R^N based on local models (Powell 1985). A local model is characterized by a center; moreover, the influence of a model is a function of a metric (distance) from the corresponding model's center.

A number of RBF networks were introduced in the literature including RBF networks for the design of neural networks (Broomhead and Lowe 1988). A RBF algorithm was presented for fast learning in networks of locally tuned neurons (Moody and Darken 1989). Placement of RBF centers can have a significant effect on performance (Poggio and Girosi 1990). Bounds on error functions in function approximation problems are functions of RBF centers. Moreover, a reduction in error was shown by moving the RBF centers (Panchapakesan et al. 2002). RBF networks were also used in classification applications (Bianchini et al. 1995; Roy et al. 1995).

RBF networks have been compared with alternative models including sigmoid perceptrons (Haykin 1994). Under minor restrictions, the functional behavior of RBF networks and fuzzy inference systems was found to be the same (Jang and Sun 1993). A comparison of RBFs with fuzzy basis functions (FBFs) in function approximation problems was shown in Kim and Mendel (1995). An advantage of the aforementioned FBFs is that they combine numeric- and linguistic- data, the latter are represented by parametric (Gaussian) fuzzy membership functions.

In the context of this work, a metric was defined in a general lattice data domain based on a positive valuation. Hence 'RBF design' can be pursued in a lattice including metric lattice $(F_+,\leq)$ of positive FINs. For instance, a grSOM model can be designed like a RBF network for classification. More generally, a 'metric design' of FISs can be pursued as explained next.

10.5.3 Novel FIS analysis and design

This section presents novel perspectives and tools reported lately (Kaburlasos 2002; Kaburlasos and Kehagias 2004, 2006b).

A fuzzy inference system (FIS) can be interpreted a look-up table, which is usually employed for function $f\colon R^N \to R^M$ approximation by interpolation (Tikk and Baranyi 2000). A FIS employs a local model per fuzzy rule; more specifically, a fuzzy rule's support defines the region of the input space where a model is active. Conventional FISs are based on (fuzzy) logic. This work proposes FISs based as well on (metric) topology.

It is remarkable that, even though an explicit connection was shown between *fuzzy sets* and *lattices* since the introduction of fuzzy set theory (Zadeh 1965) as well as later (Saade and Schwarzlander 1992; Gähler and Gähler 1994; Zhang and Hirota 1997; Höhle and Rodabaugh 1998), no tools were established for FIS analysis and design based on lattice theory. This work has engaged lattice theory above for introducing a metric $d_K(.,.)$ in the lattice $(F_+,\leq)$ of positive FINs. It follows that an employment of fuzzy rules (used as RBF centers) together with metric $d_K(.,.)$ can implement a function $f\colon F^N \to F^M$ involving jointly numeric-, interval-, and linguistic (fuzzy) input data beyond fuzzy rules' supports. An additional, potential benefit is the employment of standard function approximation techniques (Cheney 1982) in order to reduce the number of fuzzy rules (Kóczy and Hirota 1997; Dick and Kandel 1999).

Of particular significance is the problem of *structure identification*, that is an optimal placement of fuzzy sets on the real line for all rules involved in a FIS (Sugeno and Kang 1988); both the location and membership function shapes of corresponding fuzzy sets are induced from measurements. Note that usually simple parametric fuzzy membership functions are used, for example triangular, trapezoidal, and Gaussian functions. A solution to the structure identification problem is typically pursued using various clustering and/or supervised learning algorithms including (fuzzy) neural networks (Jang and Sun 1995; Kecman 2001; Leng et al. 2005; Mitra and Hayashi 2000; Papadakis and Theocharis 2002; Pomares et al. 2000, 2002; Setnes 2000; Takagi and Sugeno 1985). A modified Kohonen SOM algorithm for structure identification was proposed lately involving FINs with general membership functions (Kaburlasos and Papadakis 2006).

It is known that FISs can be improved 'genetically' by tuning parameterized membership functions as well as input-output scaling factors (Hoffmann 2001). Here genetic algorithms (GAs) are employed to optimally estimate integrable *mass functions* (Kaburlasos and Papadakis 2006). For instance, consider FINs X22, x2, and X12 in Fig.10-3(a).

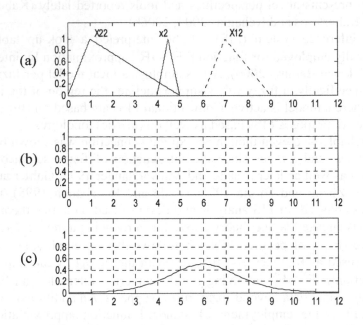

Fig.10-3 (a) Three triangular FINs X12, X22, and x2.
(b) Mass function $m_h(t) = h$, for $h=1$.
(c) Mass function $m_h(t) = h2e^{-(t-6)}/(1+e^{-(t-6)})^2$, for $h=1$.

Next, we will compute distances d_K(X22,x2) and d_K(X12,x2) using the two different mass functions shown, respectively, in Fig.10-3(b) and Fig.10-3(c). On the one hand, mass function m_h(t)= h, $h \in (0,1]$ in Fig.10-3(b) assumes that all real numbers are equally important. On the other hand, mass function m_h(t)= $h2e^{-(t-6)}/(1+e^{-(t-6)})^2$ in Fig.10-3(c) emphasizes the numbers around t=6; the corresponding positive valuation is $f_h(x)$= $h[(2/(1+e^{-(x-6)}))-1]$, namely *logistic* (or, *sigmoid*) function.

Fig.10-4(a) plots both functions d_K(X22(h),x2(h)) and d_K(X12(h),x2(h)) in solid and dashed lines, respectively, using the mass function m_h(t)= h. The area under a curve equals the corresponding distance between two FINs. It turns out d_K(X22,x2) $\approx$ 3.0 > 2.667 $\approx$ d_K(X12,x2).

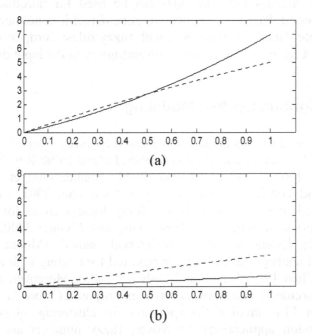

(a)

(b)

Fig.10-4 (a) Distance functions d_K(X22(h),x2(h)) and d_K(X12(h),x2(h)) are plotted in solid and dashed lines, respectively, using mass function m_h(t)= h shown in Fig.10-3(b). The area under a curve equals the corresponding distance between two FINs. It turns out d_K(X22,x2) $\approx$ 3.0 > 2.667 $\approx$ d_K(X12,x2).

(b) Distance functions d_K(X22(h),x2(h)) and d_K(X12(h),x2(h)) are plotted in solid and dashed lines, respectively, using mass function m_h(t)= $h2e^{-(t-6)}/(1+e^{-(t-6)})^2$ shown in Fig.10-3(c). The area under a curve equals the corresponding distance between two FINs. It turns out d_K(X22,x2) $\approx$ 0.328 < 1.116 $\approx$ d_K(X12,x2).

Fig.10-4(a) shows that for smaller values of h function $d_K(X12(h),x2(h))$ is larger than $d_K(X22(h),x2(h))$ and the other way around for larger values of h as expected from Fig.10-3(a) by inspection. Fig.10-4(b) plots both functions $d_K(X22(h),x2(h))$ and $d_K(X12(h),x2(h))$ in solid and dashed lines, respectively, using the mass function $m_h(t)= h2e^{-(t-6)}/(1+e^{-(t-6)})^2$. It turns out $d_K(X22,x2) \approx 0.328 < 1.116 \approx d_K(X12,x2)$.

The analysis above has demonstrated that a different mass function $m_h(t)$ can imply a different metric distance $d_K(.,.)$ between FINs. On the one hand, typical FIS applications employ (implicitly) mass function $m_h(t)=h$. On the other hand, this work has shown that alternative mass functions may imply additional flexibility in practice.

The example above has demonstrated a number of useful novelties including, first, a mass function $m_h(t)$ can be used for introducing non-linearities; second, based on a metric distance $d_K(.,.)$ it is not necessary to have the whole data domain covered with fuzzy rules; third, an input to a FIS might be a fuzzy set for dealing with ambiguity in the input data.

10.6 Ambiguous System Modeling

Ambiguity in applications is often represented by fuzzy intervals (Galichet and Foulloy 1995; Palm and Driankov 1995; Leland 1998; Roychowdhury and Pedrycz 2002; Bondia and Picó 2003; Boukezzoula et al. 2003; Fernández and Gutiérrez 2003; Foulloy and Galichet 2003; Liao et al. 2004). Of particular interest are type-2 fuzzy logic systems for handling rule uncertainties (Karnik et al. 1999; Liang and Mendel 2000; Wu and Mendel 2002). Extensions of classic interval analysis (Moore 1979) to fuzzy interval analysis have also been reported for dealing with ambiguity (Shyi-Ming Chen 1997; Moore and Lodwick 2003). Moreover, a unification of the aforementioned analyses has been proposed (Albrecht 2003).

Very often FISs involve 'interpolation' in clustering, classification, and/or regression applications. Moreover, fuzzy numbers are typically processed in FISs based on fuzzy logic. However, (conventional) FISs may not be convenient for extrapolation applications. A suitable algebra may enable extrapolation involving fuzzy numbers as explained next.

Various fuzzy arithmetics have been presented in the literature (Dubois and Prade 1980; Kaufmann and Gupta 1985; Zimmermann 1991; Filev and Yager 1997; Giachetti and Young 1997; Klir 1997). The aforementioned arithmetics usually define operations between fuzzy numbers based on fuzzy set theory's 'extension principle'. However, practical difficulties are experienced in the definition of the multiplication operation.

This work proposes a novel arithmetic involving fuzzy interval numbers (FINs) in the cone F_+ of positive FINs as detailed in section 4.5.2. Recall that a FIN may be interpreted as a fuzzy set, or it may be interpreted 'statistically', etc. A modeling problem is described next.

Based on n (N+1)-tuples of FINs for training $[X_{1,k},...,X_{N,k},Y_k]'$, $k=1,...,n$, the problem is to produce optimal estimates $\hat{Y}_k$ of Y_k, $k=1,...,n$ that minimize the mean absolute error $E_{MAE}= \sum_{k=1}^{n} d_K(Y_k, \hat{Y}_k)/n$. In particular, estimates $\hat{Y}_k$ are produced by the following autoregressive moving average (ARMA) model

$$\hat{Y}_k = T + \sum_{j=1}^{p} g_j(Y_{k-j}) + \sum_{i=1}^{N} \sum_{j=0}^{q_i} f_{i,j}(X_{i,k-j})$$

where T is a trivial FIN (threshold), g_j: $R \rightarrow R$, $j=1,...,p$ and $f_{i,j}$: $R \rightarrow R$, $i=1,...,N$, $j=0,...,q_i$ are strictly increasing functions. Model parameters including (1) the number of delays $q_1,...,q_N,p$, which take values in the set of integers $\{0,1,2,...\}$, (2) the strictly increasing functions g_j, $j=1,...,p$, $f_{i,j}$, $i=1,...,N$, $j=0,...,q_i$, which are typically in a parametric family, and (3) trivial FIN constant $T= \bigcup_{h \in (0,1]} \{[x,x]^h\}$ for $x \in R$, are to be induced from the training data $[X_{1,k},...,X_{N,k},Y_k]'$, $k=1,...,n$. Since the inputs to the aforementioned ARMA model are FINs, the model is called *granular ARMA*, or *grARMA* for short (Kaburlasos and Christoforidis 2006). Advantages of a grARMA model are summarized next.

A grARMA model can accommodate ambiguity represented by FINs. Non-linearities can be introduced either by non-linear functions g_j, $j=1,...,p$, $f_{i,j}$ or by non-linear underlying mass functions. Future work includes stability analysis regarding both the norm of FINs in a grARMA model and the corresponding FIN supports.

10.7 A Unified Treatment of Uncertainty

The same lattice theory mathematics may be given different interpretations in different contexts including probability theory, statistics, theory of evidence, fuzzy set theory, etc. as explained next.

Probabilistic techniques have been popular in computational intelligence, for instance in neural computing (Specht 1990a, 1990b; Petridis and Kehagias 1998; Anagnostopoulos et al. 2004). A number of neural networks have been proposed for probability estimation (Tråvén 1991; Lim and Harrison 1997). Lately, a granular extension of Kohonen's SOM, namely grSOM, was proposed for inducing a distribution of FINs, where a FIN represents a local probability distribution (Kaburlasos and Papadakis 2006). Note that grSOM employs metric $d_K(.,.)$ to compute a distance between two probability distributions. It turns out that metric d_K has advantages compared to the *Kullback-Leibler* (KL) distance as explained next.

The KL distance is a function for calculating proximity of two probability distributions (Duda et al. 2001). On the one hand, a disadvantage of KL-distance is that the cumulative distribution functions (CDFs) involved in the calculations need to be defined on the same elements of a set, otherwise spurious results are produced such as 0 or ∞. On the other hand, distance d_K can be used to calculate the proximity of two CDFs defined even on mutually disjoint intervals of numbers. Note also that KL-distance is not a metric because it does not satisfy laws D2 (Symmetry) and D3 (Triangle Inequality) shown in Appendix A.

Apart from probabilistic techniques a probability measure is also used in statistical techniques. Of particular practical importance is Bayesian decison-making based on the following formula

$$P(M_j/D) = \frac{P(D/M_j)P(M_j)}{P(D)}, \text{ where}$$

M_j, $j=1,...,J$ is a model, which may produce datum D; probability $P(D/M_j)$, namely *likelihood distribution*, describes the probability of observing datum D given model M_j; probability $P(M_j)$, namely *prior distribution*, represents all that is known before any data becomes available; probability $P(D)$, namely *evidence distribution*, describes the likelihood of observing datum D averaged over all models M_j, $j=1,...,J$. Bayes formula computes the posterior distribution $P(M_j/D)$, i.e. the probability of a specific model M_j, $j=1,...,J$, given datum D. An advantage of Bayesian decision-making is its capacity to fuse, rigorously, non-numeric data.

Probability theory suffers certain deficiencies in practice. For instance, in the absence of evidence, probability theory applies the 'principle of indifference' and assigns probabilities equal to 0.5 to both event and no-event leading potentially to wrong conclusions. Dempster-Shafer *theory of evidence* was introduced for overcoming decision-making problems of conventional probability theory including the aforementioned principle of indifference (Giarratano and Riley 1994). The theory of evidence originated from work by Dempster, who attempted to model uncertainty by a range of probabilities rather than by a single number (Dempster 1967). Dempster's *rule of combination* was proven especially useful in real-world applications because it can rigorously revise beliefs in the presence of new evidence. Dempster's work was later extended and refined by Shafer (Shafer 1976). From a theoretical point of view, Dempster-Shafer theory needs to satisfy fewer axioms (Liu et al. 2001). Note that a *belief function* of Dempster-Shafer theory over the lattice *frame of discernment* is a positive valuation (Kaburlasos and Petridis 2002). Note also that fuzzy extensions of the Dempster-Shafer theory have been proposed (Lucas and Araabi 1999).

An alternative to probability theory for dealing with uncertainty is fuzzy set theory, especially when uncertainty stems from ambiguity rather than from statistical occurrence. Furthermore, the fuzzy interval was used for evidence fusion (Tahani and Keller 1990). Fuzzy-logic inspired techniques were also used for improving classification by fusion (Dayou Wang et al. 1998). In this context a variety of information combination operators for data fusion have been reviewed comparatively (Bloch 1996). Synergies/connections of fuzzy set theory with probability theory and/or statistics were considered in various contexts (Ralescu and Ralescu 1984; Dubois and Prade 1986; Wonneberger 1994; Yager and Filev 1995; Kuncheva 1996; Yen and Wang 1998; Dunyak et al. 1999; Fukuda 2000).

There is an inherent relation between fuzzy set theory and lattice theory based on the lattice ordering of the fuzzy sets (Zadeh 1965). Furthermore, there is an inherent relation between probability theory and lattice theory based on the lattice ordering of events in a probability space (Birkhoff 1967). Hence, lattice theory emerges as a sound mathematical basis for a unified treatment of uncertainty including both probability theory and fuzzy set theory. Note that various authors have proposed theoretical connections between fuzzy set theory and probability theory (Goodman and Nguyen 2002; Körner and Näther 2002). This work has interpreted above a (positive) FIN as either a fuzzy set or a probabilistic distribution function.

A probability distribution, a fuzzy membership function, as well as other techniques that assign weights in various contexts (Quinlan 1986, 1990; Sun and Peterson 1999; Ishibuchi and Nakashima 2001; Gómez Sánchez et al. 2002; Kaburlasos et al. 2003; Wu and Zhang 2003) are regarded here as different interpretations of a single basic technique, which assigns a positive valuation function to elements in a lattice.

Another interesting proposal regards a probability space (Ω, S, P). It turns out that inclusion measure $k(A \subseteq B) = P(B)/P(A \cup B)$, $A, B \in S$ indicates how much a set A is included in another set B. Function $k(A,B) = k(A \subseteq B)$ could also be interpreted as the degree to which an event A implies another event B, that is the degree of truth of implication '$A \rightarrow B$' (Kaburlasos and Petridis 2002; Knuth 2005).

10.8 Machine Learning

Several of the models presented in this book can be described as machine learning (ML) models. Various connections with ML are presented next.

10.8.1 Version spaces

Perhaps no machine learning (ML) algorithm is conceptually nearest to the σ-FLN(MAP) algorithm than Find-S that outputs a *hypothesis*, from a *version space*, consistent with the training data (Mitchell 1997). Note that a *version space* is defined as a subset of hypotheses consistent with the training data; moreover, a *hypothesis* is typically a Boolean-valued expression.

There are similarities between Find-S and the σ-FLN algorithm. In particular, starting from a training datum both algorithms expand lattice intervals so as to include the training data. However, there are substantial differences between the two algorithms. For instance, Find-S typically considers a Boolean lattice of finite cardinality, whereas σ-FLN may consider a general product lattice including non-complete constituent lattices of infinite cardinalities. Moreover, a hypothesis in Find-S corresponds to a single Boolean lattice interval, whereas a hypothesis in σ-FLN corresponds to a family of lattice intervals. Finally, a fundamental difference between Find-S and σ-FLN is that the latter drives learning based on a fuzzy inclusion measure function (σ), whereas the former employs a crisp lattice order relation. Recall that σ is defined based on a positive valuation function; hence, the aforementioned fundamental difference boils down to an employment of a positive valuation function in σ-FLN but not in Find-S.

10.8.2 Instance-based learning

A number of instance-based algorithms were introduced (Mitchell 1997) including k Nearest Neighbor (kNN), RBFs, and case-based reasoning (CBR) (Aha et al. 1991; Kolodner 1993; Tsatsoulis et al. 1997; Aha 1998; Duda et al. 2001). Note that several of the aforementioned algorithms, including kNN and CBR, are 'memory-based'; the name 'lazy' is also used instead of 'memory-based' learning (Mitchell 1997; Shin et al. 2000). Unification of instance-based/CBR with rule-based induction has been proposed (Chi and Kiang 1993; Domingos 1996).

Several algorithms presented in this book including σ-FLN(MAP), FINkNN, etc. can be interpreted as instance/memory-based or CBR algorithms. However, a critical advantage with other memory-based algorithms is that the algorithms presented here can freely intermix numeric with non-numeric attributes. For instance, 'ambiguity', in a fuzzy set sense (Dubois and Prade 1980; Klir and Folger; 1988; Zimmerman 1991) can be dealt with using FINs.

A central issue in CBR is that syntactic similarity measures provide only an approximate indication of the relevance of a particular case to a particular problem. Hence, when a typical CBR algorithm attempts to reuse the retrieved cases it may uncover difficulties that were not captured by this syntactic similarity measure (Mitchell 1997). The latter problem can be alleviated in the context of this work using a (tunable) positive valuation function for tuning syntactic similarity.

10.8.3 Classification and regression trees

Classification and Regression-Trees (CARTs) as well as decision trees are popular techniques for inducing descriptive decision-making knowledge (rules) from the training data (Quinlan 1986, 1990, 1992, 1996; Breiman et al. 1998; Ruggieri 2002). The aforementioned trees typically employ an *entropy* (probabilistic) function as an objective function. It well known that maintenance and execution time of large and complex trees might be a substantial problem.

The lattice intervals computed in a product lattice by σ-FLN(MAP) in the context of this work can be interpreted as a refined decision tree. More specifically, inequalities $a_i \leq x_i \leq b_i$, i=1,...,N in constituent lattices $(L_i, \leq_i)$ i=1,...,N are implemented in parallel, one inequality per tree node. A significant advantage compared to a generic decision tree is that instead of a single inequality a pair of inequalities $a_i \leq x_i \leq b_i$, i=1,...,N is used here per tree node resulting in a finer partition of a general lattice data domain. Note also that using the lattice inclusion measure (σ) in the lattice $(\tau(L), \leq)$ of intervals an input datum $x \in \tau(L)$ can be assigned to a category even when x is outside all intervals/rules, that is an enhanced capacity for generalization. An additional advantage of σ-FLN(MAP) is its capacity to tune an underlying positive valuation function including a probability measure. It is remarkable that conventional decision trees frequently use positive valuation functions implicitly. For instance, an instance-weighting method has been proposed inspired by instance weight modification in boosting decision trees (Ting 2002). In terms of the theory presented here note that an instance weight function corresponds to a positive valuation function.

10.8.4 Classifier ensembles

It is known that an ensemble of classifiers may improve classification accuracy (Ishibuchi et al. 1996; Benediktsson et al. 1997; Kittler et al. 1998; Petridis and Kehagias 1998; Shimshoni and Intrator 1998; Drucker 1998; Avnimelech and Intrator 1999; Opitz and Maclin 1999; Dietterich 2000; Kuncheva et al. 2001; Petridis et al. 2001; Kehagias et al. 2003). Moreover, predictors including weights have been popular in bagging (Breiman 1996), boosting (Freund and Schapire 1999), predictive modular neural networks (Kehagias and Petridis 1997), linear opinion pools (Benediktsson and Swain 1992), fuzzy inference systems (Ishibuchi and Nakashima 2001), and elsewhere (Christensen et al. 2003). This work has proposed the Voting σ-FLNMAP model. Similarities and differences with popular classifier ensembles are delineated in the following.

Both bagging and boosting may employ different classifiers, e.g. CARTs, etc., whereas Voting σ-FLNMAP employs a single classifier, that is σ-FLNMAP, which operates by computing intervals. However, the wide scope of σ-FLNMAP is guaranteed by its lattice domain applicability. Bagging uses bootstrap replicates of the training data set whereas an ensemble of σ-FLNMAP voters uses different random permutations of the same training data set. The number (n_V) of different permutations (voters) may be estimated optimally during the training phase. An improved classification accuracy by σ-FLNMAP is attributed to noise canceling effects of learning intervals under different training data permutations.

AdaBoost algorithm for boosting (Freund and Schapire 1999) computes a whole population of weights over the training set, whereas σ-FLNMAP computes a much smaller set of weights, i.e. one weight per constituent lattice. Note that a set of weights (probabilities) is also maintained over a set of predictors in predictive modular neural networks (Kehagias and Petridis 1997). Consensus theoretic classification methods (Benediktsson and Swain 1992), which consider weight selections for different data sources, are regarded as the nearest to the FLN assuming that each data source corresponds to a different constituent lattice. However, applications of consensus methods are typically restricted in space R^N.

10.8.5 High dimensional techniques

In many modeling applications a popular approach is to search for an optimum model in a family of models. Furthermore, it was shown above, regarding fuzzy inference systems (FISs), that a general FIS can implement as many as $\aleph_2 = 2^{\aleph_1}$ functions, where $\aleph_1$ is the cardinality of the set R of real numbers; whereas, polynomials, ARMA models, RBF networks, MLPs, etc. can model 'only' as many as $\aleph_1$ different functions, where $\aleph_1 < \aleph_2$. Similar arguments extend to classification as explained next.

In many classification applications the objective is to compute an optimum among N-dimensional (parametric) surfaces so as to separate the training data while retaining a capacity for generalization. For instance, Support Vector Machines (SVM) employ a non-linear function $f: \mathsf{R}^N \rightarrow \mathsf{R}^M$ in order to transform a non-linearly separable classification problem in space R^N to a linearly separable classification problem in space R^M for M>>N; in particular, for a 2-categories classification problem, a hyperplane is optimally placed in R^M (Schölkopf and Smola 2002). Other authors have also considered similar techniques, which employ a map to more dimensions (Pao 1989; Looney 1997). It might be interesting to point out that both spaces R^N and R^M have identical cardinalities equal to $\aleph_1$. In other words, any increase in the number of dimensions does not change the total number of points in the space. An inherent theoretical limitation of the aforementioned techniques is shown next.

A total number of $\aleph_1$ hyperplanes exist in R^M, hence a SVM (as well as similar techniques) may theoretically accomplish as many as $\aleph_1$ different data partitions in a 2-categories classification problem. Recall that $\aleph_1$ is far less than the total number of $\aleph_2 = 2^{\aleph_1} > \aleph_1$ data partitions of the power set 2^{R^M} of R^M. In the aforementioned sense SVMs, as well as similar techniques for classification, are *inherently restrictive*. The same limitations extend to regression applications (Vapnik 1988; Chuang et al. 2002).

Now consider a classifier that learns hyperboxes in R^N (Carpenter et al. 1992; Simpson 1992, 1993; Georgiopoulos et al. 1994; Kaburlasos and Petridis 2000). An aforementioned classifier can potentially effect up to $\aleph_2$ data partitions by learning individual points in R^N by trivial hyperboxes; moreover, generalization may be possible beyond a hyperbox. It is understood that due to both the finite computer word-length and the finite computer memory only a finite number of subsets in R^N can be approximated in practice. Nevertheless, a valid theoretical bound can be useful as a bound to our practical expectations regarding a particular classifier.

10.9 Database Processing Techniques

This section presents connections of selected database processing paradigms with techniques presented in this book.

10.9.1 Granular computing

The term (*information*) *granule* was introduced in fuzzy set theory for denoting a clump of values drawn together by indistinguishability, similarity, proximity or functionality (Zadeh 1997; Hirota and Pedrycz 1999; Zadeh 1999; Bortolan and Pedrycz 2002; Pedrycz 2002). Granules have been used for data mining, knowledge extraction as well as in linguistic modeling applications (Pedrycz and Vasilakos 1999; Pedrycz and Vukovich 2001; Pedrycz and Bargiela 2002).

By definition granules are lattice ordered. A FIN used by model grSOM as well as a lattice interval used by model σ-FLN(MAP) may be interpreted as an information granule. In the aforementioned sense both neural models σ-FLN(MAP) and grSOM are inherently granular and can be used for granular computing (Zhang et al. 2000).

10.9.2 Rough sets

Rough sets are a useful alternative for dealing with uncertainty. They employ a basic idea from Frege according to which an object can have a nonempty boundary (Polkowski 2002). It turns out that rough sets can be used for set approximation (Pawlak 1991). Moreover, rough set theory has been useful in knowledge discovery applications (Ziarko 1999). The utility of rough sets in computational intelligence applications has been demonstrated (Pal et al. 2003). Moreover, comparative studies of fuzzy rough sets have been presented (Radzikowska and Kerre 2002).

Rough sets approximate a crisp concept (set) $A \subseteq X$ using partitions of the universe X. Note that the family of all partitions of a set, also called equivalence classes, is a lattice when partitions are ordered by refinement (Rota 1997). In particular, rough sets approximate A using *lower* and *upper* approximations. Therefore, lattice theory emerges implicitly in rough set theory. It follows that various lattice tools presented here based on a positive valuation function can potentially be used with rough sets.

10.9.3 Formal concept analysis

Formal concept analysis (FCA) is a branch of applied lattice theory. More specifically, based on the theory of complete lattices, FCA pursues a mathematization of *concept* and *concept hierarchy* towards conceptual data analysis and knowledge processing (Ganter and Wille 1999). Of particular importance in FCA is the notion *Galois connection* defined next.

Definition 10-2   Let φ: $P \rightarrow Q$ and ψ: $Q \rightarrow P$ be maps between two posets $(P, \leq)$ and $(Q, \leq)$. Such a pair of maps is called a *Galois connection* if

G1. $p_1 \leq p_2 \Rightarrow \varphi(p_1) \geq \varphi(p_2)$.

G2. $q_1 \leq q_2 \Rightarrow \psi(q_1) \geq \psi(q_2)$.

G3. $p \leq \psi(\varphi(p)) \Rightarrow q \leq \varphi(\psi(q))$.

A FCA algorithm is typically applied on N-tuples of (non-)numeric data and, in conclusion, a Hasse diagram of concepts is induced (Ganter and Wille 1999). Especially successful have been FCA applications for structured text representation towards clustering and classification (Carpineto and Romano 1996; Girard and Ralambondrainy 1997; Priss 2000; Doyen et al. 2001; Krohn et al. 1999; Bloehdorn et al. 2005).

A FCA algorithm is similar in spirit to a FLN algorithm; in particular, both employ complete lattices for knowledge acquisition. Nevertheless, FCA induces a lattice from the data, whereas FLN assumes a lattice ordering in the data from the outset. FCA usually considers data sets of finite cardinality, whereas FLN may consider a lattice of either finite or infinite cardinality. Furthermore, only FLN employs a (real) positive valuation function towards introducing tunable non-linearities.

It is interesting to point out that the isomorphism shown in section 3.4 between lattices $L^{\partial} \times L$ and $L \times L$ can be interpreted as a *dual* Galois connection for introducing an inclusion measure in the lattice $\tau(L)$ of intervals.

10.10 Mathematical Connections

Lattice theory is of interest in applied mathematics. In particular, in the context of computational intelligence, lattice theory was used to study (fuzzy) topological spaces (Kerre and Ottoy 1989; Georgiou and Papadopoulos 1999). This section outlines how lattice theory could further be instrumental in various mathematically inspired techniques.

10.10.1 Inclusion measures

An inclusion measure σ: $L \times L \rightarrow [0,1]$, where $(L, \leq)$ is a lattice, can be used to quantify set inclusion as explained in section 3.3. Recall that alternative definitions were proposed in the literature for quantifying a degree of inclusion of a (fuzzy) set into another one. However, on the one hand, the latter definitions typically involve only overlapping (fuzzy) sets otherwise the corresponding inclusion index equals zero. On the other hand, inclusion measure σ: L×L→[0,1] is more general because it applies to any lattice – not only to a lattice of (fuzzy) sets.

An inclusion measure σ could also be used as a *kernel* function (Schölkopf and Smola 2002). In a different context, inclusion measure σ could be used for fuzzification in a fuzzy inference system (FIS) (Takagi and Sugeno 1985). Another potential application of σ is in information retrieval tasks (Salton 1988; Carbonell et al. 2000), where an inclusion measure σ could be applied on lattice-ordered text structures towards semantic information retrieval (Pejtersen 1998).

10.10.2 'Intuitionistic' fuzzy sets

By 'intuitionistic' fuzzy sets here we mean in the sense of Atanassov (Atanassov 1999), where the degrees of membership and non-membership are defined independently. Lately, there is a growing number of publications regarding intuitionistic fuzzy sets (Grzegorzewski 2004; Gutiérrez García and Rodabaugh 2005). Moreover, there is an ongoing debate (Dubois et al. 2005), where it is shown that Atanassov's construct is isomorphic to interval-valued fuzzy sets and other similar notions.

This work does not adhere to any FIN interpretation. Rather, the interest here is in lattice-based data representations. We point out that, under any interpretation, a number of tools presented in this book can be useful.

10.10.3 Category theory

Category theory is the mathematical theory of structure and it is increasingly used for representing the hierarchical structure of knowledge and semantics (Goguen and Burstall 1992). Category theory was proposed as an alternative foundation to set theory. In particular, a primitive notion in set theory is *membership* $x \in y$, whereas a primitive notion in category theory is a *morphism f*: $a \rightarrow b$ between two objects 'a' and 'b' in a category.

Category theory was proposed for modeling semantics, the later can be represented as a hierarchy of concepts or symbolic descriptions, in the connection weights of cognitive neural systems (Healy and Caudell 2004). In the aforementioned context lattice theory emerges naturally. In particular, a lattice is a special case of a category in which the morphisms are relations $a \leq b$. Morphisms *f*: $a \rightarrow b$ might be especially useful in the context of this work for knowledge representation as explained at the end of section 5.1. Furthermore, it would be interesting to extend the utility of the tools presented here to category theory towards a tunable design.

PART IV: CONCLUSION

Part IV discusses practical implementation issues.
It also summarizes the contribution of this work.

11 Implementation Issues

This chapter discusses practical implementation issues

A model is implemented on a physical device. A CI model, in particular, is typically implemented on a digital computer. Hence, from a computational aspect, a typical CI model is a *Turing machine* whose formal definition is shown below.

11.1 Turing Machine Implementations

Entscheidungsproblem is the German term for the 'decision problem', that is 'logic cannot decide all mathematical assertions'.

In 1900 Hilbert proposed that logic could completely prove the truth or falsity of any mathematical assertion. Russell and Whitehead in *Principia Mathematica* accepted Hilbert's principle. However, Gödel showed that the Entscheidungsproblem was unsolvable by logic (Gödel 1931). Moreover, Turing showed that the Entscheidungsproblem could not be solved because the *halting problem* of *Turing machines* was itself unsolvable as explained in the following.

The *Turing machine* (Turing 1936) allows for unbounded 'external' storage (tapes) in addition to the finite information represented by the current 'internal' state (control) of the system. At every step, the machine reads the tape symbol $\alpha \in \{0,1,\text{blank}\}$ from under an access head, checks the state of the control $s \in \{1,2,\ldots,|S|\}$ and executes three operations: (1) Writes a new binary symbol $\beta \in \{0,1\}$ under the head, (2) Moves the head one letter to either right or left ($m \in \{L,R\}$), (3) Changes the state of the control to $s' \in \{1,2,\ldots,|S|\}$. The transition of the machine can be summarized by a function $f: (\alpha,s) \rightarrow (\beta,m,s')$. When the control reaches a special state, called the 'halting state', the machine stops (Siegelmann 1995).

The *halting problem* is a decision problem, which is described informally as follows: Given an algorithm and an initial input, determine whether the algorithm ever halts (completes), or it runs forever.

Vassilis G. Kaburlasos: *Implementation Issues*, Studies in Computational Intelligence (SCI) **27**, 175–178 (2006)
www.springerlink.com

Based on Gödel's ideas, the 'Church-Turing thesis' has considered *logic, Turing machines, effective function computation* and *lamda calculus* as equivalent mechanisms of problem solving (Wegner and Goldin 2003). There are various formulations of the Church-Turing thesis; one of them states that every *effective*, i.e. 'mechanical', computation can be carried out by a Turing machine. In all, a Turing machine pursues an *algorithmic* problem solving. Furthermore, a Turing machine can be implemented on electromechanical /electronic /optical /quantum etc. devices.

In the late 1940s, von Neumann designed the first modern electronic computer, which executed sequentially stored instructions (Aspray 1990). Currently, the most widely employed form of computing is carried out on silicon electronics based on Boolean logic. In particular, computations are typically carried out on digital Integrated Circuits (ICs), where the dominant switching device is the metal-oxide-semiconductor (MOS) transistor. Since the 1960s the IC industry has followed a steady path of constantly shrinking device geometries. More specifically, IC improvements have approximately doubled logic circuit density and increased performance by about 40% while quadrupling memory capacity every two to three years; that is commonly referred to as 'Moore's Law'. However, there is great uncertainty about the ability to continue similar hardware scaling beyond years 2015-2020 (Plummer and Griffin 2001).

Conventional electronics are based on the charge of the electron. Attempts to use an electron's other fundamental property, spin, gave rise to a new, rapidly evolving field, known as *spintronics* (Wolf and Treger 2003). Nevertheless, digital electronics is not unrivaled for all computing tasks. For instance due to its processing speed, optical computing is more suitable for fast Fourier transforms. A synergistic coexistence of optical- with electronics- computing was also considered (Caulfield 1998). Moreover, optical, electronic, as well as magnetic manipulation techniques have demonstrated the potential to locally control nuclear and electron spins towards quantum computing (Sih et al. 2003).

Computational intelligence (CI) models are currently implemented as Turing machines (Dote and Ovaska 2001; Reyneri 2003). Nevertheless, note that CI models may require lots of computing resources. Therefore, quantum computing (Hirvensalo 2004) might be promising. Hence, a quantum computer implementation of a fuzzy logic controller has been proposed to speed up inference in many dimensions (Rigatos and Tzafestas 2002). However, the latter quantum computer remains a Turing machine. It turns out that computing beyond Turing could pursue more efficient model implementations as explained next.

11.2 Beyond Turing Machine Implementations

Hypercomputing is a term used for denoting 'computing beyond Turing'. A machine for hypercomputing is called super-Turing machine. Different approaches to hypercomputing have been proposed as explained next.

A Turing machine is a closed system that does not accept input during its operation. One way towards hypercomputing is interactive computation (Wegner and Goldin 2003). Another way towards hypercomputing has been proposed for solving the 'halting problem' using *analogue X-machines* based on the notion of continuity (Stannett 1990). By definition, a Turing machine involves $\aleph_0$; in particular, a Turing machine involves a finite number of symbols (as well as of controls), moreover its operation proceeds at discrete time steps. Hence, one way towards hypercomputing involves $\aleph_1$. Note that neural networks with real number synaptic weights can perform computations provably uncomputable by a Turing machine; moreover, when the synaptic weights of the aforementioned neural networks are replaced by rational numbers, a network's computational capacity is reduced to that of a Turing machine (Siegelmann 1995).

From a practical implementation aspect, apart from digital Integrated Circuits (ICs), useful computations can be carried out using other devices such as analog ICs (Gilbert 2001) as well as optical buffers (Chang-Hasnain et al. 2003). An all-optical (analog) neural network has already been presented based on coupled lasers; moreover, a number of functions have been implemented to demonstrate the practicality of the aforementioned network (Hill et al. 2002). Holographic systems is a promising technology for massive data storage and instantaneous retrieval especially for accessing image-based data (Hesselink et al. 2004). It appears that hypercomputing can be pursued by analog soft computing models.

A practical advantage of hypercomputing based, for instance, on analog computers is in modeling phenomena that Turing machines cannot model. Another advantage may be sustaining an increase of computing power beyond years 2015-2020 by modeling the human mind as explained next.

Beyond analog (or digital) computing circuitry the human brain (hardware) can compute interactively. In particular, the mind (software) can process information interactively. The following chapter proposes a 'closed loop system model' for interactive computation. However, a practical problem remains, namely the 'representation problem', that is the problem of how to store and process a general lattice element. Potential solutions are investigated in the following.

Both the Boolean algebra, implemented on silicon electronics, and the collection of images on a plane, used in optical computing, imply a mathematical lattice (Kaburlasos and Petridis 2000). Alternative mathematical lattices may be considered in applications. Furthermore, an interesting proposal for computing with disparate lattice elements might be *programmable stream processors*, which effectively meet the demand for flexibility in media processing applications (Kapasi et al. 2003). However, it might not be necessary to seek a hardware implementation of general lattice elements; in particular, a lattice element could be implemented in software (Vahid 2003).

In conclusion, lattice theory proposes a conceptual shift in computing based on general lattice elements towards human intelligence modeling.

12 Discussion

This chapter discusses the contribution of this work in perspective

A (hopefully) inspiring cross-fertilization was proposed in computational intelligence (CI) – soft computing (SC) based on lattice theory. In particular, unified modeling and knowledge-representation were proposed based on tools applicable in a lattice data domain.

A merit of this book is in presenting wide-scope, effective algorithms for clustering, classification and regression using a unifying mathematical notation. Moreover, novel perspectives and tunable tools were presented. Highlights of our contribution are summarized next.

12.1 Contribution Highlights

After a relevant literature review it was argued that the art of modeling, from physical world oriented increasingly becomes human oriented. Computational intelligence (CI) or, equivalently, soft computing (SC), was presented as a useful coalition of methodologies whose enabling technology is a computer. The work of several research groups around the world on domain specific applications of lattice theory was scanned. It was also shown how lattice theory emerges, either explicitly or implicitly, in CI.

Valid mathematical results and perspectives were presented including novel ones. The employment of lattice theory here was uniquely characterized by positive valuation functions. More specifically, a positive valuation implied two useful functions, namely *inclusion measure* and *metric distance*, both of which have been used in lattices of intervals.

The mathematics here was grounded on various lattices of practical interest including the Euclidean space R^N, hyperboxes, propositional (Boolean) logic, probability spaces, fuzzy numbers, hyperspheres, structured data (graphs), disparate data, etc. Emphasis was given on (novel) fuzzy interval numbers (FINs) based, rigorously, on generalized interval analysis.

Vassilis G. Kaburlasos: *Discussion*, Studies in Computational Intelligence (SCI) **27**, 179–182 (2006)
www.springerlink.com   © Springer-Verlag Berlin Heidelberg 2006

It was shown how lattice theory proposes tunable tools for both knowledge representation and computing based on semantics. The modeling problem was formulated as a lattice function approximation problem. Furthermore, learning was pursued by a number of genetically tunable intelligent (learning) algorithms for clustering, classification and regression applicable in a lattice data domain. Successful applications of the aforementioned algorithms were demonstrated comparably in several problems involving numeric and/or non-numeric data.

Connections were detailed with established computational intelligence (CI) paradigms including Carpenter's fuzzy adaptive resonance theory (fuzzy-ART), min-max neural networks, Kohonen's self-organizing map (KSOM), fuzzy inference systems (FISs), ambiguous system modeling, various machine learning techniques, granular computing, formal concept analysis, and other. Emphasis was given on novel perspectives for improved FIS analysis and design. In all, lattice theory emerged as a sound foundation with a potential to promote technological development in CI.

Implementation of novel algorithms on conventional Turing machines was discussed; furthermore, the potential for super-Turing machine implementation was considered. Supportive material is shown below in three Appendices including definitions, mathematical proofs, and geometric interpretations on the plane, respectively.

The presentation in this book has been concise. For more details the interested reader may refer to cited journals and other publications. We point out that lattice theory is not a panacea; however, lattice theoretic analysis may imply substantial advantages. Some of them are summarized next.

The algorithms proposed in the context of this work have demonstrated, comparatively, very good results in practical applications. The latter was attributed to the capacity of (1) an inclusion measure function, and (2) a metric function in a lattice data domain including, potentially, numeric and/or non-numeric data. Furthermore, generalized interval analysis has enabled introducing tunable non-linearities in space R^N.

A major disadvantage of lattice theory is that the scientific community currently ignores it, at large.

The mathematical rigor involved in the employment of lattice theory here, apart from its utility in analysis and design, can further be practically useful by implementing mathematical principles in hardware, in order to achieve a predictable generalization involving disparate data.

Future plans include the development of improved machine learning, granular SOM, fuzzy neural networks, ambiguous system modeling, and FIS techniques. Models presented here such as the σ-FLN(MAP) have a capacity to build sophisticated (internal) representations of the (external) world. The latter capacity can be especially useful in robot applications

(Kröse et al. 2004). An inclusion measure (σ) as well as a metric (d) in a general lattice is potentially useful for database search applications (Wilson and Martinez 1997; Chávez et al. 2001; Berry et al. 1999). Moreover, application to both computational developmental psychology (Shultz 2003) and epigenetic robotics might be promising.

Authors have claimed that one of the basic functions of human neural networks is to make continuous predictions and then correct the stored worldview by feedback from the senses (Hawkins and Blakeslee 2004). Moreover, it was claimed that the development of novel system theories and gaining an understanding of biological systems are highly complementary endeavors (Doya et al. 2001). In the aforementioned context a far-reaching potential of this work is delineated next.

12.2 Future Work Speculations

Interdisciplinary cross-fertilization is of interest in science and technology. In the following we describe 'computation' in terms of system theory.

The closed loop system model in Fig.12-1 may model a Turing machine. In particular, a Turing machine corresponds to operation at discrete times without interaction (i.e. n=0). When output y equals a 'halting state' then computing stops. Otherwise, a non-zero error e drives the controller and computing proceeds. The operation of a Turing machine is described at discrete times (i.e. in $\aleph_0$) using binary digits.

One way towards hypercomputing is computing at continuous time (i.e. in $\aleph_1$) using real numbers. Another way is to consider (interactively) intention indicated as well by signal 'n' in Fig.12-1. Hence, the closed loop system in Fig.12-1 may model a super-Turing machine.

In general, signals r, y, e, and n take values in a product lattice $(L,\leq)$ including as well both binary numbers and real numbers. Design of the computing machine in Fig.12-1 may be pursued by conventional system modeling and feedback control techniques, where 'stability analysis' may correspond conditionally to the 'halting problem' of computing theory.

Fundamental, regarding a system theoretic presentation of computing, is the notion of a derivative. In particular, function $f: (\alpha,s) \rightarrow (\beta,m,s')$ of a Turing machine may be interpreted as a derivative. In the context of Fig.12-1, a derivative defined in a lattice $(L,\leq)$ could enable an employment of difference/differential equations towards hypercomputing modeling. A derivative in L could be defined as explained next.

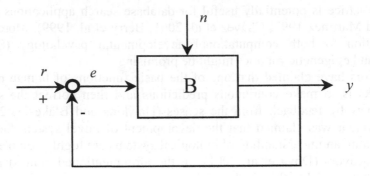

Fig.12-1 A (super-)Turing machine modeled as a closed loop system. Block B includes both a tape, with lattice (L,≤) elements on it, and a controller. The latter computes a function f: $(\alpha,s){\rightarrow}(\beta,m,s')$, where lattice (L,≤) elements 'α' and 'β' are read-from and written-on the material (tape), respectively; 's' and 's'' are states; 'm' denotes a motion. Reference signal (r) equals the 'halting state(s)'. Output signal y is the next state s'. If $y{\subseteq}r$ then computing stops; otherwise, a non-zero error drives the controller. In general, signals r, y, e, and n take values in lattice (L,≤). In particular, signals r and n may accommodate, interactively, intention.

The known derivative of a (real) function at a point x_0 is a real number. The latter includes both a magnitude and a sign. Since the lattice (R,≤) of real numbers is a chain, it follows that the sign in a derivative indicates one of two possible directions, i.e. either towards +∞ of towards -∞. Regarding a lattice (L,≤) a magnitude of change of rate can be specified by a metric based on a positive valuation function. However, regarding direction in a general lattice, the situation is more complicated since there are more than two directions through a general lattice element x_0. A possible solution for defining a derivative in a lattice (L,≤) at $x_0{\in}L$ is by considering more than one chains through x_0 as will be shown in a future publication.

A far-reaching potential is this: If human mental states could be represented in a lattice then the analysis proposed above could provide a sound theoretical basis for the development of dynamic models of the mind as well for a calculus of thought in the sense of Leibnitz (Davis 1998).

Epilogue

The biggest advances will come not from doing more and bigger and faster
of what we are already doing, but from finding new starting points.

(Winograd 1997)

Can electro-mechanical analog systems be modeled in the same terms as
electronics, discrete event systems, digital logic, linguistics, human cognition, etc.? This book suggested 'yes' based on mathematical lattice theory.

The emphasis here was on computational intelligence – soft computing.
It was shown how rigorous analysis and design can be pursued in soft
computing by conventional hard computing (HC) analytical methods. In
addition, lattice theory may suggest computing with tunable semantics.
Moreover, non-Turing computation could be pursued by accommodating
interactively human intention beyond $\aleph_0$ to $\aleph_1$ as well.

The work presented here is not conclusive. Rather, as the book title suggests, it should be regarded in perspective.

Appendix A: Useful Mathematical Definitions

A.1 Various Definitions

Definition A-1 A *relation between sets* A and B is a subset R of $A \times B$.

The *domain* of a relation is the set of all first coordinates of its elements. The *range* is the set of all second coordinates.

Definition A-2 A relation $R \subseteq A \times A$ is called an *equivalence relation on A* if it is: (a) Reflexive: aRa for all $a \in A$, (b) Symmetric: aRb implies bRa, and (c) Transitive: aRb and bRc imply aRc.

Definition A-3 A *function f* from A into B is a relation between A and B such that, for each $a \in A$, there is exactly one $b \in B$ such that $(a,b) \in f$.

For a function we write $f(a)=b$ to mean $(a,b) \in f$. Another term for function is *map(ping)*. The notation '$f: A \rightarrow B$' is interpreted as 'f is a function from the set A into the set B such that A is the domain of f and the range of f is a subset of B'.

Recall some well-known facts regarding cardinalities.
- Given a set U the *power set* of U is the family of all subsets of U and is denoted by $Pow(U)$.
- The set of *natural numbers* $\{1,2,...\}$ is denoted by N; the set of real numbers $(-\infty,\infty)$ by R; the set of real non-negative numbers $[0,\infty)$ by R_0^+.
- The *cardinality* of a set U, denoted by $card(U)$, is informally defined to be the number of elements U contains.

Of particular interest here are sets with infinite cardinalities (Kamke 1950; Stoll 1979). The following notation will be used

$$\aleph_0 = card(\mathsf{N}), \quad \aleph_1 = card(\mathsf{R}) = card(P(\mathsf{N})), \quad \aleph_2 = card(P(\mathsf{R})).$$

The use of subscripts in $\aleph_0$, $\aleph_1$, $\aleph_2$ is simply a matter of convenience and does not imply that $\aleph_1$ is the immediate successor of $\aleph_0$ or that $\aleph_2$ is the immediate successor of $\aleph_1$. However, it is well known that $\aleph_0 < \aleph_1 < \aleph_2$.

Definition A-4   A *metric* in a set A is a non-negative real function d: $A \times A \to R_0^+$ which, for all $x, y, z \in A$, satisfies:

   D0. $d(x,y) = 0 \Rightarrow x = y$
   D1. $d(x,x) = 0$
   D2. $d(x,y) = d(y,x)$ (*Symmetry*)
   D3. $d(x,z) \leq d(x,y) + d(y,z)$ (*Triangle Inequality*)

If only conditions D1, D2, and D3 are satisfied, then d is called a *pseudometric*. A *metric space* (A,d) is a set A provided with a metric d.

Let $(A_1, d_1), \ldots, (A_N, d_N)$ be metric spaces. Then the Cartesian product $A = A_1 \times \ldots \times A_N$ has the following Minkowski metrics

$$d_p(\mathbf{x}, \mathbf{y}) = [d_1(x_1, y_1)^p + \ldots + d_N(x_N, y_N)^p]^{1/p}, \text{ for } p \geq 1, \text{ and}$$

$$d_\infty(\mathbf{x}, \mathbf{y}) = \max \{d_1(x_1, y_1), \ldots, d_N(x_N, y_N)\},$$

where $\mathbf{x} = (x_1, \ldots, x_N)$ and $\mathbf{y} = (y_1, \ldots, y_N)$ are elements in A.

Condition $d(x,y) = 0$ in a pseudo-metric space M implies an equivalence relation $\sim$ in M, such that $x \sim y \Leftrightarrow d(x,y) = 0$. The *quotient* (set) of M with respect to equivalence relation $\sim$ is denoted by $M^* = M/\sim$. It follows that M^* is a metric space under $d(.,.)$.

A.2 Elements from Topology

We say that a non-empty set E of real numbers is *bounded above* (*below*) if there is a number b (a) such that $x \leq b$ ($a \leq x$), $\forall x \in E$. If a set E of real numbers is bounded both above and below it is called *bounded*.

Definition A-5   A set G in R is called *open* if, for each $x \in G$, there is a positive real number ε such that every y with $\|x-y\| < \varepsilon$ belongs to G.

Definition A-6   An *open interval* (a,b) is the set $\{x \in R: a < x < b\}$.

It has been shown by Cantor that every open set of real numbers is the union of a countable collection of mutually disjoint open intervals.

Definition A-7   A *closed interval* $[a,b]$ is the set $\{x \in R: a \leq x \leq b\}$.

Definition A-8   A point x is called a *cluster point* (or, equivalently, an *accumulation point*) of a set A if, for every $\varepsilon > 0$, there is a y in A, $y \neq x$, such that $\|x-y\| < \varepsilon$.

It can be shown that a set F in R is *closed* if and only if it contains every cluster point of F.

We say that a collection C of sets *covers* a set A if $A \subset \bigcup \{G: G \in C\}$. The collection C is then called a *cover* of A. If C contains only open sets, we call C an *open cover*. If C contains only a finite number of sets, we call C a *finite cover*. If C is a cover of A, then a subcollection C^* of C is called a *subcover* of A if C^* is also a cover of A.

A.3 Elements from Algebra

Definition A-9 An *algebra A* is a pair $[S, F]$, where S is a non-empty set, and F is a specified set of *operations* f_a, each mapping a power $S^{n(a)}$ of S into S, for some appropriate non-negative *finite* integer $n(a)$.

Otherwise stated, each operation f_a assigns to every $n(a)$-ple $(x_1, \ldots, x_{n(a)})$ of elements of S, a value $f_a(x_1, \ldots, x_{n(a)})$ in S, the result of performing the operation f_a on the sequence $x_1, \ldots, x_{n(a)}$. If $n(a)=1$, the operation f_a is called *unary*; if $n(a)=2$, it is called *binary*; if $n(a)=3$, it is called *ternary*, etc.

Definition A-10 A *congruence relation* on an algebra $A = [S, F]$ is an equivalence relation θ of A such that, for $f_a \in F$, $x_i \equiv y_i$ (mod θ), $i=1, \ldots, n(a)$, implies $f_a(x_1, \ldots, x_{n(a)}) \equiv f_a(y_1, \ldots, y_{n(a)})$ (mod θ) (Substitution Property).

Theorem A-11 Any pseudometric lattice (L,$\leq$) is a pseudometric space, in which joins and meets are uniformly continuous. The relation $d(x,y)=0$ is a congruence relation, mapping L isometrically and lattice-epimorphically onto a metric lattice.

Definition A-12 Let G be a non-empty set. Suppose that for any elements $a, b \in G$ there exists a uniquely determined element $c \in G$: $c = ab$. We call G a *group* if (i) $a(bc) = (ab)c$, (ii) there exists an element e (called the *identity element* or *unit element* of G) suct that $ae = ea = a$, for $a \in G$, and (iii) for any $a \in G$ there exists an element $x \in G$ such that $ax = xa = e$.

The element x in condition (iii) is called *inverse* of a, denoted by a^{-1}. The uniqueness of both the identity element e and the inverse a^{-1} follows readily. A group G satisfying the commutative law ($ab = ba$ for any $a,b \in G$) is called an *Abelian group*.

Definition A-13   A set K having at least two elements is called a *field* if two operations, called addition (+) and multiplication (.), are defined in K and satisfy the following three axioms.
(1) K is an Abelian group with respect to the addition (the identity element of this group is denoted by 0 and called the *zero element of K*).
(2) The set K^* of all non-zero elements of K is an Abelian group with respect to multiplication (the identity element of K^* is denoted by 1 and is called the *unity element* of K).
(3) The distributive law $a(b+c) = ab+ac$ holds.

Definition A-14   Assume a set L and a field K satisfying the following two requirements: (i) Given an arbitrary pair (a,b) of elements in L, there exists a unique element $a+b$ (called the *sum* of a,b) in L; (ii) given an arbitrary element α in K and an arbitrary element a in L, there exists a unique element αa (called the *scalar multiple* of a by α) in L. The set L is called a *linear space over K* (or *vector space over K*) if the following eight conditions are satisfied: (i) $(a+b)+c = a+(b+c)$, (ii) there exists an element $0 \in L$, called the *zero element* of L, such that $a+0 = 0+a = a$, for all $a \in L$, (iii) for any $a \in L$, there exists an element $x= -a \in L$ satisfying $a+x=x+a=0$, (iv) a+b = b+a, (v) $\alpha(a+b) = \alpha a + \alpha b$, (vi) $(\alpha\beta)a = \alpha(\beta a)$, (vii) $(\alpha+\beta)a = \alpha a + \beta a$, (viii) $1a = a$ (where 1 is the unity element of K).

An element of K is called a *scalar*, and an element of L is called a *vector*. If K is the field of real numbers R a linear space over $K=R$ is called *real linear space*.

n-tuples in K: K^n denotes the set of all sequences ($a_1,...,a_n$) of n elements in a field. Defining two operations by $(a_1,...,a_n)+(b_1,...,b_n) = (a_1+b_1,...,a_n+b_n)$ and $\lambda(a_1,...,a_n) = (\lambda a_1,..., \lambda a_n)$, $\lambda \in K$, the set K^n forms a linear space over K. An element of K^n is called an *n-tuple* in K, and a_i is called the *i*th component of ($a_1,...,a_n$).

Definition A-15   *Norm* in a real linear space L is a non-negative real function $\|.\|: L \rightarrow R_0^+$ which, for $x,y \in L$ and $\alpha \in R$, satisfies:
N1. $\|x\| = 0 \Leftrightarrow x = 0$.
N2. $\|\alpha x\| = |\alpha|.\|x\|$.
N3. $\|x + y\| \leq \|x\| + \|y\|$.
$\|x\|$ is called the *norm of vector x*, and L is called *normed linear space*.

Appendix B: Mathematical Proofs

B.1 Chapter 3 Proofs

Theorem 3-25   The existence of a positive valuation function $v: L \rightarrow R$ in a lattice $(L, \leq)$ is a sufficient condition for inclusion measure functions

(a) $k(x,u) = \dfrac{v(u)}{v(x \vee u)}$, and   (b) $s(x,u) = \dfrac{v(x \wedge u)}{v(x)}$.

Proof

Here it is assumed that $v(O)=0$ for a positive valuation; note that if $v(O) \neq 0$ then another positive valuation function v^+ with $v^+(O)=0$ can always be defined as $v^+(x) = v(x) - v(O)$, for all x. The truth of the theorem will be shown by proving the truth of conditions C0-C3 of definition 3-24.

C0. (a) $k(x,O) = \dfrac{v(O)}{v(x \vee O)} = \dfrac{v(O)}{v(x)} = 0$, for $x \neq O$.

(b) $s(x,O) = \dfrac{v(x \wedge O)}{v(x)} = \dfrac{v(O)}{v(x)} = 0$, for $x \neq O$.

C1. (a) If $x=O$ then we define $k(O,O)=1$;

otherwise, $k(x \leq x) = \dfrac{v(x)}{v(x \vee x)} = \dfrac{v(x)}{v(x)} = 1$.

(b) If $x=O$ then we define $s(O,O)=1$;

otherwise, $s(x \leq x) = \dfrac{v(x \wedge x)}{v(x)} = \dfrac{v(x)}{v(x)} = 1$.

C2. (a) In any lattice $(L, \leq)$ the operation of join $(\vee)$ is *monotone* (Birkhoff 1967), that is $u \leq w \Rightarrow x \vee u \leq x \vee w$. Furthermore, provided a positive, and hence a monotone, valuation v in a lattice $(L, \leq)$ a distance function is defined by $d(u,w) = v(u \vee w) - v(u \wedge w)$. Hence, $d(x \vee u, x \vee w) = v(x \vee w) - v(x \vee u) \geq 0$ and $d(u,w) = v(w) - v(u) \geq 0$.
It is further known (Birkhoff 1967) that if $v(.)$ is an monotone valuation then $d(x \vee u, x \vee w) + d(x \wedge u, x \wedge w) \leq d(u,w)$. Hence in our case with an initial assumption $u \leq w$, it follows

$d(x \lor u, x \lor w) \leq d(x \lor u, x \lor w) + d(x \land u, x \land w) \leq d(u, w) \Rightarrow v(x \lor w) - v(x \lor u)$
$\leq v(w) - v(u) \Rightarrow v(x \lor w) \leq v(w) - v(u) + v(x \lor u)$.

Assuming $w \neq O$ it follows

$$\frac{v(u)}{v(w)} v(x \lor w) \leq \frac{v(u)}{v(w)} [v(w) - v(u) + v(x \lor u)] = \frac{v(w) - v(u)}{v(w)} v(u) +$$

$$\frac{v(u)}{v(w)} v(x \lor u) \leq \frac{v(w) - v(u)}{v(w)} v(x \lor u) + \frac{v(u)}{v(w)} v(x \lor u) = v(x \lor u) \Rightarrow$$

$$\frac{v(u)}{v(x \lor u)} \leq \frac{v(w)}{v(x \lor w)} \Rightarrow k(x, u) \leq k(x, w).$$

(b) $u \leq w \Rightarrow x \land u \leq x \land w \Rightarrow v(x \land u) \leq v(x \land w) \Rightarrow \dfrac{v(x \land u)}{v(x)} \leq \dfrac{v(x \land w)}{v(x)}$

$\Rightarrow s(x, u) \leq s(x, w)$.

C3. (a) $x \land u < x \Rightarrow u < x \lor u \Rightarrow v(u) < v(x \lor u) \Rightarrow k(x, u) = \dfrac{v(u)}{v(x \lor u)} < 1$.

(b) $x \land u < x \Rightarrow v(x \land u) < v(x) \Rightarrow s(x, u) = \dfrac{v(x \land u)}{v(x)} < 1$.

■

Proposition 3-26 If $v_1, \ldots, v_N$ are valuations in lattices $(L_1, \leq), \ldots, (L_N, \leq)$, respectively, then function $v: L_1 \times \ldots \times L_N \to R$ given by $v = v_1 + \ldots + v_N$ is a valuation in the product lattice $(L, \leq)$, where $L = L_1 \times \ldots \times L_N$.

Proof
Let $\mathbf{x} = (x_1, \ldots, x_N)$ and $\mathbf{y} = (y_1, \ldots, y_N)$ be elements in the product lattice $(L, \leq)$. Then,

$v(\mathbf{x}) + v(\mathbf{y}) = v(x_1, \ldots, x_N) + v(y_1, \ldots, y_N) = v_1(x_1) + \ldots + v_N(x_N) + v_1(y_1) + \ldots + v_N(y_N) =$
$\quad = [v_1(x_1) + v_1(y_1)] + \ldots + [v_N(x_N) + v_N(y_N)] =$
$\quad = [v_1(x_1 \land y_1) + v_1(x_1 \lor y_1)] + \ldots + [v_N(x_N \land y_N) + v_N(x_N \lor y_N)] =$
$\quad = [v_1(x_1 \land y_1) + \ldots + v_N(x_N \land y_N)] + [v_1(x_1 \lor y_1) + \ldots + v_N(x_N \lor y_N)] =$
$\quad = v(x_1 \land y_1, \ldots, x_N \land y_N) + v(x_1 \lor y_1, \ldots, x_N \lor y_N) =$
$\quad = v(\mathbf{x} \land \mathbf{y}) + v(\mathbf{x} \lor \mathbf{y})$.

■

Proposition 3-28 Let $(L, \leq)$ be a crisp lattice and $\sigma: L \times L \to (0, 1]$ be an inclusion measure on $(L, \leq)$. Then $(L, \leq, \sigma)$ is a fuzzy lattice.

Proof
We need to prove equivalence $x \leq y \Leftrightarrow \sigma(x, y) = 1$.

(1) In the one direction let $x \leq y$.

From conditions C1 and C2 of definition 3-24 it follows

$x \leq y \Rightarrow \sigma(x,x) \leq \sigma(x,y) \Rightarrow 1 \leq \sigma(x,y) \Rightarrow \sigma(x,y) = 1.$

(2) In the other direction let $\sigma(x,y) = 1$.

There are three mutually exclusive cases (i) $x \leq y$, (ii) $y < x$, (iii) $x \parallel y$.

In the following we reject cases (ii) and (iii).

(ii) Let $y < x$. Then, based on condition C3 of definition 3-24, it follows

$y < x \Rightarrow x \wedge y < x \Rightarrow \sigma(x,y) < 1$ - contradiction.

(iii) Let $x \parallel y$. The latter implies '$x \wedge y < x < x \vee y$'.AND.'$x \wedge y < y < x \vee y$'.

Based on conditions C3 and C3' of definition 3-24 it follows

$$x \parallel y \Rightarrow \left\{ \begin{array}{l} x \wedge y < x < x \vee y \Rightarrow x \wedge y < x \\ x \wedge y < y < x \vee y \Rightarrow y < x \vee y \end{array} \right\} \Rightarrow \sigma(x,y) < 1 \text{ - contradiction.}$$

Therefore we have to accept case (i) $x \leq y$.

■

Theorem 3-29   Let $(L,\leq)$ be a complete lattice. Then the set $\tau(L)$ of intervals in $(L,\leq)$ is a complete lattice with largest and least intervals denoted by $[O,I]$ and $[I,O]$, respectively. The lattice ordering relation $[a,b] \leq [c,d]$ is equivalent to '$c \leq a$ and $b \leq d$'. For two intervals $[a,b]$, $[c,d]$ their *lattice join* is given by $[a,b] \vee [c,d] = [a \wedge c, b \vee d]$. Moreover their *lattice meet* is given by either $[a,b] \wedge [c,d] = [a \vee c, b \wedge d]$ if $a \vee c \leq b \wedge d$, or $[a,b] \wedge [c,d] = [I,O]$ otherwise.

Proof

The truth of this proposition will be shown in four steps.

(Step-1)   The collection $\tau(L)$ of intervals (including the empty set/interval) is *partially ordered*. The ordering relation $[a,b] \leq [c,d]$ is equivalent to '$c \leq a$ and $b \leq d$'. An eligible notation for the empty interval/set is $[I,O]$.

(Step-2)   Any two intervals $[a,b]$, $[c,d]$ in $\tau(L)$ have a least upper bound; in particular, it is $[a,b] \vee [c,d] = [a \wedge c, b \vee d]$.

(Step-3)   Any two intervals $[a,b]$, $[c,d]$ in $\tau(L)$ have a greatest lower bound; in particular, it is either $[a,b] \wedge [c,d] = [a \vee c, b \wedge d]$ if $a \vee c \leq b \wedge d$, or it is $[a,b] \wedge [c,d] = [I,O]$ otherwise.

(Step-4)   Lattice $(\tau(L),\leq)$ is a *complete lattice*, that is any subset of $\tau(L)$ has both a least upper bound and a greatest lower bound in $\tau(L)$.

The *monotone properties* '$x \leq y \Rightarrow x \wedge z \leq y \wedge z$, and $x \leq y \Rightarrow x \vee z \leq y \vee z$' will be used extensively in the proofs below.

Proof of Step-1:

A non-empty interval $\Delta \in \tau(L)$ has been defined as the set $[a,b] := \{x \in L: a \leq x \leq b\}$. Therefore, the collection $\tau(L)$ of intervals including the empty set is *partially ordered* under the conventional set-inclusion relation.

For non-empty intervals $[a,b]$, $[c,d] \in \tau(L)$ the partial ordering relation $[a,b] \leq [c,d]$ is equivalent to $c \leq a \leq b \leq d$. Aiming at introducing the standard interval notation for the empty set we 'relaxed' the aforementioned equivalence relation in $\tau(L)$ and we have replaced it by '$[a,b] \leq [c,d] \Leftrightarrow c \leq a$ and $b \leq d$'. All the intervals in $\tau(L)$ including the empty interval whose interval notation is, say $[e_1, e_2]$, are required to satisfy the aforementioned 'relaxed' equivalence relation.

On the one hand, the non-empty intervals already satisfy the 'relaxed' equivalence relation. On the other hand for the empty interval, the 'relaxed' equivalence relation implies $[e_1, e_2] \leq [a,b] \Leftrightarrow a \leq e_1$ and $e_2 \leq b$, for all $[a,b] \in \tau(L)$. Therefore, we propose the following interval notation for the empty interval: $[e_1, e_2] = [I, O]$.

We show now that denoting the empty interval as $[I,O]$ complies with standard lattice theoretical considerations. First, we consider a non-empty interval say $\Delta = [a,b]$ where $a,b \in L$ and $a \leq b$. Since $(L, \leq)$ is a complete lattice both the least upper bound of the set Δ, denoted by $\vee \Delta$, and the greatest lower bound of Δ, denoted by $\wedge \Delta$, exist in L. In particular it holds $\wedge \Delta = a$ and $\vee \Delta = b$. Hence a non-empty interval can be denoted as $\Delta = [\wedge \Delta, \vee \Delta]$. Second, we examine whether equality $\Delta = [\wedge \Delta, \vee \Delta]$ holds for the empty interval $[I,O]$. We cite from Davey and Priestley (1990, Remark 2.2): 'the greatest lower bound of the empty set in a complete lattice $(L, \leq)$ is I, while the least upper bound of the empty set is O'. Hence, the empty interval in $\tau(L)$ can be denoted by $[I,O]$. That is, we have reconfirmed the validity of our proposed interval notation for the empty set.

Proof of Step-2:

Consider interval $[a \wedge c, b \vee d]$. It is both $a \wedge c \leq a$ and $b \leq b \vee d$ hence $[a,b] \leq [a \wedge c, b \vee d]$. Likewise, $[c,d] \leq [a \wedge c, b \vee d]$. Therefore $[a \wedge c, b \vee d]$ is an upper bound of both $[a,b]$ and $[c,d]$. We show now that $[a \wedge c, b \vee d]$ is the least upper bound of $[a,b]$ and $[c,d]$. Towards this end assume another upper bound interval of both $[a,b]$ and $[c,d]$, say interval $[\beta, \gamma]$. The latter assumption implies both (1) $[a,b] \leq [\beta, \gamma] \Leftrightarrow \beta \leq a$, $b \leq \gamma$, and (2) $[c,d] \leq [\beta, \gamma] \Leftrightarrow \beta \leq c$, $d \leq \gamma$. Employing the aforementioned *lattice monotone properties*, we conclude both $\beta \leq a \wedge c$ and $b \vee d \leq \gamma$, hence $[a \wedge c, b \vee d] \leq [\beta, \gamma]$. In conclusion $[a,b] \vee [c,d] = [a \wedge c, b \vee d]$.

Proof of Step-3:

We first show the following equivalence:
$$[a,b] \cap [c,d] \neq \emptyset \Leftrightarrow a \vee c \leq b \wedge d, \text{ where}$$
$\cap$ is the set-intersection operator and $\emptyset$ denotes the empty set. In the one direction, assume that $[a,b]$ and $[c,d]$ intersect each other. Then $\exists \gamma \in L$: $a \leq \gamma \leq b$ and $c \leq \gamma \leq d$. Employing the aforementioned *monotone properties* we conclude both $a \vee c \leq \gamma$ and $\gamma \leq b \wedge d$; therefore $a \vee c \leq b \wedge d$. In the other direction, assume $a \vee c \leq b \wedge d$. Then $\exists \gamma \in L$: $a \leq a \vee c \leq \gamma \leq b \wedge d \leq b \Rightarrow \gamma \in [a,b]$, and $c \leq a \vee c \leq \gamma \leq b \wedge d \leq d \Rightarrow \gamma \in [c,d]$; therefore $[a,b] \cap [c,d] \neq \emptyset$.

We now resume calculation of $[a,b] \wedge [c,d]$. On the one hand, if **not** $a \vee c \leq b \wedge d$ then by employing the above equivalence we conclude $[a,b] \cap [c,d] = \emptyset$, hence we infer $[a,b] \wedge [c,d] = \emptyset$. On the other hand, if $a \vee c \leq b \wedge d$ then consider interval $[a \vee c, b \wedge d]$. It is both $a \leq a \vee c$ and $b \wedge d \leq b$, hence $[a \vee c, b \wedge d] \leq [a,b]$. Likewise $[a \vee c, b \wedge d] \leq [c,d]$. Therefore $[a \vee c, b \wedge d]$ is a lower bound of both $[a,b]$ and $[c,d]$. We show now that $[a \vee c, b \wedge d]$ is the greatest lower bound of $[a,b]$ and $[c,d]$. Towards this end assume another lower bound interval of both $[a,b]$ and $[c,d]$, say interval $[\beta,\gamma]$. The latter assumption implies both (1) $[\beta,\gamma] \leq [a,b] \Leftrightarrow a \leq \beta$, $\gamma \leq b$, and (2) $[\beta,\gamma] \leq [c,d] \Leftrightarrow c \leq \beta$, $\gamma \leq d$. Employing the aforementioned *lattice monotone properties* we conclude both $a \vee c \leq \beta$ and $\gamma \leq b \wedge d$, therefore we infer $[\beta,\gamma] \leq [a \vee c, b \wedge d]$. In conclusion $[a,b] \wedge [c,d] = [a \vee c, b \wedge d]$.

Proof of Step-4:

Considering jointly the results of Steps 1-3 above, and applying the definition of a lattice we conclude that $(\tau(L), \leq)$ is a lattice.

Let $\{[a_i,b_i]\}_{i \in I}$ denote a collection of intervals in $\tau(L)$, where I is an index set. We show now that $\vee \{[a_i,b_i]\}_{i \in I}$ exists in $\tau(L)$; in particular it is $\vee \{[a_i,b_i]\}_{i \in I} = [\wedge \{a_i\}_{i \in I}, \vee \{b_i\}_{i \in I}]$, where both $\wedge \{a_i\}_{i \in I}$ and $\vee \{b_i\}_{i \in I}$ exist in the complete lattice $(L, \leq)$. We show first that $[\wedge \{a_i\}_{i \in I}, \vee \{b_i\}_{i \in I}]$ is an upper bound of all $[a_i,b_i]$, $i \in I$, and then we show that $[\wedge \{a_i\}_{i \in I}, \vee \{b_i\}_{i \in I}]$ is their least upper bound. The details of the proof the same as in the Proof of Step-2, and they will not be repeated. In conclusion
$$\vee \{[a_i,b_i]\}_{i \in I} = [\wedge \{a_i\}_{i \in I}, \vee \{b_i\}_{i \in I}]$$
Likewise, by the same arguments as in the Proof of Step-3 we can show that $\wedge \{[a_i,b_i]\}_{i \in I}$ exists in $\tau(L)$. In particular, $\wedge \{[a_i,b_i]\}_{i \in I} = [\vee \{a_i\}_{i \in I}, \wedge \{b_i\}_{i \in I}]$ if $\vee \{a_i\}_{i \in I} \leq \wedge \{b_i\}_{i \in I}$, otherwise $\wedge \{[a_i,b_i]\}_{i \in I} = [I,O]$.

In conclusion, $(L, \leq)$ is a complete lattice.   ∎

Proposition 3-31   It holds $diag([a,b]) = d(a,b) = \max\limits_{x,y \in [a,b]} d(x,y)$.

Proof
It holds $a \le b \Rightarrow a \wedge b = a$ and $a \vee b = b$. Therefore,
$$diag([a,b]) = v(b) - v(a) = v(a \vee b) - v(a \wedge b) = d(a,b).$$
Furthermore, given $x,y \in [a,b]$ it follows that $a \le x \le b \Rightarrow a \wedge y \le x \wedge y \le b \wedge y$
$\Rightarrow a \le x \wedge y \le y \Rightarrow v(a) \le v(x \wedge y) \Rightarrow -v(x \wedge y) \le -v(a)$. Moreover, $a \le x \le b \Rightarrow$
$a \vee y \le x \vee y \le b \vee y \Rightarrow \quad y \le x \vee y \le b \Rightarrow v(x \vee y) \le v(b)$.
    Hence, $v(x \vee y) - v(x \wedge y) \le v(b) - v(a) \Rightarrow d(x,y) \le d(a,b)$.   ∎

Proposition 3-36   The collection F_c of families which represent a class $c = \bigcup\limits_j w_j$ has a greatest element, namely the *quotient* of class c denoted by
$Q(F_c) = Q(\{w_j\})$.

Proof
Let $\{w_i\}$ be a connected family of lattice intervals. A *maximal expansion* of $\{w_i\}$ is defined to be another family $\{q_i\}$ in the set F_c of families such that $\{w_i\} < \{q_i\}$. We will describe a method for constructing an ever (strictly) larger maximal expansion of a family. The aforementioned construction terminates in a finite number of steps and a global maximum will be reached, that is quotient $Q(F_c) = Q(\{w_i\})$ as explained next.

    The truth of Proposition 3-36 will be shown, first, in case family $\{w_i\}$ contains exactly two connected constituent intervals, say w_1 and w_2. To construct maximal expansion of family $\{w_1, w_2\}$, assume it is $w_1 = [w_{11}, ..., w_{1L}]$ and $w_2 = [w_{21}, ..., w_{2L}]$, where w_{1i} and w_{2i}, $i=1,...,L$ are intervals in each constituent lattice and let L be the total number of constituent lattices. The maximal expansion 'along' the first constituent lattice is determined by specifying the maximum interval 'max($w_{11} \wedge w_{21}$)' which contains $w_{11} \wedge w_{21}$ and consists of elements of w_{11} or w_{21}. Note that the latter is a trivial problem framed within the first constituent lattice. Hence, the corresponding maximal expansion implies interval [max($w_{11} \wedge w_{21}$), $w_{12} \wedge w_{22}, ..., w_{1L} \wedge w_{2L}$]. In the sequel consider the maximal expansions 'along' the remaining of the constituent lattices, these are at the most another $L-1$ maximal expansions. Finally the set-union me(w_1, w_2) of all the maximal expansions 'along' all constituent lattices has to be the maximum element in the set F_c that is the quotient $Q(F_c)$ or me(w_1, w_2) $= Q(F_c)$.

    To show the truth of the latter statement above, consider any interval u which contains only elements of $w_1 \cup w_2$. If u contains exclusively elements of w_1 OR exclusively elements of w_2 then it will be $u \leq w_1$ or $u \leq w_2$ respectively, hence $u \leq \mathrm{me}(w_1,w_2)$. On the other hand, suppose that $u=[u_1,\ldots,u_L]$ contains exclusive elements of $w_1=[w_{11},\ldots,w_{1L}]$ AND exclusive elements of $w_2=[w_{21},\ldots,w_{2L}]$. This implies that for at least one constituent lattice interval u_i, $i=1,\ldots,L$ it will be $w_{1i} \wedge w_{2i} < u_i$. But such a strict inequality can be true for at most one constituent lattice interval; otherwise u would contain elements that do not belong in neither w_1 nor w_2. Because of the way set $\mathrm{me}(w_1,w_2)$ was constructed it can be inferred that $u \leq \mathrm{me}(w_1,w_2)$. In conclusion, $\mathrm{me}(w_1,w_2)$ is the maximum family in $\{w_i\}$, $i \in \{1,2\}$, that is the quotient $Q(\{w_1,w_2\})=\mathrm{me}(w_1,w_2)$.

Now consider a third interval w_3 such that $\{w_3\} \cup Q(\{w_1,w_2\})$ is connected. Assume maximal expansions $\mathrm{me}(w_1,w_3)$ and $\mathrm{me}(w_2,w_3)$. Then any interval u containing only elements of one of w_1, w_2, w_3, w_1 and w_2, w_2 and w_3, w_3 and w_1 will be included in $\mathrm{me}(w_1,w_2) \cup \mathrm{me}(w_2,w_3) \cup \mathrm{me}(w_3,w_1)$. In addition, and in order to consider intervals containing exclusive elements of w_1 AND w_2 AND w_3, if any, the following maximal expansions have to be considered: $\mathrm{me}(w_3, \mathrm{me}(w_1,w_2))$, $\mathrm{me}(w_1, \mathrm{me}(w_2,w_3))$, and $\mathrm{me}(w_2, \mathrm{me}(w_3,w_1))$. After the corresponding simplifications the result is the quotient $Q(\{w_1,w_2,w_3\})$. Apparently the problem becomes a combinatorial one and the truth of Proposition 3-36 follows, in general, by mathematical induction.

∎

B.2 Chapter 4 Proofs

Proposition 4-3 Let f_h: R→R be a *strictly increasing* real function. Then function v_h: M^h→R given by $v_h([a,b]^h)=f_h(b) - f_h(a)$ is a *positive valuation* function in lattice $(M^h,\leq)$.

Proof
We show that function $v_h(.)$ satisfies the two conditions for a positive valuation of definition 3-22. First, let $[a,b]^h$ and $[c,d]^h$ be generalized intervals in M^h. It follows

$v_h([a,b]^h) + v_h([c,d]^h) = [f_h(b) - f_h(a)] + [f_h(d) - f_h(c)] =$
$\quad = [f_h(b) + f_h(d)] - [f_h(a) + f_h(c)] =$
$\quad = [f_h(b \vee d) + f_h(b \wedge d)] - [f_h(a \vee c) + f_h(a \wedge c)] =$
$\quad = [f_h(b \vee d) - f_h(a \wedge c)] + [f_h(b \wedge d) - f_h(a \vee c)] =$
$\quad = v_h([a \wedge c, b \vee d]^h) + v_h([a \vee c, b \wedge d]^h) =$
$\quad = v_h([a,b]^h \vee [c,d]^h) + v_h([a,b]^h \wedge [c,d]^h).$

Hence, $v_h(.)$ is a valuation function.

Second, let $[a,b]^h < [c,d]^h$ for generalized intervals $[a,b]^h$ and $[c,d]^h$ in M^h. There are three cases:

1. Both $[a,b]^h$ and $[c,d]^h$ are in M_+^h. It follows ('$c \leq a$' and '$b < d$') or ('$c < a$' and '$b \leq d$').

2. Both $[a,b]^h$ and $[c,d]^h$ are in M_-^h. It follows ('$b \leq d$' and '$c < a$') or ('$b < d$' and '$c \leq a$').

3. $[a,b]^h$ is in M_-^h and $[c,d]^h$ is in M_+^h. Hence, $b < a$ and $c \leq d$.

All three cases above imply the following strict inequality

$$f_h(b) + f_h(c) < f_h(a) + f_h(d).$$

Therefore, $[a,b]^h < [c,d]^h \Rightarrow f_h(b) + f_h(c) < f_h(a) + f_h(d) \Rightarrow$
$\Rightarrow f_h(b) - f_h(a) < f_h(d) - f_h(c) \Rightarrow v_h([a,b]^h) < v_h([c,d]^h)$.

Hence, $v_h(.)$ is a positive valuation function.   ∎

Proposition 4-5 Let $x = [a,b]^h$ be a generalized interval in M^h, $h \in (0,1]$. Then $\|x\| = h(|a| + |b|)$ is a norm in linear space M^h.

Proof

The truth of this proposition will be shown by proving the truth of conditions N1-N3 of definition A-15 in Appendix A. Assume $y = [c,d]^h \in M^h$.

(a) $\|x\| = h(|a| + |b|) \geq 0$, for all x.
    $\|x\| = 0 \Leftrightarrow |a| + |b| = 0 \Leftrightarrow a = 0 = b \Leftrightarrow x = [0,0]^h$.

(b) $\|\alpha x\| = h(|\alpha a| + |\alpha b|) = |\alpha| h(|a| + |b|) = |\alpha| \|x\|$.

(c) $\|x + y\| = \|[a,b]^h + [c,d]^h\| = h(|a+b| + |c+d|) \leq h(|a|+|b|+|c|+|d|) = h(|a|+|b|) + h(|c|+|d|) = \|x\| + \|y\|$.   ∎

Proposition 4-8 The set F of FINs is a *partially ordered set* because the ordering relation $\leq$ satisfies the following conditions

(a) *reflexive*: $F_1 \leq F_1$,

(b) *antisymmetric*: $F_1 \leq F_2$ and $F_2 \leq F_1 \Rightarrow F_1 = F_2$, and

(c) *transitive*: $F_1 \leq F_2$ and $F_2 \leq F_3 \Rightarrow F_1 \leq F_3$, for F_1, F_2, F_3 in F.

Proof

(a) Let $F_1 \in F$. Then $F_1(h) \leq F_1(h)$ for all h in $(0,1]$, hence $F_1 \leq F_1$.

(b) Let $F_1, F_2 \in F$. First, $F_1 \leq F_2$ is equivalent to $F_1(h) \leq F_2(h)$ for all h in $(0,1]$. Second, $F_2 \leq F_1$ is equivalent to $F_2(h) \leq F_1(h)$ for all h in $(0,1]$.

Since M^h is a lattice, $F_1(h) \leq F_2(h)$ and $F_2(h) \leq F_1(h)$ jointly imply $F_1(h)$ $= F_2(h)$ for all h in $(0,1]$. In conclusion, $F_1 = F_2$.

(c) Let $F_1, F_2, F_3 \in \mathsf{F}$. Then $F_1 \leq F_2$ is equivalent to $F_1(h) \leq F_2(h)$, and $F_2 \leq F_3$ is equivalent to $F_2(h) \leq F_3(h)$ for h in $(0,1]$. Since M^h is a lattice, $F_1(h) \leq F_2(h)$ and $F_2(h) \leq F_3(h)$ jointly imply $F_1(h) \leq F_3(h)$ for all h in $(0,1]$. In conclusion, $F_1 \leq F_3$.

Thus, the proposition has been proven.

■

Proposition 4-9   Let F_1 and F_2 be FINs. A *pseudometric* $d_K \colon \mathsf{F} \times \mathsf{F} \to \mathsf{R}_0^+$ is given by $d_K(F_1,F_2) = \int_0^1 d_h(F_1(h), F_2(h)) dh$, where $d_h(F_1(h), F_2(h))$ is a metric between generalized intervals $F_1(h)$ and $F_2(h)$.

Proof

From both definition $d_K(F_1,F_2) = \int_0^1 d_h(F_1(h), F_2(h)) dh$ and the fact that d_h is a metric, using standard properties of integrals it follows $d_K(F_1,F_1) = 0$, $d_K(F_1,F_2) = d_K(F_1,F_2)$, and $d_K(F_1,F_2) \leq d_K(F_1,G) + d_K(G,F_2)$.

Nevertheless, from $d_K(F_1,F_2) = \int_0^1 d_h(F_1(h), F_2(h)) dh = 0$ we cannot conclude that $F_1 = F_2$, because we may have $d_h(F_1(h), F_2(h)) \neq 0$ for an isolated point h_0 or, more generally, on a set of measure zero. Hence we have proved that d_K is a pseudometric, but not necessarily a metric.

■

Proposition 4-11   Let F_1 and F_2 be positive FINs. An *inclusion measure* function $\sigma_K \colon \mathsf{F}_+ \times \mathsf{F}_+ \to [0,1]$ is defined by $\sigma_K(F_1,F_2) = \int_0^1 k(F_1(h), F_2(h)) dh$, where $k(F_1(h), F_2(h))$ is an inclusion measure function between positive generalized intervals $F_1(h)$ and $F_2(h)$.

Proof

It is shown below that function $\sigma_K(F_1,F_2) = \int_0^1 k(F_1(h), F_2(h))dh$ satisfies conditions C1-C3 of definition 3-24.

C1. $\sigma_K(F_1,F_1) = \int_0^1 k(F_1(h), F_1(h))dh = \int_0^1 1dh = 1, \forall F_1 \in \mathsf{F}_+$.

C2. Let $F_1 \leq F_2$. It follows $F_1(h) \leq F_2(h)$ hence $k(X(h), F_1(h)) \leq k(X(h), F_2(h))$

$\Rightarrow \int_0^1 k(X(h), F_1(h))dh \leq \int_0^1 k(X(h), F_2(h))dh \Rightarrow \sigma_K(X,F_1) \leq \sigma_K(X,F_2)$.

C3. Let $F_1 \wedge F_2 < F_1$. Both FINs F_1 and F_2 have continuous membership functions hence $F_1(h) \wedge F_2(h) < F_1(h) \Rightarrow k(F_1(h), F_2(h)) < 1$ on a set of non-zero measure. Therefore, $\sigma_K(F_1,F_2) = \int_0^1 k(F_1(h), F_2(h))dh < 1$.

Thus, the proposition has been proven.   ∎

B.3 Chapter 10 Proofs

Proposition 10-1   It is $card(\mathsf{F}_n) = \aleph_1$, where $\aleph_1$ is the cardinality of the set R of real numbers.

Proof

The case of fuzzy numbers with continuous membership was proved in Kaburlasos (2002). The general case was delineated in Kaburlasos and Kehagias (2004) and will be detailed in Kaburlasos and Kehagias (2006a).

For continuous membership functions equality $card(\mathsf{F}_n) = \aleph_1$ follows from (1) $card(\mathsf{F}_n) \geq \aleph_1$, and (2) $card(\mathsf{F}_n) \leq \aleph_1$.

(1) Let T be the set of triangular fuzzy sets. Apparently $\mathsf{T} \subseteq \mathsf{F}_n$ therefore $card(\mathsf{T}) \leq card(\mathsf{F}_n)$.

A triangular FIN can be defined by a triple (a, b, c), where 'a' and 'c' define a triangle's basis, moreover 'b' defines a triangle's top.

It follows $card(\mathsf{T}) = \aleph_1^3 = (2^{\aleph_0})^3 = 2^{3\aleph_0} = 2^{\aleph_0} = \aleph_1$. In conclusion, $card(\mathsf{F}_n) \geq \aleph_1 = card(\mathsf{T})$.

(2) F_n is a subset of the set C of continuous functions, hence $card(\mathsf{F}_n) \leq card(\mathsf{C})$.

From standard Fourier series analysis it is known that a continuous function $f \in C$ can be expressed as $f(t) = \sum_{i=0}^{\infty} c_i e^{j w_0 t}$, where c_i is a Fourier complex coefficient. It follows $card(c_i) = \aleph_1$, $i \in \{0,1,2,...\}$. Therefore $card(C) = \aleph_1^{\aleph_0} = (2^{\aleph_0})^{\aleph_0} = 2^{\aleph_0 \aleph_0} = 2^{\aleph_0} = \aleph_1$. In conclusion $card(F_n) \leq \aleph_1 = card(C)$. Thus, the proposition has been proven.

The general case is based on the following proposition.

Proposition 10-1a   The cardinality $card(\mathbf{I}_{[a,b]})$ of the set $\mathbf{I}_{[a,b]}$ of non-decreasing functions on the closed interval $[a,b]$ equals $card(\mathbf{I}_{[a,b]}) = \aleph_1$.

The proof of Proposition 10-1a is based on the following lemmata.

Lemma 1: Every function $f \in \mathbf{I}_{[a,b]}$ can have at most a countable number of discontinuities.

Lemma 2: Take any countable set $X = \{x_1, x_2, ...\} \subseteq [a,b]$ and any function $f \in \mathbf{C}_{[a,b] \backslash X}$. Then f is specified by its values on $(\mathbf{Q} \cap [a,b]) \cup X$.

Lemma 3: Take any countable set $X = \{x_1, x_2, ...\} \subseteq [a,b]$. Then $card((\mathbf{Q} \cap [a,b]) \cup X) = \aleph_0$.

Lemma 4: Take any countable set $X = \{x_1, x_2, ...\} \subseteq [a,b]$. Then $card(\mathbf{C}_{[a,b] \backslash X}) = \aleph_1$.

Lemma 5: $card(\tilde{\mathbf{C}}_{[a,b]}) = \aleph_1$.

The proof of Proposition 10-1a follows from Lemma 5.

It follows (dually) that the cardinality of the set of non-increasing functions on the closed interval $[a,b]$ equals $\aleph_1$. Based on the LR-representation of fuzzy numbers, where a L-function (R-function) is a non-decreasing (non-increasing) function, it follows that the cardinality of the set F_n of fuzzy numbers equals $\aleph_1$.

■

Appendix C: Geometrical Interpretations

This appendix shows geometric interpretations on the plane

This Appendix includes insightful geometric interpretations on the plane and brief discussions regarding FLN models. In particular, functions $v(x)=x$ and $\theta(x)=1-x$ have been selected in each constituent lattice for a positive valuation function and an isomorphic function, respectively.

C.1 Inclusion Measure

This section demonstrates the utility an inclusion measure.

Example E1: Illustrating the utility of the inclusion measure k

In both Fig.C-1(a) and (b) it holds $[0.5,0.6]\times[0.3,0.4]= u \leq w= [0.4,0.9]\times[0.2,0.8]$. Inequality $\sigma(x\leq u) \leq \sigma(x\leq w)$ is demonstrated underneath for boxes x and x' in Fig.C-1(a) and (b), respectively. In Fig.C-1(a) for $x= [0.15,0.2]\times[0.15,0.2]$ it follows

$$k(x\leq u)= \frac{v(u)}{v(x \vee u)} = \frac{v(0.5,0.6,0.7,0.4)}{v((0.85,0.2,0.85,0.2) \vee (0.5,0.6,0.7,0.4))} =$$

$$= \frac{v(0.5,0.6,0.7,0.4)}{v(0.85,0.6,0.85,0.4)} = \frac{2.2}{2.7} \approx 0.8148.$$

Likewise,

$$k(x\leq w)= \frac{v(w)}{v(x \vee w)} = \frac{v(0.6,0.9,0.8,0.8)}{v((0.85,0.2,0.85,0.2) \vee (0.6,0.9,0.8,0.8))} =$$

$$= \frac{v(0.6,0.9,0.8,0.8)}{v(0.85,0.9,0.85,0.8)} = \frac{3.1}{3.4} \approx 0.9118.$$

That is, $k(x\leq u) \leq k(x\leq w)$ as expected from the Consistency Property.

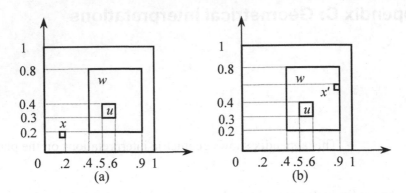

Fig.C-1   The Consistency Property guarantees that when u is inside w then x is included, in an inclusion measure (σ) sense, more in w than it is in u.

In Fig.C-1(b) a box x' is selected inside w but outside u, in particular x'= [0.85,0.9,0.55,0.6]. It follows

$$k(x'\leq u)= \frac{v(u)}{v(x'\vee u)} = \frac{v(0.5,0.6,0.7,0.4)}{v((0.15,0.9,0.45,0.6)\vee(0.5,0.6,0.7,0.4))} \approx 0.8148$$

$$k(x'\leq w)= \frac{v(w)}{v(x'\vee w)} = \frac{v(0.6,0.9,0.8,0.8)}{v((0.15,0.9,0.45,0.6)\vee(0.6,0.9,0.8,0.8))} \approx 1.0.$$

It is interesting to demonstrate that a similar consistency of inequalities is not retained by the lattice metric (distance) function d. For instance, in Fig.C-1(a), it holds
$d(x,u) = v(x\vee u)-v(x\wedge u) = 1.1 < 1.6 = v(x\vee w)-v(x\wedge w) = d(x,w).$
whereas in Fig.C-1(b) it holds
$d(x',u) = v(x'\vee u)-v(x'\wedge u) = 1.1 > 1 = v(x\vee u)-v(x\wedge u)= d(x',w).$

In conclusion, the lattice distance function d is not consistent in the sense of the lattice inclusion measure σ (Consistency Property C2) of definition 3-24.

Example E2: Dealing with 'missing' and 'don't care' values

In place of a 'missing' data value ('?') the least interval O= [1,0] is employed, whereas in place of a 'don't care' data value ('*') the greatest interval I= [0,1] is used.

Consider the two boxes $[0.6,0.7] \times [0.5,0.6] = u \le w = [0.5,0.9] \times [0.4,0.8]$ within the unit square shown in Fig.C-2. Three intervals x, x_m, and x_d are shown in Fig.C-2(a) and (b). The corresponding degrees of inclusion in boxes u and w are calculated next.

Fig.C-2(a) assumes point $x = [0.3,0.3] \times [0.3,0.3]$. Hence,

$$k(x \le u) = \frac{v(u)}{v(x \vee u)} = \frac{v(0.4,0.7,0.5,0.6)}{v((0.7,0.3,0.7,0.3) \vee (0.4,0.7,0.5,0.6))} = \frac{2.2}{2.7} \approx 0.8148.$$

$$k(x \le w) = \frac{v(w)}{v(x \vee w)} = \frac{v(0.5,0.9,0.6,0.8)}{v((0.7,0.3,0.7,0.3) \vee (0.5,0.9,0.6,0.8))} = \frac{2.8}{3.1} \approx 0.9032.$$

Fig.C-2(a) also assumes a 'missing' value $x_m = [0.3,0.3] \times [?]$. Hence,

$$k(x_m \le u) = \frac{v(u)}{v(x_m \vee u)} = \frac{v(0.4,0.7,0.5,0.6)}{v((0.7,0.3,0,0) \vee (0.4,0.7,0.5,0.6))} = \frac{2.2}{2.5} \approx 0.8800.$$

$$k(x_m \le w) = \frac{v(w)}{v(x_m \vee w)} = \frac{v(0.5,0.9,0.6,0.8)}{v((0.7,0.3,0,0) \vee (0.5,0.9,0.6,0.8))} = \frac{2.8}{3.0} \approx 0.9333.$$

Fig.C-2(b) assumes a 'don't care' value $x_d = [0.3,0.3] \times [*]$. Hence,

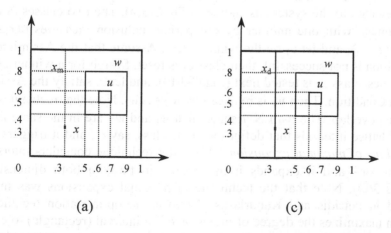

(a)   (c)

Fig.C-2   Demonstrating the capacity of inclusion measure k for handling points as well as 'missing' and 'don't care' data. (a) $x = [0.3,0.3] \times [0.3,0.3]$ is an point (atom); $x_m = [0.3,0.3] \times [?]$ includes a 'missing' data component, (b) $x_d = [0.3,0.3] \times [*]$ includes a 'don't care' data component.

$$k(x_d \leq u) = \frac{v(u)}{v(x_d \vee u)} = \frac{v(0.4,0.7,0.5,0.6)}{v((0.7,0.3,1,1) \vee (0.4,0.7,0.5,0.6))} = \frac{2.2}{3.4} \approx 0.6471.$$

$$k(x_d \leq w) = \frac{v(w)}{v(x_d \vee w)} = \frac{v(0.5,0.9,0.6,0.8)}{v((0.7,0.3,1,1) \vee (0.5,0.9,0.6,0.8))} = \frac{2.8}{3.6} \approx 0.7778.$$

In all the above cases it is $k(x \leq u) \leq k(x \leq w)$, for a datum x, as expected from the Consistency Property (C2) of definition 3-24. Note also that the existence of 'missing' data increases the degree of membership in a class, whereas the existence of 'don't care' data decreases the degree of membership in a class; more specifically both $k(x_d \leq u) \leq k(x \leq u) \leq k(x_m \leq u)$, and $k(x_d \leq w) \leq k(x \leq w) \leq k(x_m \leq w)$ hold above.

C.2 FLN and the Technique of Maximal Expansions

Example E3: Illustrating the technique of Maximal Expansions

This example presents geometric interpretations on the plane regarding learning and decision-making by σ-FLN. It also demonstrates the technique of Maximal Expansions on the plane.

Assume that the σ-FLN has already stored two distinct classes $c_1 = \{w_1\}$ and $c_2 = \{w_2\}$ and let a new input x_1, that is a rectangle in the general case, be presented to the system as shown in Fig.C-3(a). The two classes c_1 and c_2 compete with one another by comparing inclusion measures $\sigma(x_1 \leq c_1)$ and $\sigma(x_1 \leq c_2)$, and let c_1 be the winner class. Assume that the Assimilation Condition is not successful, then class c_1 is reset. Search for a winner class continues, class c_2 is tested next (Fig.C-3(b)), and let c_2 satisfy the Assimilation Condition. Then w_2 is replaced by $w'_2 = x \vee w_2$. Note that rectangles w_1 and w'_2 overlap. The σ-FLN assumes that w_1 and w'_2 are in the same family of lattice intervals that defines a single class, say c_1, and it triggers the *technique of maximal expansions*. The latter technique considers intersection $w_1 \wedge w'_2$ and it expands it maximally, in turn, in both dimensions (Fig.C-3(c)). Note that the technique of maximal expansions was introduced in Petridis and Kaburlasos (1998) as an optimization technique, which maximizes the degree of inclusion of an interval (rectangle) to class $c_1 \in C_{\tau(L)}$. The aforementioned optimization/maximization can be achieved by representing class c_1 by its quotient $Q(c_1)$.

Hence, a single class c_1 emerges represented by four rectangles $c_1=\{w_1,w'_2,w_3,w_4\}$ shown in Fig.C-3(c). In particular, rectangle w_1 is specified by its four corners 1-2-3-4, rectangle w'_2 is specified by its corners 5-6-7-8, rectangle w_3 is specified by 9-2-10-8, and rectangle w_4 by 11-6-12-4. The collection of the maximal rectangles for a given family of overlapping intervals, as shown in both Fig.C-3(c) and (d), is *the quotient* of the corresponding class. The degree of inclusion of a new input x_2 in class c_1, as shown in Fig.C-3(d), is given by $max\{\sigma(x_2\leq w_1),\ \sigma(x_2\leq w'_2),\ \sigma(x_2\leq w_3),\ \sigma(x_2\leq w_4)\}$. Note that input x_2 could be a trivial interval, that is an atom in the unit-square $[0,1]\times[0,1]$.

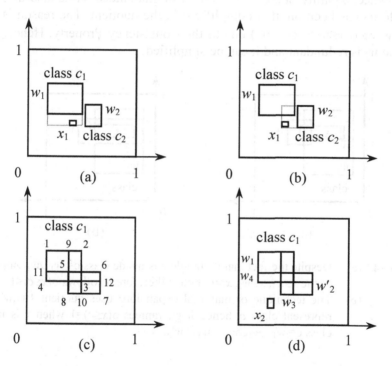

Fig.C-3   Illustrating the technique of 'maximal expansions'.
  (a) Two classes $c_1=\{w_1\}$ and $c_2=\{w_2\}$ compete over input x_1. Class c_1 is the winner because $\sigma(x_1\leq c_1)$ is larger that $\sigma(x_1\leq c_2)$, but it is reset because it does not satisfy the Assimilation Condition.
  (b) Class c_2 is selected as the new winner that satisfies the Assimilation Condition.
  (c) Rectangle w_2 is replaced by $w'_2=x_1\vee w_2$. Overlapping rectangles w_1 and w'_2 define one class, which is enhanced by inserting rectangles w_3 and w_4 produced by the technique of maximal expansions.
  (d) A new input rectangle x_2 appears and the learning cycle resumes.

Example E4: Utility of the technique of maximal expansions

The utility of the technique of maximal expansions is shown in Fig.C-4. Consider class $c = \{w_1, w_2, w_3\}$ and let rectangle x consist solely of points of w_1, w_2 and w_3 (Fig.C-4(a)). Then it is reasonable to expect $\sigma(x \leq c) = 1$. But this is not the case in Fig.C-4(a) because $w_1 < x \lor w_1$, $w_2 < x \lor w_2$ and $w_3 < x \lor w_3$, therefore $\sigma(x \leq c) = k(x \leq c) < 1$. The technique of maximal expansions comes to restore the expected equality relation by replacing class $c = \{w_1, w_2, w_3\}$ by quotient family $\{w_1, w'_2\}$ (Fig.C-4(b)). Any rectangle containing solely points of class c is contained in at least one of the quotient members w_1, w'_2; hence, equality $\sigma(x \leq c) = k(x \leq c) = 1$ is guaranteed. Note also that rectangle w_3 has been omitted (simplified) in the quotient. The reason is that $w_3 < w_1 \Rightarrow \sigma(x \leq w_3) \leq \sigma(x \leq w_1)$ due to the Consistency Property. Hence, rectangle w_3 is redundant and it can be simplified.

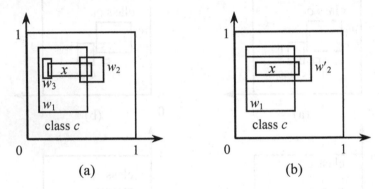

$$(a) \qquad\qquad\qquad (b)$$

Fig.C-4 (a) Despite the fact that rectangle x is inside class $c = w_1 \cup w_2 \cup w_3$ none of the $x \leq w_1$, $x \leq w_2$, $x \leq w_3$ holds; therefore it follows $\sigma(x \leq c) < 1$.

(b) The technique of maximal expansions uses quotient $\{w_1, w'_2\}$ to represent class c, hence it guarantees $\sigma(x \leq c) = 1$ when x is inside class $c = w_1 \cup w_2 \cup w_3 = w_1 \cup w'_2$.

The technique of maximal expansions is further detailed in Fig.C-5 where rectangles $w_1 = [0.15, 0.35] \times [0.05, 0.3]$, $w_2 = [0.30, 0.40] \times [0.15, 0.25]$, $w_3 = [0.55, 0.85] \times [0.15, 0.45]$, and an input $x = [0.47, 0.47] \times [0.20, 0.20]$ are shown. Two classes c_1 and c_2 are specified as $c_1 = w_1 \cup w_2$ and $c_2 = w_3$.

Using Euclidean distances of x from the nearest edge of the three rectangles it follows 'distance between x and w_1' $= |0.47 - 0.35| = 0.12$, 'distance between x and w_2' $= |0.47 - 0.40| = 0.07$, and 'distance between x and w_3' $= |0.47 - 0.55| = 0.08$. Hence it makes sense to classify x in class-c_1. However, based on inclusion measure k in Fig.C-5(a) and without the technique of maximal expansions, input x is classified in class $c_2 = w_3$ because

$$k(x \leq w_1) = \frac{v(w_1)}{v(x \vee w_1)} = \frac{2.45}{2.57} \approx 0.953, \quad k(x \leq w_2) = \frac{v(w_2)}{v(x \vee w_2)} = \frac{2.2}{2.27} \approx 0.969,$$

and $k(x \leq w_3) = \dfrac{v(w_3)}{v(x \vee w_3)} = \dfrac{2.6}{2.68} \approx 0.970.$

The above 'counter-intuitive' classification decision can be mended using the quotient $\{w_1, w'_2\}$ with $w'_2 = [0.15, 0.40] \times [0.15, 0.25]$ produced by the technique of maximal expansions. Winner now will be class $c_1 = w_1 \cup w'_2$ over class $c_2 = w_3$, because

$$k(x \leq w_1) \approx 0.953, \quad k(x \leq w'_2) = \frac{2.35}{2.42} \approx 0.971, \text{ and } k(x \leq w_3) \approx 0.970.$$

Hence the technique of maximal expansions can make the difference in pattern recognition problems since it may imply a common sense decision. It should be pointed out that the technique of maximal expansions does not change the definition of class c, but rather it only changes the actual representation of class c by employing the unique family $Q(c)$, which includes any other family representing class c.

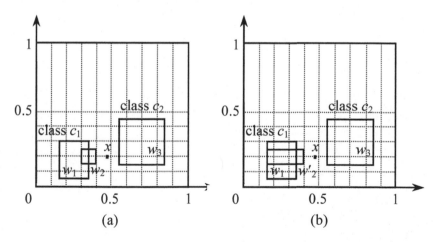

(a)   (b)

Fig.C-5 (a) Trivial interval x is included more in class-c_2 than in class-c_1.
(b) The technique of maximal expansions is an optimization technique which replaces family $\{w_1, w_2\}$ representing class $c_1 = w_1 \cup w_2$ by the quotient family $Q(c) = \{w_1, w'_2\}$ in order to maximize the degree of inclusion of a box to class c_1. As a result, in this example, x is now included more in class-c_1 than it is in class-c_2.

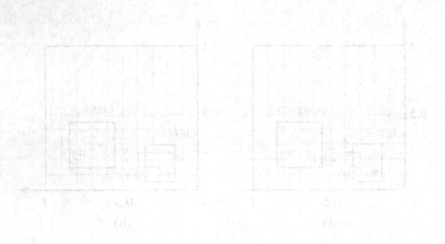

References

Adam NR, Atluri V, Bertino E, Ferrari E (2002) A content-based authorization model for digital libraries. IEEE Trans Knowledge Data Engineering 14(2): 296-315.

Agouris P, Stefanidis A (2003) Efficient summarization of spatiotemporal events. Comm ACM 46(1): 65-66.

Aha DW (1998) The omnipresence of case-based reasoning in science and application. Knowledge-Based Systems 11(5-6): 261-273.

Aha DW, Kibler D, Albert MK (1991) Instance-based learning algorithms. Machine Learning 6(1): 37-66.

Ait-Kaci H, Boyer R, Lincoln P, Nasr R (1989) Efficient implementation of lattice operations. ACM Trans Programming Languages Systems 11(1): 115-146.

Ajmal N, Thomas KV (1994) Fuzzy lattices. Info Sciences, 79(3-4): 271-291.

Albrecht RF (2003) Topological interpretation of fuzzy sets and intervals. Fuzzy Sets Systems 135(1): 11-20.

Alefeld G, Herzberger J (1983) Introduction to interval computation. Academic Press, New York, NY.

Amari S, Takeuchi A (1978) Mathematical theory on formation of category detecting nerve cells. Biological Cybernetics 29(3): 127-136.

An N, Jin J, Sivasubramaniam A (2003) Toward an accurate analysis of range queries on spatial data. IEEE Trans Knowledge Data Engin 15(2): 305-323.

Anagnostopoulos GC, Georgiopoulos M (2002) Category regions as new geometrical concepts in Fuzzy-ART and Fuzzy-ARTMAP. Neural Networks 15(10): 1205-1221.

Anagnostopoulos I, Anagnostopoulos C, Loumos V, Kayafas E (2004) Classifying Web pages employing a probabilistic neural network. IEE Proc - Software 151(3): 139-150.

Andresen SL (2002) John McCarthy: Father of AI. IEEE Intel Syst 17(5): 84-85.

Angluin D, Frazier M, Pitt L (1992) Learning conjunctions of horn clauses. Machine Learning 9: 147-164.

Antsaklis PJ (2000) Special issue on hybrid systems: Theory and applications. Proc of the IEEE 88(7): 879-887.

Arbib MA (ed.) (2003) The handbook of brain theory and neural networks, 2nd ed. The MIT Press, Cambridge, MA.

Aspray W (1990) John von Neumann and the origins of modern computing. The MIT Press, Cambridge, MA.

Atanassov KT (1999) Intuitionistic fuzzy sets: Theory and applications. Physica-Verlag, Heidelberg, Germany.

Athanasiadis IN, Kaburlasos VG, Mitkas PA, Petridis V (2003) Applying machine learning techniques on air quality data for real-time decision support. In: Proc 1st Intl. NAISO Symposium on Information Technologies in Environmental Engineering (ITEE 2003), Gdansk, Poland.

Athanasiadis IN, Mitkas PA (2004) An agent-based intelligent environmental monitoring system. Managenent of Environmental Quality: An International Journal 15(3): 238-249.

Avnimelech R, Intrator N (1999) Boosted mixture of experts: An ensemble learning scheme. Neural Computation 11(2): 483-497.

Bains S (2003) Intelligence as physical computation. Artificial Intelligence and Simulation of Behavior Journal 1(3): 225-240.

Baker D, Brett PN, Griffiths MV, Reyes L (1996) A mechatronic drilling tool for ear surgery: A case study of some design characteristics. Mechatronics 6(4): 461-477.

Bandemer H, Gottwald S (1996) Fuzzy sets, fuzzy logic, fuzzy methods with applications. John Wiley & Sons, Chichester, England.

Baraldi A, Alpaydin E (2002) Constructive feedforward ART clustering networks. IEEE Trans Neural Networks, 13(3): 645-677.

Baralis E, Ceri S, Paraboschi S (1998) Compile-time and runtime analysis of active behaviors. IEEE Trans Knowledge Data Engineering 10(3): 353-370.

Baum EB, Haussler D (1989) What size net gives valid generalization?. Neural Computation 1: 151-160.

Benediktsson JA, Swain PH (1992) Consensus theoretic classification methods. IEEE Trans Systems, Man, Cybernetics 22(4): 688-704.

Benediktsson JA, Sveinsson JR, Ersoy OK, Swain PH (1997) Parallel consensual neural networks. IEEE Trans Neural Networks 8(1): 54-64.

Ben-Yacoub S, Abdeljaoued Y, Mayoraz E (1999) Fusion of face and speech data for person identity verification. IEEE Trans Neural Networks 10(5): 1065-1074.

Bengio Y, Buhmann JM, Embrechts MJ, Zurada JM (2000) Introduction to the special issue on neural networks for data mining and knowledge discovery. IEEE Trans Neural Networks 11(3): 545-549.

Berners-Lee T (2003) The future of the World Wide Web. The Royal Society, UK, http://www.w3.org/

Berners-Lee T, Hendler J, Lassila O (2001) The Semantic Web. Scientific American, 284(5): 34-43.

Berretti S, Del Bimbo A, Pala P (2000) Retrieval by shape similarity with perceptual distance and effective indexing, IEEE Trans Multimedia, 2(4): 225-239.

Berretti S, Del Bimbo A, Vicario E (2001) Efficient matching and indexing of graph models in content-based retrieval. IEEE Trans Pattern Analysis Machine Intelligence 23(10): 1089-1105.

Berry MW, Dumais ST, Obrien GW (1995) Using linear algebra for intelligent information-retrieval. SIAM Review 37(4): 573-595.

Berry MW, Drmac Z, Jessup ER (1999) Matrices, vector-spaces, and information-retrieval. SIAM Review 41(2): 335-362.

Bertsekas DP, Nedić A, Ozdaglar AE (2003) Convex analysis and optimization. Athena Scientific, Belmont, MA.

Berztiss AT (1973) A backtrack procedure for isomorphism of directed graphs. J ACM 20(3): 365-377.

Bianchini M, Frasconi P, Gori M (1995) Learning without local minima in radial basis function networks. IEEE Trans Neural Networks 6(3): 749-756.

Bienenstock EL, Cooper LN, Munro PW (1982) Theory for the development of neuron selectivity: Orientation specificity and binocular interaction in visual cortex. J Neuroscience 2(1): 32-48.

Birkhoff G (1967) Lattice theory. American Mathematical Society, Colloquium Publications, 25, Providence, RI.

Birkhoff G, von Neumann J (1936) The logic of quantum mechanics. Annals of Mathematics 37(4): 823-843.

Bishop CM (1997) Neural networks for pattern recognition. Oxford University Press, New York, NY.

Bishop CM, Svensén M, Williams CKI (1998) GTM: The generative topographic mapping. Neural Computation 10(1): 215-234.

Bloch I (1996) Information combination operators for data fusion: A comparative review with classification. IEEE Trans Systems, Man, Cybernetics - A 26(1): 52-67.

Bloch I, Maitre H (1995) Fuzzy mathematical morphologies: A comparative study. Pattern Recognition 28(9): 1341-1387.

Bloehdorn S, Cimiano P, Hotho A, Staab S (2005) An ontology-based framework for text mining. LDV Forum - GLDV Journal for Computational Linguistics and Language Technology 20(1): 87-112.

Blume M, Esener S (1995) Optoelectronic fuzzy ARTMAP processor. Optical Computing 10: 213-215.

Blumer A, Ehrenfeucht A, Haussler D, Warmuth MK (1989) Learnability and the Vapnik-Chervonenkis dimension. J ACM 36(4): 929-965.

Bobrowski L, Bezdek JC (1991) C-means clustering with the l_1 and l_∞ norms. IEEE Trans Systems, Man, Cybernetics 21(3): 545-554.

Boinski S, Cropp SJ (1999) Disparate data sets resolve squirrel-monkey (Saimiri) taxonomy: Implications for behavioral ecology and biomedical usage. Intl J Primatology 20(2): 237-256.

Bollmann-Sdorra P, Wong SKM, Yao YY (1993) A measure-theoretic axiomatization of fuzzy sets. Fuzzy Sets Systems 60(3): 295-307.

Bondia J, Picó J (2003) Analysis of linear systems with fuzzy parametric uncertainty. Fuzzy Sets Systems 135(1): 81-121.

Bonissone PP, Chen YT, Goebel K, Khedkar PS (1999) Hybrid soft computing systems: Industrial and commercial applications. Proc of the IEEE 87(9): 1641-1667.

Bortolan G, Pedrycz W (2002) Fuzzy descriptive models: An interactive framework of information granulation. IEEE Trans Fuzzy Systems 10(6): 743-755.

Boukezzoula R, Galichet S, Foulloy L (2003) Nonlinear internal model control: Application of inverse model based fuzzy control. IEEE Trans Fuzzy Systems 11(6): 814-829.

Box G, Jenkins FM (1976) Time series analysis: Forecasting and control, 2nd ed. Holden-Day, Oakland, CA.

Braga-Neto UM (2001) Connectivity in image processing and analysis: Theory, multiscale extensions and applications. PhD thesis, The Johns Hopkins University, Baltimore, USA.

Breiman L (1996) Bagging Predictors. Machine Learning 24 (2): 123-140.

Breiman L, Friedman JH, Olshen RA, Stone CJ (1998) Classification and regression trees. Chapman & Hall /CRC, Boca Raton, FL.

Broomhead DS, Lowe D (1988) Multivariate functional interpolation and adaptive networks. Complex Systems 2: 321-355.

Browne A, Sun R (2001) Connectionist inference models. Neural Networks 14(10): 1331-1355.

Brunelli R, Poggio T (1993) Face recognition: Features versus templates, IEEE Trans Pattern Analysis Machine Intelligence 15(10): 1042-1052.

Buchanan BG, Shortliffe EH (1984) Rule-Based Expert Systems: The MYCIN Experiments of the Stanford Heuristic Programming Project. Addison-Wesley, Reading, MA.

Buede DM, Girardi P (1997) A target identification comparison of Bayesian and Dempster-Shafer multisensor fusion. IEEE Trans Systems, Man, Cybernetics - A 27(5): 569-577.

Burillo P, Frago N, Fuentes R (2000) Inclusion grade and fuzzy implication operators. Fuzzy Sets Systems 114(3): 417-429.

Burt PJ (2002) A pyramid-based front-end processor for dynamic vision applications. Proc of the IEEE 90(7): 1188-1200.

Caines PE, Wei YJ (1998) Hierarchical hybrid control systems: A lattice theoretic formulation. IEEE Trans Automatic Control 43(4): 501-508.

Cano Izquierdo JM, Dimitriadis YA, Gómez Sánchez E, López Coronado J (2001) Learning from noisy information in FasArt and FasBack neuro-fuzzy systems. Neural Networks 14(4-5): 407-425.

Carbonell J, Yang Y, Cohen W (2000) Special issue of machine learning on information retrieval Introduction. Machine Learning 39(2-3): 99-101.

Card HC, Rosendahl GK, McNeill DK, McLeod RD (1998) Competitive learning algorithms and neurocomputer architecture. IEEE Trans Computers 47(8): 847-858.

Carozza M, Rampone S (2005) An incremental regression method for graph structured data. Neural Networks 18(8): 1087-1092.

Carpenter GA, Grossberg S (1987a) A massively parallel architecture for a self-organizing neural pattern recognition machine. Computer Vision, Graphics, and Image Processing 37: 54-115.

Carpenter GA, Grossberg S (1987b) ART2: Self-organization of stable category recognition codes for analog input patterns. Applied Optics 26(23): 4919-4930.

Carpenter GA, Grossberg S, Rosen DB (1991a) ART 2-A: An adaptive resonance algorithm for rapid category learning and recognition. Neural Networks 4(4): 493-504.

Carpenter GA, Grossberg S, Rosen DB (1991b) Fuzzy ART: Fast stable learning and categorization of analog patterns by an adaptive resonance system. Neural Networks 4(6): 759-771.

Carpenter GA, Grossberg S, Markuzon N, Reynolds JH, Rosen DB (1992) Fuzzy ARTMAP: A neural network architecture for incremental supervised learning of analog multidimensional maps. IEEE Trans Neural Networks 3(5): 698-713.

Carpineto C, Romano G (1996) A lattice conceptual clustering system and its application to browsing retrieval. Machine Learning 24(2): 95-122.

Carrasco RC, Forcada ML (2001) Simple strategies to encode tree automata in sigmoid recursive neural networks. IEEE Trans Knowledge Data Engineering 13(2): 148-156.

Cassandras CG, Pepyne DL, Wardi Y (2001) Optimal control of a class of hybrid systems. IEEE Trans Automatic Control 46(3): 398-415.

Castano S, de Antonellis V (2001) Global viewing of heterogeneous data sources. IEEE Trans Knowledge Data Engineering 13(2): 277-297.

Castro JL, Delgado M (1996) Fuzzy systems with defuzzification are universal approximators. IEEE Trans Systems, Man, Cybernetics - B 26(1): 149-152.

Castro JL, Mantas CJ, Benitez JM (2002) Interpretation of artificial neural networks by means of fuzzy rules. IEEE Trans Neural Networks 13(1): 101-116.

Castro J, Georgiopoulos M, Demara R, Gonzalez A (2005) Data-partitioning using the Hilbert space filling curves: Effect on the speed of convergence of Fuzzy ARTMAP for large database problems. Neural Networks 18(7): 967-984.

Caudell TP (1992) Hybrid optoelectronic adaptive resonance theory neural processor, ART1. Applied Optics 31(29): 6220-6229.

Caulfield HJ (1998) Perspectives in optical computing. Computer 31(2): 22-25.

Cercone N, An A, Chan C (1999) Rule-induction and case-based reasoning: Hybrid architectures appear advantageous. IEEE Trans Knowledge Data Engineering 11(1): 166-174.

Chae SB (1995) Lebesgue integration, 2nd ed. Springer-Verlag, New York, NY.

Chaib-draa B (2002) Causal maps: Theory, implementation, and practical applications in multiagent environments. IEEE Trans Knowledge Data Engineering 14(6): 1201-1217.

Chakrabarty K (2001) On fuzzy lattice. In: Lecture Notes in Computer Science, W Ziarko and Y Yao (eds.) LNAI 2005: 238-242. Springer-Verlag, Berlin.

Chakrabarti S, Dom BE, Kumar SR, Raghavan P, Rajagopalan S, Tomkins A, Gibson D, Kleinberg J (1999) Mining the Web's link structure. Computer 32(8): 60-67.

Chandrasekaran B, Goel A (1988) From numbers to symbols to knowledge structures: Artificial intelligence perspectives on the classification task. IEEE Trans Systems, Man, Cybernetics 18(3): 415-424.

Chandrasekaran B, Johnson TR, Smith JW (1992) Task-structure analysis for knowledge modeling. Comm ACM 35(9): 124-137.

Chandrasekaran B, Josephson JR, Benjamins VR (1999) What are ontologies, and why do we need them?. IEEE Intelligent Systems 14(1): 20-26.

Chang CH, Hsu CC (1999) Enabling concept-based relevance feedback for information retrieval on the WWW. IEEE Trans. Knowledge Data Engineering 11(4): 595-609.

Chang PT, Lee ES (1994) Fuzzy linear regression with spreads unrestricted in sign. Computers Math Applic 28(4): 61-70.

Chang SK, Znati T (2001) Adlet: An active document abstraction for multimedia information fusion. IEEE Trans Knowledge Data Engineering 13(1): 112-123.

Chang-Hasnain CJ, Ku PC, Kim J, Chuang SL (2003) Variable optical buffer using slow light in semiconductor nanostructures. Proc of the IEEE 91(11): 1884-1897.

Chatzis V, Pitas I (1995) Mean and median of fuzzy numbers. In: Proc IEEE Workshop Nonlinear Signal Image Processing, Neos Marmaras, Greece, pp 297-300.

Chatzis V, Pitas I (1999) Fuzzy scalar and vector median filters based on fuzzy distances. IEEE Trans Image Processing 8(5): 731-734.

Chatzis V, Pitas I (2000) A generalized fuzzy mathematical morphology and its application in robust 2-D and 3-D object representation. IEEE Trans Image Processing 9(10): 1798-1810.

Chatzis V, Bors AG, Pitas I (1999) Multimodal decision-level fusion for person authentication. IEEE Trans Systems, Man, Cybernetics - A 29(6): 674-680.

Chávez E, Navarro G, Baeza-Yates R, Marroquín JL (2001) Searching in metric spaces, ACM Computing Surveys 33(3): 273-321.

Chella A, Gaglio S, Pirrone R (2001) Conceptual representations of actions for autonomous robots. Robotics and Autonomous Systems 34(4): 251-263.

Chen SM (2002) Weighted fuzzy reasoning using weighted fuzzy Petri nets. IEEE Trans Knowledge Data Engineering 14(2): 386-397.

Chen TP, Chen H (1995) Approximation capability to functions of several variables, nonlinear functionals, and operators by radial basis function neural networks. IEEE Trans. Neural Networks 6(4): 904-910.

Chen JH, Chen CS (2002) Fuzzy kernel perceptron. IEEE Trans Neural Networks, 13(6): 1364-1373.

Chen JX, Nakano A (2003) High-dimensional data acquisition, computing, and visualization. Computing in Science and Engineering 5(2): 12-13.

Chen JQ, Xi YG (1998) Nonlinear system modeling by competitive learning and adaptive fuzzy inference system. IEEE Trans Systems, Man, Cybernetics - C 28(2): 231-238.

Chen X, Zhang X (2003) A popularity-based prediction model for Web prefetching. Computer 36(3): 63-70.

Chen S, Cowan CFN, Grant PM (1991) Orthogonal least squares learning algorithm for radial basis function networks. IEEE Trans Neural Networks 2(2): 302-309.

Chen M, Ghorbani AA, Bhavsar VC (2005) Incremental communication for adaptive resonance theory networks. IEEE Transactions on Neural Networks 16(1): 132-144.

Cheney EW (1982) Introduction to Approximation Theory, 2nd ed. Chelsea, New York, NY.

Chi RTH, Kiang MY (1993) Reasoning by coordination: An integration of case-based and rule-based reasoning systems. Knowledge-Based Systems 6(2): 103-113.

Christensen SW, Sinclair I, Reed PAS (2003) Designing committees of models through deliberate weighting of data points. J Machine Learning Research, 4(Apr): 39-66.

Chuang CC, Su SF, Jeng JT, Hsiao CC (2002) Robust support vector regression networks for function approximation with outliers. IEEE Trans Neural Networks, 13(6): 1322-1330.

Cios KJ, Pedrycz W, Swiniarski RM (1998) Data mining methods for knowledge discovery. Kluwer, Boston, MA.

Citterio C, Pelagotti A, Piuri V, Rocca L (1999) Function approximation – A fast-convergence neural approach based on spectral analysis. IEEE Trans Neural Networks 10(4): 725-740.

Cofer DD, Garg VK (1996) Supervisory control of real-time discrete-event systems using lattice theory. IEEE Trans Automatic Control 41(2): 199-209.

Cohen WW (1998) Hardness results for learning first-order representations and programming by demonstration. Machine Learning 30(1): 57-88.

Conklin D, Fortier S, Glasgow J (1993) Knowledge discovery in molecular databases. IEEE Trans Knowledge Data Engineering 5(6): 985-987.

Cook DJ, Holder LB (2000) Graph-based data mining. IEEE Intelligent Systems 15(2): 32-41.

Cook DJ, Holder LB, Su S, Maglothin R, Jonyer I (2001) Structural mining of molecular biology data. IEEE Engineering in Medicine and Biology Magazine 20(4): 67-74.

Cornelis C, van der Donck C, Kerre E (2003) Sinha-Dougherty approach to the fuzzification of set inclusion revisited. Fuzzy Sets Systems 134(2): 283-295.

Corridoni JM, Del Bimbo A (1998) Structured representation and automatic indexing of movie information content. Pattern Recognition 31(12): 2027-2045.

Cotter NE (1990) The Stone-Weierstrass theorem and its application to neural networks. IEEE Trans Neural Networks 1(4): 290-295.

Cottrell M, Ibbou S, Letrémy P (2004) SOM-based algorithms for qualitative variables. Neural Networks 17(8-9): 1149-1167.

Cover TM (1965) Geometrical and statistical properties of systems of linear inqualities with applications in pattern recognition. IEEE Trans Electronic Computers 14(3): 326-334.

Cripps A, Kaburlasos VG, Nguyen N, Papadakis SE (2003) Improved experimental results using Fuzzy Lattice Neurocomputing (FLN) classifiers. In: Proc Intl Conf Machine Learning; Models, Technologies and Applications (MLMTA'03), Las Vegas, NV, pp 161-166.

Cristianini N, Shawe-Taylor J (2000) Support vector machines. Cambridge Univ. Press, Cambridge, UK.

Culler D, Estrin D, Srivastava M (2004) Overview of sensor networks. Computer 37(8): 41-49.

Cybenko G (1989) Approximation by superpositions of a sigmoid function. Math Contr Signals Syst 2: 303-314.

Dagher I, Georgiopoulos M, Heileman GL, Bebis G (1999) An ordering algorithm for pattern presentation in fuzzy ARTMAP that tends to improve generalization performance. IEEE Trans Neural Networks 10(4): 768-778.

Dayou Wang, Keller JM, Carson CA, McAdo-Edwards KK, Bailey CW (1998) Use of fuzz-logic-inspired features to improve bacterial recognition through classifier fusion. IEEE Trans Systems, Man, Cybernetics - B 28(4): 583-591.

Davenport MP, Titus AH (2004) Multilevel category structure in the ART-2 network. IEEE Trans Neural Networks 15(1): 145-158.

Davey BA, Priestley HA (1990) Introduction to Lattices and Order. Cambridge University Press, Cambridge, Great Britain.

Davis R (1998) What are intelligence? And why?. AI Magazine 19(1): 91-111.

Davoren JM, Nerode A (2000) Logics for hybrid systems. Proc of the IEEE 88(7): 985-1010.

Dawant BM, Jansen BH (1991) Coupling numerical and symbolic methods for signal interpretation. IEEE Trans Systems, Man, Cybernetics 21(1): 115-124.

Dawant BM, Garbay C (1999) Special topic section on biomedical data fusion. IEEE Trans Biomedical Engineering 46(10): 1169-1170.

De Baets B, Kerre EE (1993) The generalized modus ponens and the triangular fuzzy data model. Fuzzy Sets Systems 59(3): 305-317.

de Jesus JD, Calvo JJV, Fuente AI (2000) Surveillance system based on data fusion from image and acoustic array sensors. IEEE Aerospace and Electronic Systems Magazine 15(2): 9-16.

Dempster AP (1967) Upper and lower probabilities induced by multivalued mappings. Annals of Math Stat 38: 325-329.

Dey D, Sarkar S, De P (2002) A distance-based approach to entity reconciliation in heterogeneous databases. IEEE Trans Knowledge Data Engineering 14(3): 567-582.

Diamond P (1988) Fuzzy least squares. Information Sciences 46(3): 141-157.

Diamond P, Kloeden P (1994) Metric spaces of fuzzy sets: Theory and applications. World Scientific, Singapore.

Diamond P, Körner R (1997) Extended fuzzy linear models and least squares estimates. Computers Math Applic 33(9): 15-32.

Díaz P, Aedo I, Panetsos F (2001) Modeling the dynamic behavior of hypermedia applications. IEEE Trans Software Engineering 27(6): 550-572.

Dick S, Kandel A (1999) Comment on "Combinatorial rule explosion eliminated by a fuzzy rule configuration". IEEE Trans Fuzzy Systems 7(4): 475-477.

Dickerson JA, Kosko B (1996) Fuzzy function approximation with ellipsoidal rules. IEEE Trans Systems, Man, Cybernetics - B 26(4): 542-560.

Dietterich TG (2000) An experimental comparison of three methods for constructing ensembles of decision trees: Bagging, boosting, and randomization. Machine Learning 40(2): 139-157.

Dietterich TG, Lathrop RH, Lozano-Perez T (1997) Solving the multiple-instance problem with axis-parallel rectangles. Artificial Intelligence 89(1-2): 31-71.

Domingos P (1996) Unifying instance-based and rule-based induction, Machine Learning 24(2): 141-168.

Doob JL (1994) Measure Theory. Springer-Verlag, New York, NY.

Dote Y, Ovaska SJ (2001) Industrial applications of soft computing: A review. Proc of the IEEE 89(9): 1243-1265.

Dougherty ER, Sinha D (1995) Computational gray-scale mathematical morphology on lattices (a comparator-based image algebra) part II: image operators. Real-Time Imaging 1: 283-295.

Downs J, Harrison RF, Kennedy RL, Cross SS, (1996) Application of the fuzzy ARTMAP neural-network model to medical pattern-classification tasks. Artificial Intelligence in Medicine 8(4): 403-428.

Doya K, Kimura H, Kawato M (2001) Neural mechanisms of learning and control. IEEE Control Syst Mag, 21(4): 42-54.

Doyen AL, Duquenne V, Nuques S, Carlier M (2001) What can be learned from a lattice analysis of a laterality questionnaire?. Behavior Genetics 31(2): 193-207.

Drucker H. (1998) Boosting using neural networks. Springer-Verlag, New York.

Drucker H, Wu D, Vapnik VN (1999) Support Vector Machines for Spam Categorization. IEEE Trans Neural Networks 10(5): 1048-1054.

Dubois D, Prade H (1980) Fuzzy Sets and Systems - Theory and Applications. Academic Press, San Diego, CA.

Dubois D, Prade H (1986) Fuzzy sets and statistical data. European J Operational Research 25: 345-356.

Dubois D, Gottwald S, Hajek P, Kacprzyk J, Prade H (2005) Terminological difficulties in fuzzy set theory - The case of "intuitionistic fuzzy sets". Fuzzy Sets Systems 156(3): 485-491.

Duch W, Setiono R, Zurada JM (2004) Computational intelligence methods for rule-based data understanding. Proc of the IEEE, 92(5): 771-805.

Duda RO, Hart PE, Stork DG (2001) Pattern classification, 2nd ed. John Wiley & Sons, New York, NY.

Dunyak J, Saad IW, Wunsch D (1999) A theory of independent fuzzy probability for system reliability. IEEE Trans Fuzzy Systems 7(3): 286-294.

Duric Z, Gray WD, Heishman R, Li F, Rosenfeld A, Schoelles MJ, Schunn C, Wechsler H (2002) Integrating perceptual and cognitive modeling for adaptive and intelligent human-computer interaction. Proc of the IEEE 90(7): 1272-1289.

Edmonds EA (1980) Lattice fuzzy logics. Intl J Man-Machine Studies 13(4): 455-465.

Edwards D (2000) Introduction to graphical modelling, 2nd ed. Springer, New York, NY.

Edwards PJ, Peacock AM, Renshaw D, Hannah JM, Murray AF (2002) Minimizing risk using prediction uncertainty in neural network estimation fusion and its application to papermaking. IEEE Trans Neural Networks, 13(3): 726-731.

Ehrenfeucht A, Haussler D, Kearns M, Valiant L (1989) A general lower bound on the number of examples needed for learning. Information and Computation 82(3): 247-261.

El-Alfy AE, El-Gamal AF, Haggag MH, El-Allmi ME (2003) Integration of quantitative and qualitative knowledge for online decision support. Intl J Intelligent Computing and Information Sciences 3(1): 62-74.

Elkan C (1994) The paradoxical success of fuzzy logic. IEEE Expert 9(4): 3-8.

Erdogmus D, Principe JC (2002) Generalized information potential criterion for adaptive system training. IEEE Trans Neural Networks, 13(5): 1035-1044.

Evans B, Fisher D (1994) Overcoming process delays with decision tree induction. IEEE Expert 9(1): 60-66.

Faloutsos C, Jagadish HV, Manolopoulos Y (1997) Analysis of the n-dimensional quadtree decomposition for arbitrary hyperrectangles. IEEE Trans Knowledge Data Engineering 9(3): 373-383.

Fan J, Xie W, Pei J (1999) Subsethood measure: New definitions. Fuzzy Sets Systems 106(2): 201-209.

Fang Y, Rousseau R (2001) Lattices in citation networks: An investigation into the structure of citation graphs. Scientometrics 50(2): 273-287.

Feller W (1968) An introduction to probability theory and its applications. John Wiley & Sons, New York, NY.

Fernández F, Gutiérrez J (2003) A Tagaki-Sugeno model with fuzzy inputs viewed from multidimensional interval analysis. Fuzzy Sets Systems 135(1): 39-61.

Filev DP, Yager RR (1997) Operations on fuzzy numbers via fuzzy reasoning. Fuzzy Sets Systems 91(2): 137-142.

Fine TL (1999) Feedforward neural network methodology. Springer-Verlag, New York.

Fogel DB (1999) Evolutionary computation: Toward a new philosophy of machine intelligence, 2nd ed. IEEE Press, Piscataway, NJ.

Fogel DB, Fukuda T, Guan L (1999) Special issue on computational intelligence. Proc of the IEEE 87(9): 1415-1422.

Foulloy L, Galichet S (2003) Fuzzy control with fuzzy inputs. IEEE Trans Fuzzy Systems 11(4): 437-449.

Frasconi P, Gori M, Sperduti A (1998) A general framework for adaptive processing of data structures. IEEE Trans Neural Networks 9(5): 768-786.

Freund Y, Schapire RE (1999) A short introduction to boosting. J Japan Soc Artificial Intelligence 14(5): 771-780.

Fritzke B (1994) Growing cell structures - A self-organizing network for unsupervised and supervised learning. Neural Networks 7(9): 1441-1460.

Fuessel D, Isermann R (2000) Hierarchical motor diagnosis utilizing structural knowledge and a self-learning neuro-fuzzy scheme. IEEE Trans Industrial Electronics 47(5): 1070-1077.

Fuhr N, Rölleke T (1997) A probabilistic relational algebra for the integration of information retrieval and database systems. ACM Trans Information Systems 15(1): 32-66.

Fujihara H, Simmons DB, Ellis NC, Shannon RE (1997) Knowledge conceptualization tool. IEEE Trans Knowledge Data Engineering 9(2): 209-220.

Fukuda T (2000) On fuzzy statistical analysis for vague random data associated with human feelings. In: Proc 9th IEEE Intl Workshop Robot and Human Interactive Communication (RO-MAN), pp 126-131.

Fukuda T, Michelini R, Potkonjak V, Tzafestas S, Valavanis K, Vukobratovic M (2001) How far away is "artificial man". IEEE Robotics & Automation Mag 8(1): 66-73.

Gabrys B, Bargiela A (2000) General fuzzy min-max neural network for clustering and classification. IEEE Trans Neural Networks 11(3): 769-783.

Gähler S, Gähler W (1994) Fuzzy real numbers. Fuzzy Sets Systems 66(2): 137-158.

Gaines BR (1978) Fuzzy and probability uncertainty logics. Information and Control 38: 154-169.

Galichet S, Foulloy L (1995) Fuzzy controllers: Synthesis and equivalences. IEEE Trans Fuzzy Systems 3(2): 140-148.

Ganter B, Wille R (1999) Formal concept analysis. Springer, Heidelberg, Germany.

Garg A, Pavlović V, Rehg JM (2003) Boosted learning in dynamic Bayesian networks for multimodal speaker detection. Proc of the IEEE 91(9): 1355-1369.

Georgiopoulos M, Huang J, Heileman GL (1994) Properties of learning in ARTMAP. Neural Networks 7(3): 495-506.

Georgiopoulos M, Fernlund H, Bebis G, Heileman GL (1996) Order of search in fuzzy ART and fuzzy ARTMAP: Effect of the choice parameter. Neural Networks 9(9): 1541-1559.

Georgiopoulos M, Dagher I, Heileman GL, Bebis G (1999) Properties of learning of a fuzzy ART variant. Neural Networks 12(6): 837-850.

Georgiou DN, Papadopoulos BK (1999) Convergences in fuzzy topological spaces. Fuzzy Sets Systems 101(3): 495-504.

Giachetti RE, Young RE (1997) A parametric representation of fuzzy numbers and their arithmetic operators. Fuzzy Sets Systems 91(2): 185-202.

Giarratano J, Riley G (1994) Expert Systems - Principles and Programming, 2nd ed. PWS Publishing Company, Boston, MA.

Gierz G, Hofmann KH, Keimel K, Lawson JD, Mislove M, Scott DS (1980) A compendium of continuous lattices. Springer-Verlag, Berlin, Germany.

Gilbert B (2001) Analog at milepost 2000: A personal perspective. Proc of the IEEE 89(3): 289-304.

Giles CL, Omlin CW, Thornber KK (1999) Equivalence in knowledge representation: Automata, recurrent neural networks, and dynamical fuzzy systems. Proc of the IEEE 87(9): 1623-1640.

Girard R, Ralambondrainy H (1997) Computing a concept lattice from structured and fuzzy data. In: Proc 6th IEEE Intl Conf Fuzzy Systems, vol. 1, pp. 135-142.

Gödel K (1931) On formally undecidable propositions in principia mathematica and related systems. Monatshefte fur Mathematik und Physic 38 (in German); English translation in Davis M (ed.) (1965) The Undecidedable. Raven Press.

Goguen JA (1967) L-fuzzy sets. J Math Analysis and Applications 18: 145-174.

Goguen JA, Burstall RM (1992) Institutions: Abstract model theory for specification and programming. J Association of Computing Machinery 39(1): 95-146.

Goldberg DE (1989) Genetic algorithms in search, optimization and machine learning. Addison-Wesley, New York, NY.

Goldfarb L (1992) What is a distance and why do we need the metric model for pattern learning?. Pattern Recognition 25(4): 431-438.

Goldfarb L, Deshpande S (1997) What is a symbolic measurement process?. In: Proc IEEE Intl Conf Systems, Man, Cybernetics, pp 4139-4145.

Goldfarb L, Golubitsky O, Korkin D (2000) What is a structural representation?. Tech rep TR00-37, Faculty of Computer Science, Univ Brunswick, Canada.

Gómez Sánchez E, Dimitriadis YA, Cano Izquierdo JM, Lopez Coronado J (2002) μARTMAP: Use of mutual information for category reduction in Fuzzy ARTMAP. IEEE Trans Neural Networks 13(1): 58-69.

Gonzalez A, Perrez R (2001) Selection of relevant features in a fuzzy genetic learning algorithm. IEEE Trans Systems, Man, Cybernetics - B 31(3): 417-425.

González J, Rojas I, Pomares H, Ortega J, Prieto A (2002) A new clustering technique for function approximation. IEEE Trans Neural Networks 13(1): 132-142.

Goodman IR, Nguyen HT (2002) Fuzziness and randomness. In: Statistical Modeling, Analysis and Management of Fuzzy Data, C Bertoluzza, MA Gil, and DA Ralescu (eds.) 87: 3-21. Physica-Verlag, Heidelberg, Germany.

Gori M, Scarselli F (1998) Are multilayer perceptrons adequate for pattern recognition and verification?. IEEE Trans. Pattern Analysis Machine Intelligence 20(11): 1121-1132.

Gori M, Scarselli F, Tsoi AC (1998) On the closure of the set of functions that can be realized by a given multilayer perceptron. IEEE Trans Neural Networks 9(6): 1086-1098.

Goshko BM (1997) Algebraic-logical model of an information-retrieval system with an interface component. Cybernetics Systems Analysis 33(2): 168-170.

Goutsias J, Heijmans HJAM (2000) Nonlinear multiresolution signal decomposition schemes. I. morphological pyramids. IEEE Trans Image Processing 9(11): 1862-1876.

Grätzer G (1971) Lattice theory. WH Freeman & Co, San Francisco, CA.

Green SJ (1999) Building hypertext links by computing semantic similarity. IEEE Trans Knowledge Data Engineering, 11(5): 713-730.

Grossberg S (1969) Embedding fields: A theory of learning with physiological implications. J of Mathematical Psychology 6(2): 209-239.

Grossberg S (1976a) Adaptive pattern classification and universal recoding, I: Parallel development and coding of neural feature detectors. Biological Cybernetics 23: 121-134.

Grossberg S (1976b) Adaptive pattern classification and universal recoding, II: Feedback, expectation, olfaction, and illusions. Biological Cybernetics 23: 187-202.

Grzegorzewski P (2004) Distances between intuitionistic fuzzy sets and/or interval-valued fuzzy sets based on the Hausdorff metric. Fuzzy Sets Systems 148(2): 319-328.

Guadarrama S, Muñoz S, Vaucheret C (2004) Fuzzy Prolog: A new approach using soft constraints propagation. Fuzzy Sets Systems 144(1): 127-150.

Guglielmo EJ, Rowe NC (1996) Natural-language retrieval of images based on descriptive captions. ACM Trans Information Systems 14(3): 237-267.

Gutiérrez García J, Rodabaugh SE (2005) Order-theoretic, topological, categorical redundancies of interval-valued sets, grey sets, vague sets, interval-valued "intuitionistic" sets, "intuitionistic" fuzzy sets and topologies. Fuzzy Sets Systems 156(3): 445-484.

Güting RH, Zicari R, Choy DM (1989) An algebra for structured office documents. ACM Trans Information Systems 7(2): 123-157.

Hagenbuchner M, Sperduti A, Tsoi AC (2003) A self-organizing map for adaptive processing of structured data. IEEE Trans Neural Networks, 14(3): 491-505.

Hall DL, Llinas J (1997) An introduction to multisensor data fusion. Proc of the IEEE 85(1): 6-23.

Halmos PR (1960) Naive Set Theory. Van Nostrand Co., New York, NY.

Halmos P, Givant S (1998) Logic as algebra. The Mathematical Association of America, The Dolciani Mathematical Expositions, no 21.

Hammer B, Villmann T (2002) Generalized relevance learning vector quantization. Neural Networks 15(8-9): 1059-1068.

Hammer B, Micheli A, Sperduti A, Strickert M (2004) A general framework for unsupervised processing of structured data. Neurocomputing 57: 3-35.

Harel D, Rumpe B (2004) Meaningful modeling: What's the semantics of semantics?. Computer 37(10): 64-72.

Harnad S (2002) Symbol grounding and the origin of language. In: Computationalism – New Directions, M Scheutz (ed.) 143-158. The MIT Press, Cambridge, MA.

Hartemink AJ, Gifford DK, Jaakkola TS, Young RA (2002) Bayesian methods for elucidating genetic regulatory networks. IEEE Intel Syst 17(2): 37-43.

Hathaway RJ, Bezdek JC, Pedrycz W (1996) A parametric model for fusing heterogeneous fuzzy data. IEEE Trans Fuzzy Systems 4(3): 270-281.

Hathaway RJ, Bezdek JC, Hu YK (2000). Generalized fuzzy c-means clustering strategies using L_p norm distances. IEEE Trans Fuzzy Systems 8(5): 576-582.

Haussler D (1989) Learning conjunctive concepts in structural domains. Machine Learning 4(1): 7-40.

Hawkins J, Blakeslee S (2004) On intelligence. Times Books, New York, NY.

Haykin S (1994) Neural Networks. Macmillan, New Jersey.

Haykin S, de Freitas N (2004) Special issue on sequential state estimation. Proc of the IEEE 92(3): 399-400.

Healy MJ (1999) A topological semantics for rule extraction with neural networks. Connection Science 11(1): 91-113.

Healy MJ, Caudell TP (1997) Acquiring rule sets as a product of learning in a logical neural architecture. IEEE Trans Neural Networks 8(3): 461-474.

Healy MJ, Caudell TP (1998) Guaranteed two-pass convergence for supervised and inferential learning. IEEE Trans Neural Networks 9(1): 195-204.

Healy MJ, Caudell TP (2004) Neural networks, knowledge, and cognition: A mathematical semantic model based upon category theory. University of New Mexico Dept Electrical and Computer Engineering Technical Report: EECE-TR-04-020 available at https://repository.unm.edu/handle/1928/33.

Hearst MA (1997) Banter on Bayes: Debating the usefulness of Bayesian approaches to solving practical problems. IEEE Expert 12(6): 18-21.

Hearst MA, Levy AY, Knoblock C, Minton S, Cohen W (1998) Information integration. IEEE Intelligent Systems 13(5): 12-24.

Heijmans HJAM, Goutsias J (2000) Nonlinear multiresolution signal decomposition schemes. II. morphological wavelets. IEEE Trans Image Processing 9(11): 1897-1913.

Heilpern S (1997) Representation and application of fuzzy numbers. Fuzzy Sets Systems 91(2): 259-268.

Heisele B, Verri A, Poggio T (2002) Learning and vision machines. Proc of the IEEE 90(7): 1164-1177.

Herbrich R (2002) Learning kernel classifiers: Theory and algorithms. The MIT Press, Cambridge, MA.

Hesselink L, Orlov SS, Bashaw MC (2004) Holographic data storage systems. Proc of the IEEE 92(8): 1231-1280.

Hettich S, Blake CL, Merz CJ (1998) UCI repository of machine learning databases [http://www.ics.uci.edu/~mlearn/MLRepository.html]. Univ California, Irvine, CA, Dept Information and Computer Science.

Hill MT, Frietman EEE, de Waardt H, Khoe Gd, Dorren HJS (2002) All fiber-optic neural network using coupled SOA based ring lasers. IEEE Trans Neural Networks, 13(6): 1504-1513.

Hinton GE (1990) Mapping part-whole hierarchies into connectionist networks. Artificial Intelligence 46(1-2): 47-75.

Hirose A (ed.) (2003) Complex-valued neural networks – Theories and applications. World Scientific, New Jersey.

Hirota K, Pedrycz W (1999) Fuzzy computing for data mining. Proc of the IEEE 87(9): 1575-1600.

Hiruma T (2005) Photonics technology for molecular imaging. Proc of the IEEE 93(4): 829-843.

Hirvensalo M (2004) Quantum computing, 2nd ed. Springer, Heidelberg, Germany.

Hoffmann F (2001) Evolutionary algorithms for fuzzy control system design. Proc of the IEEE 89(9): 1318-1333.

Höhle U, Rodabaugh SE (1998) Mathematics of fuzzy sets: Logic, topology, and measure theory. Springer, Berlin, Germany.

Holder LB, Cook DJ (1993) Discovery of inexact concepts from structural data. IEEE Trans Knowledge Data Engineering 5(6): 992-994.

Holmes N (2001) The great term robbery. Computer 34(5): 94-96.

Hong L, Jain A (1998) Integrating faces and fingerprints for personal identification. IEEE Trans Pattern Analysis Machine Intelligence 20(12): 1295-1307.

Hongyi Li, Deklerck R, De Cuyper B, Hermanus A, Nyssen E, Cornelis J (1995) Object recognition in brain CT-scans: Knowledge-based fusion of data from multiple feature extractors. IEEE Trans Medical Imaging 14(2): 212-229.

Hopfield JJ (1982) Neural networks and physical systems with emergent collective computational abilities. Proc Nat Academy Sci, 79, pp. 2554-2558.

Horiuchi T (1998) Decision rule for pattern classification by integrating interval feature values. IEEE Trans Pattern Analysis Machine Intelligence 20(4): 440-448.

Hornik K, Stinchcombe M, White H (1989) Multilayer feedforward networks are universal approximators. Neural Networks 2(5): 359-366.

Hou JL, Sun MT, Chuo HC (2005) An intelligent knowledge management model for construction and reuse of automobile manufacturing intellectual properties. Intl J Advanced Manufacturing Technology 26(1-2): 169-182.

Hsinchun Chen, Schatz B, Ng T, Martinez J, Kirchhoff A, Chienting Lin (1996) A parallel computing approach to creating engineering concept spaces for semantic retrieval: The Illinois Digital Library Initiative project, IEEE Trans Pattern Analysis Machine Intelligence 18(8): 771-782.

Huang J, Georgiopoulos M, Heileman G (1995) Fuzzy ART properties. Neural Networks 8(2): 203-213.

Huang Q, Puri A, Liu Z (2000) Multimedia search and retrieval: New concepts, system implementation, and application. IEEE Trans Circuits and Systems for Video Technology 10(5): 679-692.

Hummel R, Manevitz L (1996) A statistical approach to the representation of uncertainty in beliefs using spread of opinions. IEEE Trans Systems, Man, Cybernetics - A 26(3): 378-384.

Hutchinson EG, Thornton JM (1996) Promotif: A program to identify and analyze structural motifs in proteins. Protein Science 5(2): 212-220.

Huynh QQ, Cooper LN, Intrator N, Shouval H (1998) Classification of underwater mammals using feature extraction based on time-frequency analysis and BCM theory. IEEE Trans Signal Processing 46(5): 1202-1207.

Isermann R (1998) On fuzzy logic applications for automatic control, supervision, and fault diagnosis. IEEE Trans Systems, Man, Cybernetics - A 28(2): 221-235.

Ishibuchi H, Nakashima T (2001) Effect of rule weights in fuzzy rule-based classification systems. IEEE Trans Fuzzy Systems 9(4): 506-515.

Ishibuchi H, Fujioka R, Tanaka H (1993) Neural networks that learn from fuzzy if-then rules. IEEE Trans Fuzzy Systems 1(2): 85-97.

Ishibuchi H, Kwon K, Tanaka H (1995) A learning algorithm of fuzzy neural networks with triangular fuzzy weights. Fuzzy Sets Systems 71(3): 277-293.

Ishibuchi H, Morisawa T, Nakashima T (1996) Voting schemes for fuzzy-rule-based classification systems. In: Proc IEEE Intl Conf Fuzzy Systems, pp 614-620.

Ishibuchi H, Nakashima T, Murata T (1999) Performance evaluation of fuzzy classifier systems for multidimensional pattern classification problems. IEEE Trans Systems, Man, Cybernetics - B 29(5): 601-618.

Ishigami H, Fukuda T, Shibata T, Arai F (1995) Structure optimization of fuzzy neural network by genetic algorithm. Fuzzy Sets Systems 71(3): 257-264.

Itô K (ed.) (1987) Encyclopedic Dictionary of Mathematics, 2nd ed. The Mathematical Society of Japan, English translation The MIT Press, Cambridge, MA.

Iyengar S, Brooks RR (1997) Multi-sensor fusion: Fundamentals and applications with software. Prentice Hall, NJ.

Jain AK, Bin Yu (1998) Document representation and its application to page decomposition. IEEE Trans Pattern Analysis Machine Intelligence 20(3): 294-308.

Jain AK, Murty MN, Flynn PJ (1999) Data clustering: A review. ACM Computing Surveys 31(3): 264-323.

Jain AK, Duin RPW, Jianchang Mao (2000) Statistical pattern recognition: A review. IEEE Trans Pattern Analysis Machine Intelligence 22(1): 4-37.

Jang JSR (1993) ANFIS: Adaptive-network-based fuzzy inference system. IEEE Trans Systems, Man, Cybernetics 23(3): 665–685.

Jang JSR, Sun CT (1993) Functional equivalence between radial basis function networks and fuzzy inference systems. IEEE Trans Neural Networks 4(1): 156-159.

Jang JSR, Sun CT (1995) Neuro-fuzzy modeling and control. Proc of the IEEE 83(3): 378-406.

Jang JSR, Sun CT, Mizutani E (1997) Neuro-fuzzy and soft computing – A computational approach to learning and machine intelligence. Prentice Hall, Upper Saddle River, NJ.

Jia-Lin Chen, Jyh-Yeong Chang (2000) Fuzzy perceptron neural networks for classifiers with numerical data and linguistic rules as inputs. IEEE Trans Fuzzy Systems 8(6): 730-745.

Jianchang Mao, Jain AK (1996) A self-organizing network for hyperellipsoidal clustering (HEC). IEEE Trans Neural Networks 7(1): 16-29.

Jilani LL, Desharnais J, Mili A (2001) Defining and applying measures of distance between specifications. IEEE Trans Software Engineering 27(8): 673-703.

Jones AS, Eis KE, Vonderhaar TH (1995) A method for multisensor-multispectral satellite data fusion. J Atmospheric and Oceanic Technology 12(4): 739-754.

Joshi A, Ramakrishman N, Houstis EN, Rice JR (1997) On neurobiological, neuro-fuzzy, machine learning, and statistical pattern recognition techniques. IEEE Trans Neural Networks 8(1): 18-31.

Jordan MI, Sejnowski TJ (eds.) (2001) Graphical models: Foundations of neural computation. The MIT Press, Cambridge, MA.

Juang C-F, Lin C-T (1998) An online self-constructing neural fuzzy inference network and its applications. IEEE Trans Fuzzy Systems 6(1): 12-32.

Kaburlasos VG (1992) Adaptive resonance theory with supervised learning and large database applications. PhD thesis, Univ Nevada, Reno, USA, Library of Congress-Copyright Office.

Kaburlasos VG (2002) Novel fuzzy system modeling for automatic control applications. In: Proc Intl Conf Technology & Automation, Thessaloniki, Greece, pp 268-275.

Kaburlasos VG (2003) Improved Fuzzy Lattice Neurocomputing (FLN) for semantic neural computing. In: Proc Intl Joint Conf Neural Networks (IJCNN'2003), Portland, OR, pp 1850-1855.

Kaburlasos VG (2004a) FINs: Lattice theoretic tools for improving prediction of sugar production from populations of measurements. IEEE Trans Systems, Man, Cybernetics - B 34(2): 1017-1030.

Kaburlasos VG (2004b) A device for linking brain to mind based on lattice theory. In: Proc Intl Conf Cognitive and Neural Systems (ICCNS 2004), Boston University, Boston, MA, p 58.

Kaburlasos VG, Christoforidis A (2006) Granular auto-regressive moving average (grARMA) model for predicting a distribution from other distributions. Real-world applications. In: Proc World Congress Computational Intelligence (WCCI) IEEE Intl Conf Fuzzy Systems (FUZZ-IEEE), Vancouver, Canada.

Kaburlasos VG, Kazarlis S (2002) σ-FLNMAP with Voting (σFLNMAPwV): A genetically optimized ensemble of classifiers with the capacity to deal with partially-ordered, disparate types of data. Application to financial problems. In: Proc Intl Conf Technol & Automation, Thessaloniki, Greece, pp 276-281.

Kaburlasos VG, Kehagias A (2004) Novel analysis and design of fuzzy inference systems based on lattice theory. In: Proc IEEE Intl Conf Fuzzy Systems (FUZZ-IEEE), Budapest, Hungary, pp 281-286.

Kaburlasos VG, Kehagias A (2006a) Novel fuzzy inference system (FIS) analysis and design based on lattice theory, part I: Working principles. Intl J General Systems (in press).

Kaburlasos VG, Kehagias A (2006b) Novel fuzzy inference system (FIS) analysis and design based on lattice theory. IEEE Trans Fuzzy Systems (in press).

Kaburlasos VG, Papadakis SE (2006) Granular self-organizing map (grSOM) for structure identification. Neural Networks (in press).

Kaburlasos VG, Petridis V (1997) Fuzzy lattice neurocomputing (FLN): A novel connectionist scheme for versatile learning and decision making by clustering. Intl J Computers and Their Applications 4(3): 31-43.

Kaburlasos VG, Petridis V (1998) A unifying framework for hybrid information processing. In: Proc ISCA Intl Conf Intelligent Systems (ICIS'98), Paris, France, pp 68-71.

Kaburlasos VG, Petridis V (2000) Fuzzy Lattice Neurocomputing (FLN) models. Neural Networks 13(10): 1145-1170.

Kaburlasos VG, Petridis V (2002) Learning and decision-making in the framework of fuzzy lattices. In: New Learning Paradigms in Soft Computing, LC Jain and J Kacprzyk (eds.) 84: 55-96. Physica-Verlag, Heidelberg, Germany.

Kaburlasos V, Petridis V, Allotta B, Dario P (1997) Automatic detection of bone breakthrough in orthopedics by fuzzy lattice reasoning (FLR): The case of drilling in the osteosynthesis of long bones. In: Proc Mechatronical Computer Systems for Perception and Action (MCPA'97), Pisa Italy, pp 33-40.

Kaburlasos VG, Petridis V, Brett P, Baker D (1999) Estimation of the stapes-bone thickness in stapedotomy surgical procedure using a machine-learning technique. IEEE Trans Information Technology in Biomedicine 3(4): 268-277.

Kaburlasos VG, Spais V, Petridis V, Petrou L, Kazarlis S, Maslaris N, Kallinakis A (2002) Intelligent clustering techniques for prediction of sugar production. Mathematics and Computers in Simulation 60(3-5): 159-168.

Kaburlasos VG, Papadakis SE, Kazarlis S (2003) A genetically optimized ensemble of σ-FLNMAP neural classifiers based on non-parametric probability distribution functions. In: Proc Intl J Conf Neural Networks (IJCNN'03), Portland, OR, pp 426-431.

Kaburlasos VG, Chatzis V, Tsiantos V, Theodorides M (2005) granular Self-Organizing Map (grSOM) neural network for industrial quality control. In:

Proc of SPIE – Mathematical Methods in Pattern and Image Analysis, San Diego, CA, vol. 5916, pp. 59160J: 1-10.

Kalman RE (1960) A new approach to linear filtering and prediction problems. Trans ASME J Basic Eng, 82: 35-45.

Kamke E (1950) Theory of sets. Dover, New York, NY.

Kandel A, Zhang YQ, Miller T (1995) Knowledge representation by conjunctive normal forms and disjunctive normal forms based on n-variable-m-dimensional fundamental clauses and phrases. Fuzzy Sets Systems 76(1): 73-89.

Kapasi UJ, Rixner S, Dally WJ, Khailany B, Ahn JH, Mattson P, Owens JD (2003) Programmable stream processors. Computer 36(8): 54-62.

Karayiannis NB, Bezdek JC (1997) An integrated approach to fuzzy learning vector quantization and fuzzy c-means clustering. IEEE Trans Fuzzy Systems 5(4): 622-628.

Karayiannis NB, Randolph-Gips MM (2003) Soft learning vector quantization and clustering algorithms based on non-Euclidean norms: Multinorm algorithms. IEEE Trans Neural Networks, 14(1): 89-102.

Karnik NN, Mendel JM, Liang Q (1999) Type-2 fuzzy logic systems. IEEE Trans Fuzzy Systems 7(6): 643-658.

Kaufmann A, Gupta MM (1985) Introduction to Fuzzy Arithmetic – Theory and Applications. Van Nostrand Reinhold, New York, NY.

Kearns MJ, Vazirani UV (1994) An introduction to computational learning theory. The MIT Press, Cambridge, MA.

Kearns M, Li M, Valiant L (1994) Learning Boolean formulas, Journal of the Association of Computing Machinery 41(6): 1298-1328.

Kecman V (2001) Learning and Soft Computing. The MIT Press, Cambridge, MA.

Kehagias A (2002) An example of L-fuzzy join space. Rend Circ Mat Palermo 51: 503-526.

Kehagias A (2003) L-fuzzy join and meet hyperoperations and the associated L-fuzzy hyperalgebras. Rend Circ Mat Palermo 52: 322-350.

Kehagias A, Konstantinidou M (2003) L-fuzzy valued inclusion measure, L-fuzzy similarity and L-fuzzy distance. Fuzzy Sets Systems 136(3): 313-332.

Kehagias A, Petridis V (1997) Predictive modular neural networks for time series classification. Neural Networks 10(1): 31-49.

Kehagias A, Petridis V, Kaburlasos VG, Fragkou P (2003) A comparison of word- and sense-based text categorization using several classification algorithms. J Intel Info Syst 21(3): 227-247.

Kelly GA (1992) The psychology of personal constructs. Routledge, London, UK.

Kent RE (2000) Conceptual knowledge markup language: An introduction. Netnomics 2(2): 139-169.

Kerre EE, Ottoy PL (1989) Lattice properties of neighbourhood systems in Chang fuzzy topological spaces. Fuzzy Sets Systems 30(2): 205-213.

Kim HM, Mendel JM (1995) Fuzzy basis functions: Comparisons with other basis functions. IEEE Trans Fuzzy Systems 3(2): 158-168.

Kim I, Vachtsevanos G (1998) Overlapping object recognition: A paradigm for multiple sensor fusion. IEEE Robotics & Automation Magazine 5(3): 37-44.

Kittler J, Hatef M, Duin RPW, Matas J (1998) On combining classifiers. IEEE Trans Pattern Analysis Machine Intelligence 20(3): 226-239.

Klir GJ (1997) Fuzzy arithmetic with requisite constraints. Fuzzy Sets Systems 91(2): 165-175.

Klir GJ, Folger TA (1988) Fuzzy sets, uncertainty, and information. Prentice-Hall, Englewood Cliffs, NJ.

Klir GJ, Yuan B (1995) Fuzzy sets and fuzzy logic: Theory and applications. Prentice Hall, Upper Saddle River, NJ.

Knuth KH (2005) Lattice duality: The origin of probability and entropy. Neurocomputing 67: 245-274.

Kobayashi I, Chang MS, Sugeno M (2002) A study on meaning processing of dialogue with an example of development of travel consultation system. Information Sciences 144(1-4): 45-74.

Kóczy LT, Hirota K (1997) Size reduction by interpolation in fuzzy rule bases. IEEE Trans. Systems, Man, Cybernetics - B 27(1): 14-25.

Kohavi Z (1978) Switching and finite automata theory, 2nd ed. McGraw-Hill, NY.

Kohonen T (1990) The self-organizing map. Proc of the IEEE 78(9) 1464-1480.

Kohonen T (1995) Self-organizing maps. Springer, Berlin, Germany.

Kohonen T, Somervuo P (1998) Self-organizing maps of symbols strings. Neurocomputing 21(1-3): 19-30.

Kohonen T, Somervuo P (2002) How to make large self-organizing maps for non-vectorial data. Neural Networks 15(8-9): 945–952.

Kohonen T, Oja E, Simula O, Visa A, Kangas J (1996) Engineering applications of the self-organizing map. Proc of the IEEE 84(10) 1358-1384.

Kohonen T, Kaski S, Lagus K, Salojärvi J, Honkela J, Paatero V, Saarela A (2000) Self organization of a massive document collection. IEEE Trans Neural Networks 11(3): 574-585.

Koiran P, Sontag ED (1998) Vapnik-Chervonenkis dimension of recurrent neural networks. Discrete Applied Mathematics 86(1): 63-79.

Kolodner J (1993) Case-based reasoning. Morgan Kaufmann, San Mateo, CA.

Körner R, Näther W (2002) On the variance of random fuzzy variables. In: Statistical Modeling, Analysis and Management of Fuzzy Data, C Bertoluzza, MA Gil, and DA Ralescu (eds.) 87: 25-42. Physica-Verlag, Heidelberg, Germany.

Kosko B (1994) Fuzzy systems as universal approximators. IEEE Trans Computers 43(11): 1329-1333.

Kosko B (1997) Fuzzy Engineering. Prentice-Hall, Upper Saddle River, NJ.

Kosko B (1998) Global stability of generalized additive fuzzy systems. IEEE Trans Systems, Man, Cybernetics - C 28(3): 441-452.

Koufakou A, Georgiopoulos M, Anagnostopoulos G, Kasparis T (2001) Cross-validation in Fuzzy ARTMAP for large databases. Neural Networks 14(9): 1279-1291.

Kourti T (2002) Process analysis and abnormal situation detection: From theory to practice. IEEE Control Syst Mag, 22(5): 10-25.

Krishnapuram R, Keller JM (1993) A possibilistic approach to clustering. IEEE Trans Fuzzy Systems 1(2): 98-110.

Krohn U, Davies NJ, Weeks R (1999) Concept lattices for knowledge management. BT Technology Journal 17(4): 108-116.

Kröse B, Bunschoten R, ten Hagen S, Terwijn B, Vlassis N (2004) Household robots look and learn. IEEE Robotics and Automation Mag, 11(4): 45-52.

Kuncheva LI (1996) On the equivalence between fuzzy and statistical classifiers. Intl J Uncertainty, Fuzziness and Knowledge-Based Systems 4(3): 245-253.

Kuncheva LI, Bezdek JC, Duin RPW (2001) Decision templates for multiple classifier fusion: An experimental comparison. Pattern Recognition 34(2): 299-314.

Kwak N, Choi CH (2002) Input feature selection for classification problems. IEEE Trans Neural Networks, 13(1): 143-159.

Lavoie P, Crespo JF, Savaria Y (1999) Generalization, discrimination, and multiple categorization using adaptive resonance theory. IEEE Trans Neural Networks 10(4): 757-767.

Lecun Y, Bottou L, Bengio Y, Haffner P (1998) Gradient-Based Learning Applied to Document Recognition. Proc of the IEEE 86(11): 2278-2324.

Lee CH, Teng CC (2000) Identification and control of dynamic systems using recurrent fuzzy neural networks. IEEE Trans Fuzzy Systems 8(4): 349-366.

Lee DL, Huei Chuang, Seamons K (1997) Document ranking and the vector-space model. IEEE Software 14(2): 67-75.

Leland RP (1998) Feedback linearization control design for systems with fuzzy uncertainty. IEEE Trans Fuzzy Systems 6(4): 492-503.

Leng G, McGinnity TM, Prasad G (2005) An approach for on-line extraction of fuzzy rules using a self-organising fuzzy neural network. Fuzzy Sets Systems 150(2): 211-243.

Leung FHF, Lam HK, Ling SH, Tam PKS (2003) Tuning of the structure and parameters of a neural network using an improved genetic algorithm. IEEE Trans Neural Networks, 14(1): 79-88.

Levine DS (2000) Introduction to Neural and Cognitive Modeling, 2nd ed. Lawrence Erlbaum, Mahwah, NJ.

Li C, Biswas G (2002) Unsupervised learning with mixed numeric and nominal data. IEEE Trans Knowledge Data Engineering 14(4): 673-690.

Liang Q, Mendel JM (2000) Interval type-2 fuzzy logic systems: Theory and design. IEEE Trans. Fuzzy Systems 8(5): 535-550.

Liao SS, Tang TH, Liu WY (2004) Finding relevant sequences in time series containing crisp, interval, and fuzzy interval data. IEEE Trans Systems, Man and Cybernetics - B 34(5): 2071-2079.

Lim CP, Harrison RF (1997) An incremental adaptive network for on-line supervised learning and probability estimation. Neural Networks 10(5): 925-939.

Lin JK (2005) Lattice theoretic aspects of inference and parameter estimation. In: Bayesian Inference Maximum Entropy Methods in Science and Engineering.

Lin CT, Lee CSG (1996) Neural fuzzy systems – A neuro-fuzzy synergism to intelligent systems. Prentice Hall, Upper Saddle River, NJ.

Lin CJ, Lin CT (1997) An ART-based fuzzy adaptive learning control network. IEEE Trans Fuzzy Systems 5(4): 477-496.

Liu CL (1969) Lattice functions, pair algebras, and finite-state machines. J ACM 16(3): 442-454.

Liu P (2002) Mamdani fuzzy system: Universal approximator to a class of random processes. IEEE Trans Fuzzy Systems 10(6): 756-766.

Liu ZQ, Yan F (1997) Fuzzy neural network in case-based diagnostic system, IEEE Transactions on Fuzzy Systems 5(2): 209-222.

Liu J, Maluf DA, Desmarais MC (2001) A new uncertainty measure for belief networks with applications to optimal evidential inferencing. IEEE Trans Knowledge Data Engineering 13(3): 416-425.

Liu J, Wong CK, Hui KK (2003) An adaptive user interface based on personalized learning. IEEE Intel Syst 18(2): 52-57.

Lloyd JW (2003) Logic for learning – Learning comprehensible theories from structured data. Springer, Berlin, Germany.

Long J (1991) Theory in human-computer interaction?. In: Proc IEE Colloquium on Theory in Human-Computer Interaction (HCI), London, UK, pp 2/1-2/6.

Long PM, Tan L (1998) PAC learning axis-aligned rectangles with respect to product distributions from multiple-instance examples. Machine Learning 30(1): 7-21.

Looney CG (1988) Fuzzy Petri nets for rule-based decision-making. IEEE Trans Systems, Man, Cybernetics 18(1): 178-183.

Looney CG (1997) Pattern recognition using neural networks. Oxford University Press, New York, NY.

Lu YZ, He M, Xu CW (1997) Fuzzy modeling and expert optimization control for industrial processes. IEEE Trans Control Systems Technology 5(1): 2-12.

Lu JJ, Nerode A, Subrahmanian VS (1996) Hybrid knowledge bases. IEEE Trans Knowledge Data Engineering 8(5): 773-785.

Lubeck O, Sewell C, Gu S, Chen X, Cai DM (2002) New computational approaches for *de Novo* peptide sequencing from MS/MS experiments. Proc of the IEEE 90(12): 1868-1874.

Lucas C, Araabi BN (1999) Generalization of the Dempster-Shafer theory: A fuzzy-valued measure. IEEE Trans Fuzzy Systems 7(3): 255-270.

Lund HH (2004) Modern artificial intelligence for human-robot interaction. Proc of the IEEE 92(11): 1821-1838.

Luxemburg WAJ, Zaanen AC (1971) Riesz Spaces. North-Holland, Amsterdam, The Netherlands.

Magdon-Ismail M, Atiya A (2002) Density estimation and random variate generation using multiplayer networks. IEEE Trans Neural Networks 13(3): 497-520.

Mamdani EH, Assilian S (1975) An experiment in linguistic synthesis with a fuzzy logic controller. Intl J Man-Machine Studies 7: 1-13.

Manning CD, Schütze H (1999) Foundations of statistical natural language processing. The MIT Press, Cambridge, MA.

Mao KZ (2002) Fast orthogonal forward selection algorithm for feature selection. IEEE Trans Neural Networks, 13(5): 1218-1224.

Maragos P (2005) Lattice image processing: A unification of morphological and fuzzy algebraic systems. J Math Imaging and Vision 22(2-3): 333-353.

Marcus GF (2001) The algebraic mind: Integrating connectionism and cognitive science. The MIT Press, Cambridge, MA.

Mardia KV, Kent JT, Bibby JM (1979) Multivariate analysis. Academic Press, London, UK.

Margaris A (1990) First order mathematical logic. Dover Publications, New York.

Martens JB (2002) Multidimensional modeling of image quality. Proc of the IEEE 90(1): 133-153.

Massey L (2003) On the quality of ART1 text clustering. Neural Networks 16(5-6): 771-778.

Masticola SP, Marlowe TJ, Ryder BG (1995) Lattice frameworks for multisource and bi-directional data flow problems. ACM Trans Programming Languages and Systems 17(5): 777-803.

Matsakis P, Keller JM, Wendling L, Marjamaa J, Sjahputera O (2001) Linguistic description of relative positions in images. IEEE Trans Systems, Man, Cybernetics - B 31(4): 573-588.

McLachlan GJ (1992) Discriminant analysis and statistical pattern recognition. Wiley & Sons, New York, NY.

Meghini C, Sebastiani F, Straccia U (2001) A model of multimedia information retrieval. J ACM 48(5): 909-970.

Mendel JM, John RIB (2002) Type-2 fuzzy sets made simple. IEEE Trans Fuzzy Systems 10(2): 117-127.

Mitaim S, Kosko B (2001) The shape of fuzzy sets in adaptive function approximation. IEEE Trans Fuzzy Systems 9(4): 637-656.

Mitchell TM (1997) Machine Learning. McGraw-Hill, New York, NY.

Mitchell TM (1999) Machine learning and data mining. Comm ACM 42(11): 31-36.

Mitra S, Hayashi Y (2000) Neuro-fuzzy rule generation: Survey in soft computing framework. IEEE Trans Neural Networks 11(3): 748-768.

Mitra S, Pal SK (1995) Fuzzy multi-layer perceptron, inferencing and rule generation. IEEE Trans Neural Networks 6(1): 51-63.

Mitra S, Pal SK (1996) Fuzzy self-organization, inferencing, and rule generation. IEEE Trans Systems, Man, Cybernetics - A 26(5): 608-620.

Mitra S, De RK, Pal SK (1997) Knowledge-based fuzzy MLP for classification and rule generation. IEEE Trans Neural Networks 8(6): 1338-1350.

Mitra S, Pal SK, Mitra P (2002) Data mining in soft computing framework: A survey. IEEE Trans Neural Networks, 13(1): 3-14.

Moody JE, Darken CJ (1989) Fast learning in networks of locally-tuned processing units. Neural Computation 1: 281-294.

Moore RE (1979) Methods and applications of interval analysis. SIAM Studies in Applied Mathematics, Philadelphia, PA.

Moore R, Lodwick W (2003) Interval analysis and fuzzy set theory. Fuzzy Sets Systems 135(1): 5-9.

Moray N (1987) Intelligent aids, mental models, and the theory of machines. Intl J Man-Machine Studies 27: 619-629.

Moreau Y, DeSmet F, Thijs G, Marchal K, DeMoor B (2002) Functional bioinformatics of microarray data: From expression to regulation. Proc of the IEEE 90(11): 1722-1743.

Motoda H, Mizoguchi R, Boose J, Gaines B (1991) Knowledge acquisition for knowledge-based systems. IEEE Expert Mag 6(4): 53-64.

Murphy RR (1998) Dempster-Shafer theory for sensor fusion in autonomous mobile robots. IEEE Trans Robotics and Automation 14(2): 197-206.

Myrtveit I, Stensrud E, Olsson UH (2001) Analyzing data sets with missing data: An empirical evaluation of imputation methods and likelihood-based methods. IEEE Trans Software Engineering 27(11): 999-1013.

Nachtegael M, Kerre EE (2001) Connections between binary, gray-scale and fuzzy mathematical morphologies. Fuzzy Set and Systems 124(1): 73-85.

Nakamura S (2002) Statistical multimodal integration for audio-visual speech processing. IEEE Trans Neural Networks, 13(4): 854-866.

Nanda S (1989) Fuzzy lattices. Bulletin Calcutta Math Soc, 81: 1-2.

Naphade MR, Huang TS (2002) Extracting semantics from audio-visual content: The final frontier in multimedia retrieval. IEEE Trans Neural Networks, 13(4): 793-810.

Nauck D, Kruse R (1999) Neuro-fuzzy systems for function approximation. Fuzzy Sets Systems 101(2): 261-271.

Nozaki K, Ishibuchi H, Tanaka H (1997) A simple but powerful heuristic method for generating fuzzy rules from numerical data. Fuzzy Sets Systems 86(3): 251-270.

O'Shaughnessy D (2003) Interacting with computers by voice: Automatic speech recognition and synthesis. Proc of the IEEE 91(9): 1272-1305.

Obrenovic Z, Starcevic D (2004) Modeling multimodal human-computer interaction. Computer 37(9): 65-72.

Omlin CW, Giles CL (1996a) Extraction of rules from discrete-time recurrent neural networks. Neural Networks 9(1): 41-52.

Omlin CW, Giles CL (1996b) Constructing deterministic finite-state automata in recurrent neural networks. J ACM 43(6): 937-972.

Omori T, Mochizuki A, Mizutani K, Nishizaki M (1999) Emergence of symbolic behavior from brain like memory with dynamic attention. Neural Networks 12(7-8): 1157-1172.

Opitz D, Maclin R (1999) Popular ensemble methods: An empirical study. J Artificial Intelligence Research 11: 169-198.

Oppenheim AV, Lim JS (1981) The importance of phase in signals. Proc of the IEEE, 69(5): 529-541.

Oppenheim AV, Willsky AS, Nawab SH (1997) Signals and systems. Prentice-Hall, Englewood-Cliffs, New Jersey.

Paccanaro A, Hinton GE (2001) Learning distributed representations of concepts using linear relational embedding. IEEE Trans Knowledge Data Engineering 13(2): 232-244.

Pal SK, Mitra S (1992) Multilayer perceptron, fuzzy sets, and classification. IEEE Trans Neural Networks 3(5): 683-697.

Pal NR, Bezdek JC, Tsao ECK (1993) Generalized clustering networks and Kohonen's self-organizing scheme. IEEE Trans Neural Networks 4(4): 549-557.

Pal SK, Talwar V, Mitra P (2002) Web mining in soft computing framework: Relevance, state of the art and future directions. IEEE Trans Neural Networks, 13(5): 1163-1177.

Pal SK, Mitra S, Mitra P (2003) Rough-fuzzy MLP: Modular evolution, rule generation, and evaluation. IEEE Trans Knowledge Data Engineering 15(1): 14-25.

Palm R, Driankov D (1995) Fuzzy inputs. Fuzzy Sets Systems 70(2-3): 315-335.

Palopoli L, Saccà D, Terracina G, Ursino D (2003) Uniform techniques for deriving similarities of objects and subschemes in heterogeneous databases. IEEE Trans Knowledge Data Engineering 15(2): 271-294.

Panchapakesan C, Palaniswami M, Ralph D, Manzie C (2002) Effects of moving the centers in an RBF network. IEEE Trans Neural Networks, 13(6): 1299-1307.

Pantic M, Rothkrantz LJM (2003) Towards an affect-sensitive multimodal human-computer interaction. Proc of the IEEE 91(9): 1370-1390.

Pao YH (1989) Adaptive pattern recognition and neural networks. Addison-Wesley, Reading, MA.

Papadakis SE, Kaburlasos VG (2005) mass-grSOM: A flexible rule extraction for classification. In: Workshop on Self-Organizing Maps (WSOM 2005), Paris, France, pp 553-560.

Papadakis SE, Theocharis JB (2002) A GA-based fuzzy modeling approach for generating TSK models. Fuzzy Sets Systems 131(2): 121-152.

Papadakis SE, Marinagi CC, Kaburlasos VG, Theodorides MK (2004) Estimation of industrial production using the granular Self-Organizing Map (grSOM). In: Proc Mediterranean Conf Control and Autom (MED'04), Kusadasi, Turkey.

Papadakis SE, Tzionas P, Kaburlasos VG, Theocharis JB (2005) A genetic based approach to the Type I structure identification problem. Informatica 16(3): 365-382.

Parrado-Hernández E, Gómez-Sánchez E, Dimitriadis YA (2003) Study of distributed learning as a solution to category proliferation in Fuzzy ARTMAP based neural systems. Neural Networks 16(7): 1039-1057.

Parsons O, Carpenter GA (2003) ARTMAP neural networks for information fusion and data mining: Map production and target recognition methodologies. Neural Networks 16(7): 1075-1089.

Parzen E (1962) On estimation of a probability density function and mode. J Math Stat 33: 1065-1076.

Passino K, Yurkovich S (1998) Fuzzy Control. Addison Wesley Longman, Reading, MA.

Paul S, Kumar S (2002) Subsethood-product fuzzy neural inference system (SuP-FuNIS). IEEE Trans Neural Networks, 13(3): 578-599.

Pawlak Z (1991) Rough sets: Theoretical aspects of reasoning about data. Kluwer, Boston, MA.

Pearl J (2000) Causality. Cambridge University Press, Cambridge. UK.

Pedrycz W (1992) Associations of fuzzy sets. IEEE Trans Systems, Man and Cybernetics 22(6): 1483-1488.

Pedrycz W (1994) Why triangular membership functions?. Fuzzy Sets Systems 64(1): 21-30.

Pedrycz W (1996) Fuzzy multimodels. IEEE Trans Fuzzy Systems 4(2): 139-148.

Pedrycz W (1998) Computational intelligence – An introduction. CRC Press, Boca Raton, FL.

Pedrycz W (2002) Granular Networks and Granular Computing. In: New Learning Paradigms in Soft Computing, LC Jain and J Kacprzyk (eds.) 84: 30-54. Physica-Verlag, Heidelberg, Germany.

Pedrycz W, Bargiela A (2002) Granular clustering: A granular signature of data. IEEE Trans Systems, Man, Cybernetics - B 32(2): 212-224.

Pedrycz W, Reformat M (1997) Rule-based modeling of nonlinear relationships. IEEE Trans Fuzzy Systems 5(2): 256-269.

Pedrycz W, Vasilakos AV (1999). Linguistic models and linguistic modeling. IEEE Trans Systems, Man, Cybernetics - B 29(6): 745-757.

Pedrycz W, Vukovich G (2001) Abstraction and specialization of information granules. IEEE Trans Systems, Man, Cybernetics - B 31(1): 106-111.

Pedrycz W, Bezdek JC, Hathaway RJ, Rogers GW (1998) Two nonparametric models for fusing heterogeneous fuzzy data, IEEE Trans Fuzzy Systems 6(3): 411-425.

Pejtersen AM (1998) Semantic information retrieval. Comm ACM 41(4): 90-92.

Peltonen J, Klami A, Kaski S (2004) Improved learning of Riemannian metrics for exploratory analysis. Neural Networks 17(8-9) 1087-1100.

Penrose R (1991) The emperor's new mind. Penguin, New York, NY.

Petridis V, Kaburlasos VG (1998) Fuzzy lattice neural network (FLNN): A hybrid model for learning. IEEE Trans Neural Networks 9(5): 877-890.

Petridis V, Kaburlasos VG (1999) Learning in the framework of fuzzy lattices. IEEE Trans Fuzzy Systems 7(4): 422-440.

Petridis V, Kaburlasos VG (2000) An intelligent mechatronics solution for automated tool guidance in the epidural surgical procedure. In: Proc 7th Annual Conf Mechatronics and Machine Vision in Practice (M2VIP'00), Hervey Bay, Australia, pp 201-206.

Petridis V, Kaburlasos VG (2001) Clustering and classification in structured data domains using Fuzzy Lattice Neurocomputing (FLN). IEEE Trans Knowledge Data Engineering 13(2): 245-260.

Petridis V, Kaburlasos VG (2003) FINkNN: A fuzzy interval number k-nearest neighbor classifier for prediction of sugar production from populations of samples. J Machine Learning Research, 4(Apr): 17-37.

Petridis V, Kehagias A (1998) Predictive modular neural networks – Applications to time series. Kluwer Academic Publishers, Norwell, MA.

Petridis V, Kehagias A, Petrou L, Bakirtzis A, Kiartzis S, Panagiotou H, Maslaris N (2001) A Bayesian multiple models combination method for time series prediction. J Intel Robotic Systems 31(1-3): 69-89.

Pindyck RS, Rubinfeld DL (1991) Econometric models and economic forecasts, 3rd ed. McGraw-Hill, NY.

Plate TA (2003) Holographic reduced representation – Distributed representation for cognitive structures. CSLI Publications, Stanford, CA.

Plummer JD, Griffin PB (2001) Material and process limits in silicon VLSI technology. Proc of the IEEE 89(3): 240-258.

Poggio T, Girosi F (1990) Networks for approximation and learning. Proc of the IEEE 78(9): 1481-1497.

Polani D (1999) On the optimization of self-organizing maps by genetic algorithms. In: Kohonen Maps, E Oja and S Kaski (eds.): 157-169. Elsevier, Amsterdam, NL.

Polikar R, Udpa L, Udpa SS, Honovar V (2001) Learn++: An incremental learning algorithm for supervised neural networks. IEEE Trans. Systems, Man, Cybernetics - C 31(4): 497-508.

Polkowski L (2002) Rough sets – Mathematical foundations. In: Advances in Soft Computing, J Kacprzyk (series ed.) Physica-Verlag, Heidelberg, Germany.

Pollack JB (1990) Recursive distributed representations. Artificial Intelligence 46(1-2): 77-105.

Pomares H, Rojas I, Ortega J, Gonzalez J, Prieto A (2000) A systematic approach to a self-generating fuzzy rule-table for function approximation. IEEE Trans. Systems, Man, Cybernetics - B 30(3): 431-447.

Pomares H, Rojas I, González J, Prieto A (2002) Structure identification in complex rule-based fuzzy systems. IEEE Trans Fuzzy Systems 10(3): 349-359.

Popov AT (1997) Convexity indicators based on fuzzy morphology. Pattern Recognition Letters 18(3): 259-267.

Powell MJD (1985) Radial basis functions for multivariable interpolation: A review. In: Proc IMA Conf Algorithms for Approximation of Functions and Data, Shrivenham, UK, pp 143-167.

Principe JC, Tavares VG, Harris JG, Freeman WJ (2001) Design and implementation of a biologically realistic olfactory cortex in analog VLSI. Proc of the IEEE 89(7): 1030-1051.

Priss U (2000) Lattice-based information retrieval. Knowledge Organization 27(3): 132-142.

Pykacz J (2000) Łukasiewicz operations in fuzzy set and many-valued representations of quantum logics. Foundations of Physics 30(9): 1503-1524.

Quinlan JR (1986) Induction of decision trees. Machine Learning 1(1): 81-106.

Quinlan JR (1990) Decision trees and decision-making, IEEE Trans Systems, Man, Cybernetics 20(2): 339-346.

Quinlan JR (1992) C4.5: Programs for machine learning. Morgan Kaufman, San Mateo, CA.

Quinlan JR (1996) Improved use of continuous attributes in C4.5. J Artificial Intelligence Research 4: 77-90.

Radzikowska AM, Kerre EE (2002) A comparative study of fuzzy rough sets. Fuzzy Sets Systems 126(2): 137-155.

Ralescu AL, Ralescu DA (1984) Probability and fuzziness. Information Sciences 34(2): 85-92.

Rauber A, Merkl D, Dittenbach M (2002) The growing hierarchical self-organizing map: Exploratory analysis of high-dimensional data. IEEE Trans Neural Networks, 13(6): 1331-1341.

Reyneri LM (2003) Implementation issues of neuro-fuzzy hardware: Going toward HW/SW codesign. IEEE Trans Neural Networks, 14(1): 176-194.

Richards D, Simoff SJ (2001) Design ontology in context – A situated cognition approach to conceptual modelling. Artificial Intelligence in Engineering 15(2): 121-136.

Riecanova Z (1998) Lattices and quantum logics with separated intervals, atomicity. Intl J Theoretical Physics 37(1): 191-197.

Riesz F (1928) Sur la décomposition des opérations fonctionnelles linéaires. In: Atti del Congresso Internazionale dei Matematici, Bologna, 3: 143-148.

Rigatos GG, Tzafestas SG (2002) Parallelization of a fuzzy control algorithm using quantum computation. IEEE Trans Fuzzy Systems 10(4): 451-460.

Ritter H (1999) Self-organizing maps on non-Euclidean spaces. In: Kohonen Maps, E Oja and S Kaski (eds.): 97-109. Elsevier, Amsterdam, NL.

Ritter GX, Urcid G (2003) Lattice algebra approach to single-neuron computation. IEEE Trans Neural Networks 14(2): 282-295.

Ritter GX, Sussner P, Diaz-de-Leon JL (1998) Morphological associative memories. IEEE Trans Neural Networks 9(2): 281-293.

Ritter GX, Diaz-de-Leon JL, Sussner P (1999) Morphological bidirectional associative memories. Neural Networks 12(6): 851-867.

Rizzi A, Panella M, Frattale Mascioli FM (2002) Adaptive resolution min-max classifiers. IEEE Trans Neural Networks, 13(2): 402-414.

Rodríguez MA, Egenhofer MJ (2003) Determining semantic similarity among entity classes from different ontologies. IEEE Trans Knowledge Data Engineering 15(2): 442-456.

Rota GC (1997) The many lives of lattice theory. Notices of the American Mathematical Society 44(11): 1440-1445.

Rowe NC (2002) Marie-4: A high-recall, self-improving Web crawler that finds images using captions. IEEE Intel Syst 17(4): 8-14.

Roy A, Govil S, Miranda R (1995) An algorithm to generate radial basis function (RBF)-like nets for classification problems. Neural Networks 8(2): 179-201.

Roychowdhury S, Pedrycz W (2002) Modeling temporal functions with granular regression and fuzzy rules. Fuzzy Sets Systems 126(3): 377-387.

Ruggieri S (2002) Efficient C4.5. IEEE Trans Knowledge Data Engineering 14(2): 438-444.

Rumelhart DE, McClelland JL, and the PDP Research Group (1986) Parallel distributed processing. The MIT Press, Cambridge, MA.

Runkler TA (1997) Selection of appropriate defuzzification methods using application specific properties. IEEE Trans Fuzzy Systems 5(1): 72-79.

Runkler TA, Bezdek JC (1999a) Function approximation with polynomial membership functions and alternating cluster estimation. Fuzzy Sets Systems 101(2): 207-218.

Runkler TA, Bezdek JC (1999b) Alternating cluster estimation: A new tool for clustering and function approximation, IEEE Trans Fuzzy Systems 7(4): 377-393.

Rus D, Subramanian D (1997) Customizing information capture and access. ACM Trans Information Systems 15(1): 67-101.

Russell B (2000) Human knowledge – Its scope and limits. Routledge, London, UK.

Rust BW (2001) Fitting nature's basic functions. Part I: Polynomials and linear least squares. Computing in Science and Engineering 3(5): 84-89.

Rutherford DE (1965) Introduction to Lattice Theory. Oliver and Boyd Ltd., Edinburgh, Great Britain.

Rutkowski L, Cpalka K (2003) Flexible neuro-fuzzy systems. IEEE Trans Neural Networks, 14(3): 554-574.

Saade JJ, Schwarzlander H (1992) Ordering fuzzy sets over the real line: An approach based on decision making under uncertainty. Fuzzy Sets Systems 50(3): 237-246.

Saade JJ, Diab HB (2000) Defuzzification techniques for fuzzy controllers. IEEE Trans Systems, Man, Cybernetics - B 30(1): 223-229.

Sabatti C, Lange K (2002) Genomewide motif identification using a dictionary model. Proc of the IEEE 90(11): 1803-1810.

Sahami M (1995) Learning classification rules using lattices. In: Proc 8th European Conf Machine Learning (ECML-95). Heraklion, Crete, Greece.

Sahami M (1998) Using machine learning to improve information access. PhD thesis, Stanford University, CA.

Salton G (1988) Automatic text processing: The transformation analysis and retrieval of information by computer. Addison-Wesley, New York, NY.

Salzberg S (1991) A nearest hyperrectangle learning method. Machine Learning 6(3): 251-276.

Samet H (1988) Hierarchical representations of collections of small rectangles. ACM Computing Surveys 20(4): 271-309.

Sánchez-Garfias FA, Díaz-de-León SJL Yáñez-Márquez C (2005) A new theoretical framework for the Steinbuch's Lernmatrix. In: Proc of SPIE – Mathematical Methods in Pattern and Image Analysis, San Diego, CA, vol. 5916, pp. 59160N: 1-9.

Saridis G (1983) Intelligent robotic control. IEEE Trans Automatic Control, 28(5): 547-557.

Sastry SS, Sztipanovits J, Bajcsy R, Gill H (2003) Special issue on modeling and design of embedded software. Proc of the IEEE 91(1): 3-10.

Scarselli F, A.C. Tsoi AC (1998) Universal approximation using feedforward neural networks: A survey of some existing methods, and some new results. Neural Networks 11(1): 15-37.

Schapire RE (1990) The Strength of Weak Learnability. Machine Learning 5(2): 197-227.

Scheutz M, ed. (2002) Computationalism – New Directions. The MIT Press Cambridge, MA.

Schölkopf B, Smola AJ (2002) Learning with kernels. The MIT Press, Cambridge, MA.

Scott D (1976) Data types as lattices. SIAM J Comput, 5(3): 522-587.

Searle JR (1980) Minds, brains, and programs. Behavioral and Brain Sciences, 3(3): 417-457.

Sebastiani F (2002) Machine learning in automated text categorization. ACM Computing Surveys 34(1): 1-47.

Seo A, Obermayer K (2004) Self-organizing maps and clustering methods for matrix data. Neural Networks 17(8-9): 1211-1229.

Serrano-Gotarredona T, Linares-Barranco B (1997) An ART1 microchip and its use in multi-ART1 systems. IEEE Trans Neural Networks 8(5): 1184-1194.

Setiono R, Leow WK, Zurada JM (2002) Extraction of rules from artificial neural networks for nonlinear regression. IEEE Trans Neural Networks, 13(3): 564-577.

Setnes M (2000) Supervised fuzzy clustering for rule extraction. IEEE Trans Fuzzy Systems 8(4): 416-424.

Setnes M, Babuska R (1999) Fuzzy relational classifier trained by fuzzy clustering. IEEE Trans Systems, Man, Cybernetics - B 29(5): 619-625.

Setnes M, Babuska R, Verbruggen HB (1998a) Rule-based modeling: Precision and transparency. IEEE Trans Systems, Man, Cybernetics - C 28(1): 165-169.

Setnes M, Babuška R, Kaymak U, van Nauta Lemke HR (1998b) Similarity measures in fuzzy rule base simplification. IEEE Trans Systems, Man, Cybernetics - B 28(3): 376-386.

Shafer GA (1976) A mathematical theory of evidence. Princeton Univ Press, Princeton, NJ.

Shannon CE (1953) The lattice theory of information. IEEE Trans Information Theory 1(1): 105-107.

Shimojima K, Fukuda T, Hasegawa Y (1995) Self-tuning fuzzy modeling with adaptive membership function, rules, and hierarchical structure based on genetic algorithms. Fuzzy Sets Systems 71(3): 295-309.

Shimshoni Y, Intrator N (1998) Classification of seismic signals by integrating ensembles of neural networks, IEEE Trans Signal Processing 46(5): 1194-1201.

Shin CK, Yun UT, Kim HK, Park SC (2000) A hybrid approach of neural network and memory-based learning to data mining. IEEE Trans Neural Networks 11(3): 637-646.

Shmulevich I, Dougherty ER, Zhang W (2002) From Boolean to probabilistic Boolean networks as models of genetic regulatory networks. Proc of the IEEE 90(11): 1778-1792.

Shultz TR (2003) Computational developmental psychology. The MIT Press, Cambridge, MA.

Shyi-Ming Chen (1997) Interval-valued fuzzy hypergraph and fuzzy partition. IEEE Trans Systems, Man, Cybernetics - B 27(4): 725-733.

Shyi-Ming Chen, Yih-Jen Horng, Chia-Hoang Lee (2001) Document retrieval using fuzzy-valued concept networks. IEEE Trans Systems, Man, Cybernetics - B 31(1): 111-118.

Siegelmann HT (1995) Computation beyond the Turing limit. Science 268: 545-548.

Sifakis J, Tripakis S, Yovine S (2003) Building models of real-time systems from application software. Proc of the IEEE 91(1): 100-111.

Sih VA, Johnston-Halperin E, Awschalom DD (2003) Optical and electronic manipulation of spin coherence in semiconductors. Proc of the IEEE 91: 752-760.

Simon HA (1955) A behavioral model of rational choice. The Quarterly Journal of Economics LXIX(February): 99-117.

Simpson PK (1992) Fuzzy min-max neural networks - part1: Classification. IEEE Trans Neural Networks 3(5): 776-786.

Simpson PK (1993) Fuzzy min-max neural networks - part2: Clustering. IEEE Trans Fuzzy Systems 1(1): 32-45.

Sinha D, Dougherty ER (1993) Fuzzification of set inclusion: Theory and applications. Fuzzy Sets Systems 55(1): 15-42.

Sinha D, Dougherty ER (1995) A general axiomatic theory of intrinsically fuzzy mathematical morphologies. IEEE Trans Fuzzy Systems 3(4): 389-403.

Solaiman B, Debon R, Pipelier F, Cauvin JM, Roux C (1999a) Information fusion, application to data and model fusion for ultrasound image segmentation. IEEE Trans Biomedical Engineering 46(10): 1171-1175.

Solaiman B, Pierce LE, Ulaby FT (1999b) Multisensor data fusion using fuzzy concepts: Application to land-cover classification using ERS-1/JERS-1 SAR composites. IEEE Trans Geoscience and Remote Sensing 37(3): 1316-1326.

Somervuo PJ (2004) Online algorithm for the self-organizing map of symbol strings. Neural Networks 17(8-9): 1231-1239.

Song M, Tarn TJ, Xi N (2000) Integration of task scheduling, action planning, and control in robotic manufacturing systems. Proc of the IEEE 88(7): 1097-1107.

Sorenson HW (ed.) (1985) Kalman filtering: Theory and application. IEEE Press, New York, NY.

Sowa JF (2000) Knowledge representation. Brooks Cole, Pacific Grove, CA.

Specht DF (1990a) Probabilistic neural networks and the polynomial Adaline as complementary techniques for classification. IEEE Trans Neural Networks 1(1): 111-121.

Specht DF (1990b) Probabilistic neural networks. Neural Networks 3(1): 109-118.

Sperduti A, Starita A (1997) Supervised neural networks for the classification of structures. IEEE Trans. Neural Networks 8(3): 714-735.

Srihari RK, Rao A, Han B, Munirathnam S, Xiaoyun Wu (2000) A model for multimodal information retrieval. In: Proc IEEE Intl Conf Multimedia and Expo (ICME), pp. 701-704.

Stannett M (1990) X-machines and the halting problem: Building a super-Turing machine. Formal Aspects of Computing 2(4): 331-341.

Stoll RR (1979) Set theory and logic. Dover, New York, NY.

Su MC, Chang HT (2000) Application of neural networks incorporated with real-valued genetic algorithms in knowledge acquisition. Fuzzy Sets Systems 112(1): 85-97.

Sugeno M, Kang GT (1988) Structure identification of fuzzy model. Fuzzy Sets Systems 28(1): 15-33.

Sugeno M, Yasukawa T (1993) A fuzzy-logic-based approach to qualitative modeling. IEEE Trans Fuzzy Systems 1(1): 7-31.

Sun R (1997) Learning, action and consciousness: A hybrid approach toward modeling consciousness. Neural Networks 10(7): 1317-1331.

Sun R, Peterson T (1999) Multi-agent reinforcement learning: Weighting and partitioning. Neural Networks 12(4-5): 727-753.

Sussner P (2003) Associative morphological memories based on variations of the kernel and dual kernel methods. Neural Networks 16(5-6): 625-632.

Sussner P, Ritter GX (1997) Decomposition of gray-scale morphological templates using the rank method. IEEE Trans Pattern Analysis and Machine Intelligence 19(6): 649-658.

Sussner P, Graña M (2003) Special issue on morphological neural networks. J Math Imaging and Vision 19(2): 79-80.

Sutton RS, Barto AG (1999) Reinforcement learning: An introduction. The MIT Press, Cambridge, MA.

Swartout W, Tate A (1999) Ontologies. IEEE Intelligent Systems. 14(1): 18-19.

Tahani H, Keller JM (1990) Information fusion in computer vision using the fuzzy integral. IEEE Trans Systems, Man, Cybernetics. 20(3): 733-741.

Takagi T, Sugeno M (1985) Fuzzy identification of systems and its applications to modeling and control. IEEE Trans Systems, Man, Cybernetics 15(1): 116-132.

Tan AH (1997) Cascade ARTMAP: Integrating neural computation and symbolic knowledge processing. IEEE Trans Neural Networks 8(2): 237-250.

Tanaka H, Lee H (1998) Interval regression analysis by quadratic programming approach. IEEE Trans Fuzzy Systems 6(4): 473-481.

Tanaka H, Uejima S, Asai K (1982) Linear regression analysis with fuzzy model. IEEE Trans Systems, Man, Cybernetics 12(6): 903-907.

Tanaka H, Sugihara K, Maeda Y (2003) Interval probability and its properties. In: Rough Set Theory and Granular Computing, M Inuiguchi, S Hirano, S Tsumote (eds.) 125: 69-78. Springer-Verlag, Heidelberg, Germany.

Taylor JG, Fragopanagos NF (2005) The interaction of attention and emotion. Neural Networks 18(4): 353-369.

Tepavcevic A, Trajkovski G (2001) L-fuzzy lattices: An introduction. Fuzzy Sets Systems 123(2): 209-216.

Thrun S (2002) Probabilistic robotics. Comm ACM 45(3): 52-57.

Tikk D, Baranyi P (2000) Comprehensive analysis of a new fuzzy rule interpolation method. IEEE Trans Fuzzy Systems 8(3): 281-296.

Ting KM (2002) An instance-weighting method to induce cost-sensitive trees. IEEE Trans Knowledge Data Engineering 14(3): 659-665.

Tomlin CJ, Mitchell I, Bayen AM, Oishi M (2003) Computational techniques for the verification of hybrid system. Proc of the IEEE 91(7): 986-1001.

Tråvén HGC (1991) A neural network approach to statistical pattern classification by 'semiparametric' estimation of probability density functions. IEEE Trans Neural Networks 2(3): 366-377.

Tsao ECK, Bezdek JC, Pal NR (1994) Fuzzy Kohonen clustering networks. Pattern Recognition 27(5): 757-764.

Tsatsoulis C, Cheng Q, Wei HY (1997) Integrating case-based reasoning and decision theory. IEEE Expert 12(4): 46-55.

Turing AM (1936) On computable numbers, with an application to the Entscheidungsproblem. Proc London Math Soc 42(2): 230-265.

Turing AM (1950) Computing machinery and intelligence. Mind LIX: 433-460.

Uehara K, Fujise M (1993) Fuzzy inference based on families of α-level sets. IEEE Trans Fuzzy Systems 1(2): 111-124.

Uykan Z (2003) Clustering-based algorithms for single-hidden-layer sigmoid perceptron. IEEE Trans Neural Networks, 14(3): 708-715.

Vahid F (2003) The softening of hardware. Computer 36(4): 27-34.

Valiant LG (1984) A theory of the learnable. Comm ACM 27(11): 1134-1142.

Vapnik V (1988) The support vector method of function estimation. In: Nonlinear Modeling: Advanced Black-Box Techniques, J Suykens, and J Vandewalle (eds.): 55-86. Kluwer Academic, Boston, MA.

Vapnik VN (1998) Statistical learning theory. John Wiley & Sons, New York.

Vapnik VN (1999) An overview of statistical learning theory. IEEE Trans Neural Networks 10(5): 988-999.

Varshney PK (1997) Special Issue on Data Fusion. Proc of the IEEE, 85(1): 3-5.

Verbeek JJ, Vlassis N, Kröse BJA (2005) Self-organizing mixture models. Neurocomputing 63: 99-123.

Vidyasagar M (1978) Nonlinear systems analysis. Prentice-Hall, Englewood Cliffs, NJ.

Vidyasagar M (1997) A theory of learning and generalization. Springer-Verlag Ltd, London, GB.

von der Malsburg C (1973) Self-organization of orientation sensitive cells in the striate cortex. Kybernetik 14: 85-100.

von Neumann J (2000) The computer and the brain, 2nd ed. Yale Univ Press, New Haven, CT.

Vulikh BZ (1967) Introduction to the theory of partially ordered vector spaces. Gronigen: Wolters-Noordhoff Scientific Publications, XV.

Vuorimaa P (1994) Fuzzy self-organizing map. Fuzzy Sets Systems 66(2): 223-231.

Wan W, Fraser D (1999) Multisource data fusion with multiple self-organizing maps. IEEE Trans Geoscience and Remote Sensing 37(3): 1344-1349.

Wang MH (2005) Extension Neural Network-Type 2 and Its Applications. IEEE Trans Neural Networks 16(6): 1352-1361.

Wang LX, Mendel JM (1992a) Fuzzy basis functions, universal approximation, and orthogonal least squares learning. IEEE Trans Neural Networks 3(5): 807-814.

Wang LX, Mendel JM (1992b) Generating fuzzy rules by learning from examples. IEEE Trans Systems, Man, Cybernetics 22(6): 1414-1427.

Wang JH, Rau JD, Liu WJ (2003) Two-stage clustering via neural networks. IEEE Trans Neural Networks, 14(3): 606-615.

Weber C, Obermayer K (2000) Structured models from structured data: Emergence of modular information processing within one sheet of neurons. In: Proc Intl Joint Conf Neural Networks (IJCNN), 4: 608-613.

Wegner P, Goldin D (2003) Computation beyond Turing machines. Comm ACM 46(4): 100-102.

Wenocur RS, Dudley RM (1981) Some special Vapnik-Chervonenkis classes. Discrete Mathematics 33: 313-318.

Werbos PJ (1993) Neurocontrol and elastic fuzzy logic: Capabilities, concepts, and applications. IEEE Trans Industrial Electronics 40(2): 170-180.

Wettschereck D, Dietterich TG (1995) An experimental comparison of the nearest-neighbor and nearest-hyperrectangle algorithms. Machine Learning 19(1): 5-27.

Wiesman F, Hasman A (1997) Graphical information retrieval by browsing meta-information. Computer Methods and Programs in Biomedicine 53(3): 135-152.

Williamson JR (1996) Gaussian ARTMAP: A neural network for fast incremental learning of noisy multidimensional maps. Neural Networks 9(5): 881-897.

Williamson JR (2001) Self-organization of topographic mixture networks using attentional feedback. Neural Computation 13: 563-593.

Wilson DR, Martinez TR (1997) Improved heterogeneous distance functions. J Artificial Intelligence Research 6: 1-34.

Winograd T (1997) The design of interaction, Beyond calculation: The next 50 years of computing. Copernicus, New York, 149-161.

Wittgenstein L (1975) Tractatus logico-philosophicus. Routledge, New York, NY.

Wolf SA, Treger DM (2003) Special issue on spintronics. Proc of the IEEE 91(5): 647-651.

Wolff JE, Florke H, Cremers AB (2000) Searching and browsing collections of structural information. Proc IEEE Advances in Digital Libraries, pp. 141-150.

Wong SKM, Ziarko W, Raghavan VV, Wong PCN (1987) On modeling of information retrieval concepts in vector spaces. ACM Trans Database Systems 12(2): 299-321.

Wonneberger S (1994) Generalization of an invertible mapping between probability and possibility. Fuzzy Sets Systems 64(2): 229-240.

Wu H, Mendel JM (2002) Uncertainty bounds and their use in the design of interval type-2 fuzzy logic systems. IEEE Trans Fuzzy Systems 10(5): 622-639.

Wu X, Zhang S (2003) Synthesizing high-frequency rules from different data sources. IEEE Trans Knowledge Data Engineering 15(2): 353-367.

Wuwongse V, Nantajeewarawat E (2002) Declarative programs with implicit implication. IEEE Trans Knowledge Data Engineering 14(4): 836-849.

Xu Y, Ruan D, Qin K, Liu J (2003) Lattice-valued logic. Springer, Berlin, Germany.

Yager RR, Filev DP (1995) Including probabilistic uncertainty in fuzzy logic controller modeling using Dempster-Shafer theory. IEEE Trans Systems, Man, Cybernetics 25(8): 1221-1230.

Yang Y (1999) An evaluation of statistical approaches to text categorization. Information Retrieval 1(1-2): 69-90.

Yang MS, Ko CH (1997) On cluster-wise fuzzy regression analysis. IEEE Trans Systems, Man, Cybernetics - B 27(1): 1-13.

Yao X (1999) Evolving artificial neural networks. Proc of the IEEE 87(9): 1423-1447.

Yao YY, Wong SKM, Wang LS (1995) A nonnumeric approach to uncertain reasoning. Intl J General Systems 23(4): 343-359.

Yen J, Wang L (1998) Application of statistical information criteria for optimal fuzzy model construction. IEEE Trans Fuzzy Systems 6(3): 362-372.

Yoshitaka A, Ichikawa T (1999) A survey on content-based retrieval for multimedia databases. IEEE Trans Knowledge Data Engineering 11(1): 81-93.

Young VR (1996) Fuzzy subsethood. Fuzzy Sets Systems 77(3): 371-384.

Yuan Bo, Wu Wangming (1990) Fuzzy ideals on a distributive lattice. Fuzzy Sets Systems 35(2): 231-240.

Zadeh LA (1965) Fuzzy sets. Inform Contr 8: 338-353.

Zadeh LA (1978) Fuzzy sets as a basis for a theory of possibility. Fuzzy Sets Systems 1(1): 3-28.

Zadeh LA (1996) Fuzzy logic = Computing with words. IEEE Trans Fuzzy Systems 4(2): 103-111.

Zadeh LA (1997) Toward a theory of fuzzy information granulation and its centrality in human reasoning and fuzzy logic. Fuzzy Sets Systems, 90(2): 111-127.

Zadeh LA (1999) From computing with numbers to computing with words. From manipulation of measurements to manipulation of perceptions. IEEE Trans Circuits Systems – I, 45(1): 105-119.

Zadeh LA (2001) Scanning the technology. Proc of the IEEE 89(9): 1242.

Zadeh LA (2004) Precisiated natural language (PNL) - Toward an enlargement of the role of natural languages in scientific theories. In: IEEE Intl Conf Fuzzy Systems (FUZZ-IEEE), Budapest, Hungary, Keynote Talk I.

Zeng XJ, Singh MG (1996) Approximation accuracy analysis of fuzzy systems as function approximators. IEEE Trans Fuzzy Systems 4(1): 44-63.

Zeng XJ, Singh MG (1997) Fuzzy bounded least-squares method for the identification of linear systems. IEEE Trans Systems, Man, Cybernetics - A 27(5): 624-635.

Zeng XJ, Singh MG (2003) Knowledge bounded least squares method for the identification of fuzzy systems. IEEE Trans Systems, Man, Cybernetics - C 33(1): 24-32.

Zhang K, Hirota K (1997) On fuzzy number lattice $(R, \leq)$. Fuzzy Sets Systems 92(1): 113-122.

Zhang YQ, Fraser MD, Gagliano RA, Kandel A (2000) Granular neural networks for numerical-linguistic data fusion and knowledge discovery. IEEE Trans Neural Networks 11(3): 658-667.

Ziarko W (1999) Discovery through rough set theory. Comm ACM 42(11): 55-57.

Zimmermann HJ (1991) Fuzzy set theory and its applications, 2nd ed. Kluwer, Boston, MA.

Zurada JM, Marks II RJ, Robinson CJ (eds.) (1994) Computational Intelligence - Imitating Life. IEEE Press, New York, NY.

Zwick R, Carlstein E, Budescu DV (1987) Measures of similarity among fuzzy concepts: A comparative analysis. Intl J Approx Reason 1(2): 221-242.

Index